ALLE ZEIT WACH 1842

Springer Series in Synergetics Editor: Hermann Haken

Synergetics, an interdisciplinary field of research, is concerned with the cooperation of individual parts of a system that produces macroscopic spatial, temporal or functional structures. It deals with deterministic as well as stochastic processes.

Volume 1 **Synergetics** An Introduction 3rd Edition
By H. Haken

Volume 2 **Synergetics** A Workshop Editor: H. Haken

Volume 3 **Synergetics** Far from Equilibrium Editors: A. Pacault and C. Vidal

Volume 4 **Structural Stability in Physics** Editors: W. Güttinger and H. Eikemeier

Volume 5 **Pattern Formation by Dynamic Systems and Pattern Recognition**
Editor: H. Haken

Volume 6 **Dynamics of Synergetic Systems** Editor: H. Haken

Volume 7 **Problems of Biological Physics** By L. A. Blumenfeld

Volume 8 **Stochastic Nonlinear Systems** in Physics, Chemistry, and Biology
Editors: L. Arnold and R. Lefever

Volume 9 **Numerical Methods in the Study of Critical Phenomena**
Editors: J. Della Dora, J. Demongeot, and B. Lacolle

Volume 10 **The Kinetic Theory of Electromagnetic Processes** By Yu. L. Klimontovich

Volume 11 **Chaos and Order in Nature** Editor: H. Haken

Volume 12 **Nonlinear Phenomena in Chemical Dynamics** Editors: C. Vidal and A. Pacault

Volume 13 **Handbook of Stochastic Methods** for Physics, Chemistry and the Natural Sciences
By C. W. Gardiner

Volume 14 **Concepts and Models of a Quantitative Sociology**
The Dynamics of Interacting Populations By W. Weidlich and G. Haag

Volume 15 **Noise-Induced Transitions** Theory and Applications in Physics, Chemistry, and Biology By W. Horsthemke and R. Lefever

Volume 16 **Physics of Bioenergetic Processes** By L. A. Blumenfeld

Volume 17 **Evolution of Order and Chaos** in Physics, Chemistry, and Biology
Editor: H. Haken

Volume 18 **The Fokker-Planck Equation** By H. Risken

Volume 19 **Chemical Oscillations, Waves, and Turbulence** By Y. Kuramoto

Volume 20 **Advanced Synergetics** By H. Haken

Volume 21 **Stochastic Phenomena and Chaotic Behaviour in Complex Systems**
Editor: P. Schuster

Volume 22 **Synergetics - From Microscopic to Macroscopic Order** Editor: E. Frehland

Volume 23 **Synergetics of the Brain** Editors: E. Başar, H. Flohr, H. Haken, and A. J. Mandell

Volume 24 **Chaos and Statistical Methods** Editor: Y. Kuramoto

Stochastic Phenomena and Chaotic Behaviour in Complex Systems

Proceedings of the Fourth Meeting of the UNESCO Working Group on Systems Analysis
Flattnitz, Kärnten, Austria, June 6–10, 1983

Editor: P. Schuster

With 108 Figures

Springer-Verlag
Berlin Heidelberg New York Tokyo 1984

Professor Dr. Peter Schuster
Institut für Theoretische Chemie und Strahlenchemie der Universität Wien,
Währinger Straße 17
A-1090 Wien, Austria

Series Editor:
Professor Dr. Dr. h. c. Hermann Haken
Institut für Theoretische Physik der Universität Stuttgart, Pfaffenwaldring 57/IV,
D-7000 Stuttgart 80, Fed. Rep. of Germany

ISBN-13:978-3-642-69593-3 e-ISBN-13:978-3-642-69591-9
DOI: 10.1007/978-3-642-69591-9

Library of Congress Cataloging in Publication Data. Main entry under title: Stochastic phenomena and chaotic behaviour in complex systems. (Springer series in synergetics ; v. 21). "Workshop was sponsored by UNESCO, Paris and the Bundesministerium für Wissenschaft und Forschung, Wien" - Pref. 1. Chaotic behavior in systems-Congresses. 2. Stochastic processes-Congresses. I. Schuster, P. (Peter), 1941-. II. Unesco. III. Austria. Bundesministerium für Wissenschaft und Forschung. IV. Series. QA402.S8475 1984 003 84-1251

Softcover reprint of the hardcover 1st edition 1984

2153/3130-543210

Preface

This book contains all invited contributions of an interdisciplinary workshop of the UNESCO working group on systems analysis of the European and North American region entitled "Stochastic Phenomena and Chaotic Behaviour in Complex Systems". The meeting was held at Hotel Winterthalerhof in Flattnitz, Kärnten, Austria from June 6-10, 1983.

This workshop brought together some 20 mathematicians, physicists, chemists, biologists, psychologists and economists from different European and American countries who share a common interest in the dynamics of complex systems and their analysis by mathematical techniques. The workshop in Flattnitz continued a series of meetings of the UNESCO working group on systems analysis which started in 1977 in Bucharest and was continued in Cambridge, U.K., 1981 and in Lyon, 1982.

The title of the meeting was chosen in order to focus on one of the current problems of the analysis of dynamical systems. A deeper understanding of the various sources of stochasticity is of primary importance for the interpretation of experimental observations. Chaotic dynamics plays a central role since it introduces a stochastic element into deterministic systems.

The workshop was sponsored by UNESCO, Paris and the Bundesministerium für Wissenschaft und Forschung, Wien. We are greatly indebted to the Bundesminister für Wissenschaft und Forschung, Dr. Herta Firnberg and to the Head of the Scientific Cooperation Bureau for the European and North American Region, Prof. Dr. J. Jaz, who made this workshop possible through their financial support. The warm hospitality of Familie Klimbacher and the staff of Hotel Winterthalerhof is gratefully acknowledged. We thank Mrs. J. Jakubetz, Mag. B. Gassner, Mag. M. Reithmaier, Dr. F. Kemler and Mr. J. König for their technical assistence in organizing the meeting and preparing the proceedings. In addition, our thanks go to Professor H. Haken for having these Proceedings included in the Springer Series in Synergetics.

Wien
November 1983

Peter Schuster

Contents

Introductory Remarks

P. Schuster
Institut für Theoretische Chemie und Strahlenchemie, Universität Wien
Währinger Straße 17, A-1090 Wien, Austria

The notions "strange attractor", "chaotic dynamics" or simply "chaos" are synonyms for a kind of complicated dynamical behaviour with two main features: (1) The long-time dynamics is characterized by irregular changes in variables which are lacking any periodicity or quasiperiodicity (2) trajectories through close by lying points diverge exponentially with time. The search for strange attractors in nonlinear dynamical systems became kind of a fashion after LORENZ [1] had published his famous work on a model differential equation for hydrodynamic flow which shows chaotic dynamics for certain choices of parameters. Indeed, most people were successful and found chaotic solutions for various nonlinear difference and differential equations provided the number of independent variables was high enough and the nonlinearities were sufficiently general. It took much longer to work out the deeper mathematical properties of chaos, its physical meaning is not very well understood yet. For some publications concerning current problems of the research on strange attractors and their properties see [2-4]. In this volume the contribution by ROESSLER deals with the problem of classification of strange attractors, MAYER presents an investigation on invariant measures for exponentially expanding maps.

Chaos introduces an interface between determinism and randomness as FORD [5] points out in a well-written recent article. Truly deterministic description of chaotic dynamics requires infinite precision in the choice of initial conditions and thus, is a scientific chimera. Therefore, chaos introduces a fundamental uncertainty which is more general than HEISENBERG's uncertainty in quantum mechanics: it concerns also macroscopic dynamics and restricts the available information for every variable whereas the quantum mechanical uncertainty operates on canonically conjugate pairs of variables only. The present state of affairs in the description of real systems with chaotic dynamics calls for a new approach towards a more complete description of experiments. Such a description takes into account explicitly the inevitable uncertainties by means of an appropriate mathematical formalism and unites thereby deterministic kinetics and the various sources of stochasticity. A mathematical tool for such an approach in available, at least in principle: almost all real dynamical systems in continuous time can be described properly by a master equation. The problem actually is the search for solutions which most often is extremely hard and successful in exceptional cases only. New methods of approximation were developed during the last decade [6,7] but they are not applicable to the complex dynamics we are basically interested in here.

Difference equations of very simple algebraic structures may have extremely complicated dynamical behaviour. The discretized logistic equation is one of the most popular examples of this kind.(For an easy to read and recent review see [8]). The very rich dynamics introduced by the choice of discrete time may also cause serious problems for the numerical approach: numerical integration inevitably is based on discrete time and can result in highly complicated spurious solutions as the contribution by PEITGEN nicely demonstrates.

How common is chaotic dynamics in systems with continuous time? In order to find an answer we divide dynamical systems into two classes:

(1) Systems which are derived from "true" equations of motion and hence in general have a non-zero kinetic energy:

$$m_k \frac{d^2x_k}{dt^2} + \gamma_k \frac{dx_k}{dt} - F_k(x_1,\ldots,x_n) = 0 \; ; \; k=1,\ldots,n \quad (1).$$

In systems of this kind chaotic dynamics leading to ergodic behaviour is very frequent. Separable Hamiltonian systems characterized by very simple dynamics are rare in reality although they make up for most examples treated in elementary texts of classical mechanics. The contribution by RICHTER and SCHOLZ dealing with chaotic dynamics found with the double pendulum is a characteristic example. Closely related are problems in hydrodynamics: chaotic solutions of the NAVIER-STOKES equation are discussed in relation to turbulence [3,9].

(2) Systems which can be derived formally from (1) by extrapolation to zero mass (or zero kinetic energy)[1]. An alternative interpret-

---

[1] Such an extrapolation is not trivial since one boundary condition is lost thereby as FRIEDRICHS [10] pointed out some time ago. It has to be performed with sufficient care.

ation of the neglect of the second derivatives is motion at very high friction, i.e. motion in a medium of high viscosity (high values of γ). Equations of this truncated type

$$\frac{dx_k}{dt} = G_k(x_1,\ldots,x_n) \;;\; k=1,\ldots,n \qquad (2)$$

are commonly considered in chemical kinetics of homogeneous solutions, i.e. when diffusion can be neglected or in well stirred reactors, and in many other disciplines like mathematical ecology, population genetics, game dynamics, dynamical economies. In such systems with zero kinetic energy strange attractors are common when G_k is sufficiently non-linear and $k \geq 3$. In almost all well studied examples the occurrence of chaotic dynamics is confined to very small regions in parameter space. Detecting these regions requires mathematical intuition and numerical skill. New methods are being developed for this purpose, one example is presented by COULLET in his contribution. Needless to say, the experimental verification of chaotic dynamics in systems of this kind is an extremely hard problem: how to distinguish "true" chaos in the solution of the differential equations and irregular fluctuations of external parameters within the range of control ? The highly precise and elegant experimental and numerical works by HUDSON, MANKIN, ROESSLER and NOYES as well as their scientific dispute serve as an illustrative demonstration of the enormous difficulties to be encountered in this new branch of chemical kinetics. OLSEN's contribution presents an example for the occurrence of a strange attractor in an enzyme-catalysed reaction. NICOLIS, MAYER-KRESS and HAUBS discuss the role of chaotic dynamics in the operation of an information processor, particularly the human brain.

The contributions in Part One deal with new general concepts in systems analysis: a dynamical information theory by HAKEN, an approach to describe optimization and adaptation in fluctuating systems by ERMOLIEV and a contribution on relaxed Markov chains by WHITTLE.

One part of these proceedings contains three papers dealing with the search for stability and instability in non-linear dynamical systems. PHILLIPSON presents a new method to construct analytical approximations to the solutions of ordinary differential equations which is shown to be useful as diagnostics of stability.

The two contributions by HOFBAUER, SIGMUND and SCHUSTER deal with biochemical and biological examples of stability analysis in non-linear dynamical systems.

The last part presents a collection of individual contributions on stochastic phenomena in complex systems from various fields, from biochemistry (SCHUSTER, SIGMUND and RODRIGUEZ-VARGAS), ecology (REJMANEK), economies and urban development (ALLEN, SANGLIER and ENGELEN) and general theory of non-linear systems (ALBRECHT, CHETVERIKOV, EBELING, FUNKE, MENDE and PESCHEL).

References

1 E.N.Lorenz: J.Atmos.Sci.20, 130 (1963)
2 M.J.Feigenbaum: Comm. Math.Phys.77, 65 (1980)
3 D.Ruelle: Bull.Am.Math.Soc.5, 29 (1981)
4 J.D.Farmer, E.Ott, J.A.Yorke: The Dimension of Chaotic Attractors, Physica D, in press (1983)
5 J.Ford: Physics Today, April 1983, 40
6 C.W.Gardiner, S.Chaturvedi: J.Statist.Phys. 17, 429 (1977)
7 N.G.van Kampen: "Stochastic Processes in Physics and Chemistry" (North Holland, Amsterdam 1983)
8 D.R.Hofstadter: Sci.Am.245(5), 19 (1981)
9 D.Ruelle, F.Takens: Comm.Math.Phys.20, 167 (1971)
10 K.O.Friedrichs: Bull.Am.Math.Soc.61, 485 (1955)

Part I

General Concepts

Some Basic Ideas on a Dynamic Information Theory

H. Haken

Institut für Theoretische Physik, Universität Stuttgart, Pfaffenwaldring 57
D-7000 Stuttgart 80, Fed. Rep. of Germany

1. Introduction

"Information" is a well defined quantity within communication theory [1], [2] in contrast to the meaning we attach to the word "information" in usual language. The price to be paid for the precise definition of information rests in the fact that information in the former sense does not have a meaning such as "useful" or "useless" or "meaningful" or "meaningless". It is rather a measure for a message being scarce.

In this paper an attempt is made to introduce a new concept which is a step towards including semantics. We are led to the basic idea by the observation that we can only attribute a meaning to a message if the response of the receiver of that message is taken into account.

2. An approach to dynamic information theory

Since we wish to introduce new concepts it may be advisable to start with rather concrete situations. Let us consider a set of messages which are specified by a string of numbers. The central problem consists in modelling the receiver. We do this by invoking modern concepts of dynamic system theory or, still more generally, by concepts of synergetics [3], [4] . We model the receiver as a dynamic system whose behavior is described by trajectories within a state space.

The point of each trajectory at time t is then determined by the vector $\underset{\sim}{q}(t)$. To be still more specific we assume that $\underset{\sim}{q}$ obeys a set of first order nonlinear differential equations of the form

$$\dot{\underset{\sim}{q}} = \underset{\sim}{N}(\underset{\sim}{q}, \alpha) + \underset{\sim}{F}(t). \tag{2.1}$$

In it $\underset{\sim}{N}$ is a vector function which depends on $\underset{\sim}{q}$ and on certain control parameters α . Later on it will turn out that stochastic forces $\underset{\sim}{F}(t)$ play an important role. We assume that the receipt of a message by the system means that the parameters α and the initial value of q_0 are set by the message. For the moment being we shall assume that these parameters are then uniquely fixed. An extension of the theory to an incomplete message is straightforward (see below). We first ignore the role of fluctuations. We assume that before the message arrives the system has been in an attractor which we shall call the neutral state. The attractor may be a resting state i.e. a point sink, but it could equally well be a limit cycle, a torus or a strange attractor, or a type of attractor still to be discovered by dynamic system theory. We shall call this attractor $\underset{\sim}{q}_0$. After the message has been received and the parameters α and the initial value $\underset{\sim}{q}$ are newly set, in principle two things may happen. Let us assume that we are allowed to wait for a certain measuring time so that the dynamic system can be assumed to be in one of its possible attractors. Then either the message has left the system in the q_0 state. In such a case the message is evidently useless or meaningless.

The other case is that the system goes into a new attractor. We first assume that this attractor is uniquely determined by the incident message. Clearly, different messages can give rise to the same attractor. In this case we will speak of redundancy of the messages.

Finally, especially in the realm of biology it has been a puzzle so far how information can be generated. This can be easily visualized, however, if we assume that the incident message provides the situation as depicted in fig. 1, which is clearly ambiguous. Two new stable points (or attractors) can be realized depending on a fluctuation within the system itself. Here the incident message contains an information in the ordinary sense of the word, which is ambiguous and the ambiguity is resolved by a fluctuation of the system. Loosely speaking, the original information is doubled because now two attractors become available.

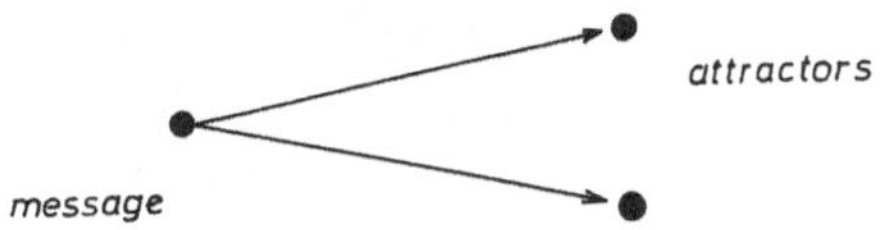

Fig. 1

In the case of biology these fluctuations are realized by mutations. On the other hand in the realm of physics we should rather speak of symmetry breaking effects.

Taking all these different processes together we may find the elementary schemes of fig. 2. Of course, when we consider the effect of different messages, more complicated schemes such as those of fig. 3 may evolve.

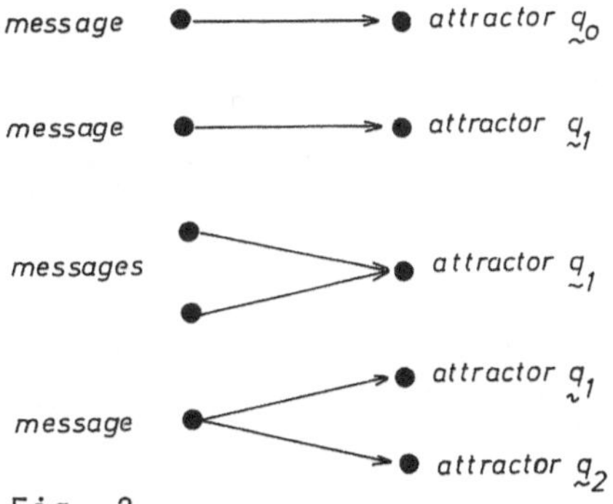

Fig. 2

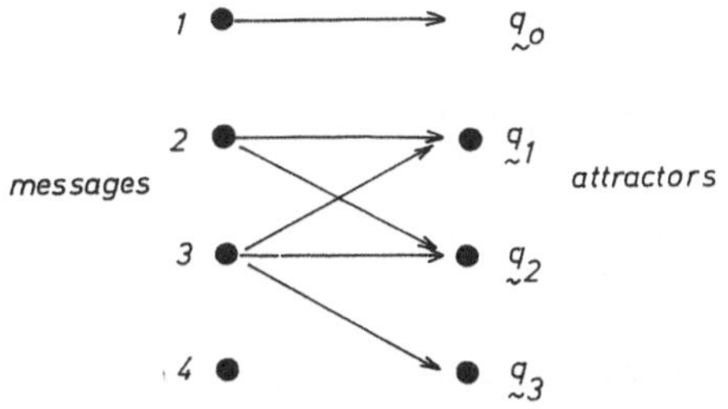

Fig. 3

We shall now treat the question in which way we can attribute values to the incident messages or, more precisely speaking, we want to define a "relative importance of the messages". To this end we first have to introduce a "relative importance" for the individual attractors. In reality, the individual attractors will be the origin of new messages which are then put into a new dynamical system and we can continue this process ad infinitum. However, for practical purposes, we have to cut this hierarchical sequence at a certain level and at this level we have to attribute values of the relative importance to the individual attractors. Since our procedure can already be clearly demonstrated if we have a one-step process, let us consider this process in detail.

Let us contribute a "relative importance" to the individual attractors where the attractor 0 with q_0 has the value 0, while the other attractors may have values $0 \leq p'_j \leq 1$, which we normalize to $\sum_j p'_j = 1$.

The fixation of p'_j depends on the task the dynamic system has to fulfill. We may think of a specific task which can be fulfilled just by a specific attractor or we may think of an ensemble of tasks whose fulfillment is of a given relative importance. Clearly the relative importance of the messages p_j does not only depend on the dynamic system but on its tasks it has to fulfill. The problem now is: What are the values p_j of the incident messages? To this end we consider the links between a message and the attractor into which the dynamical system is driven after receipt of this message. If an attractor k (including the 0 attractor) is reached after receipt of the message j we attribute to this process the matrix element $M_{jk} = 1$ (or $= 0$). If we allow for internal fluctuations of a system, a single message can drive the system via fluctuations into several different attractors which may occur with branching rates M_{jk} with $\sum_k M_{jk} = 1$. We define the "relative importance p_j" by

$$p_j = \sum_k L_{jk}\, p'_k = \sum_k \frac{M_{jk}}{\sum_{j'} M_{j'k} + \varepsilon}\, p'_k \tag{2.2}$$

where we let $\varepsilon \to 0$. (This trick serves the purpose that the ratio remains determined in the case that denominator and nominator vanish simultaneously.) We first assume that for any $p'_k \neq 0$ at least one $M_{jk} \neq 0$. One readily convinces oneself that p_j is normalized which can be shown by the steps

$$\sum_j p_j = \sum_{kj} \frac{M_{jk}}{\sum_{j'} M_{j'k} + \varepsilon}\, p'_k = \tag{2.3}$$

$$= \sum_k \left(\sum_j \frac{M_{jk}}{\sum_{j'} M_{j'k} + \varepsilon}\right) p'_k \tag{2.4}$$

$$= \sum_k p'_k \tag{2.5}$$

where the bracket in (2.4) is equal to 1.

Now consider the case that some $M_{jk'} = 0$ for $p_{k'} \neq 0$ and all j. In this case in the sums over k in (2.3) and (2.4) some coefficients of $p'_k \neq 0$ vanish and, since $\sum p'_k = 1$, we obtain $\sum_j p_j < 1$. If this inequality holds, we shall speak of an <u>information deficiency</u>.

In a more abstract way we may adopt the left hand side of (2.2) as a basic definition where we assume

$$\sum_j L_{jk} \leqq 1 \qquad (2.6)$$

where the equality sign holds in the case of absence of an information deficiency.

We note that instead of the requirement $M_{jk} = 1$ in case of a single final attractor for an incident message M_{jk} can be generalized to

$$0 < M_{jk} \leq 1. \qquad (2.7)$$

The form (2.2), left hand side, allows us to immediately write down the formulas when several systems are coupled one after the other. For instance in the two step process we immediately obtain

$$p_j = \sum_k L^{(1)}_{jk} \, p'_k = \sum_{kk'} L^{(1)}_{jk} \, L^{(2)}_{kk'} \, p''_{k'} \qquad (2.8)$$

where one can convince oneself very easily that $\Sigma p_j = 1$ provided $\Sigma p'_k = 1$ and $\sum_j L_{jk} = 1$. The individual steps read

$$\Sigma p_j = \sum_j \sum_{kk'} L_{jk} L_{kk'} p''_k = \sum_{kk'} \underbrace{\left(\sum_j L_{jk} \right)}_{= 1} L_{kk'} p''_{k'} \qquad (2.9)$$

$$= \sum_{k'} \underbrace{\sum_k L_{kk'}}_{= 1} p''_k = 1 \qquad (2.1o)$$

We may define

$$L'_{jk'} = \sum_k L^{(1)}_{jk} \, L^{(2)}_{kk'} \qquad (2.11)$$

Because the L's are positive we find

$$L'_{jk} \geq 0 \qquad (2.12)$$

and because of the normalization properties (in case of absence of information deficiency)

$$\left.\begin{aligned} \sum_j L'_{jk'} &= \sum_k \underbrace{\sum_j L^{(1)}_{jk}} \, L^{(2)}_{kk} \\ &= \sum_k L^{(2)}_{kk'} = 1 \end{aligned}\right\} \qquad (2.13)$$

we readily obtain

$$L'_{jk} \leq 1 \qquad (2.14)$$

so that L'_{jk} obeys the inequality

$$0 \leq L'_{jk} \leq 1. \qquad (2.15)$$

We mention that the recursion from p" or still higher order $p^{(n)}$ to p may depend on the paths.

3. Some conclusions

Our above approach does not only introduce the new concept of relative importance of a message but it also provides us with an algorithm to determine p_j which has some conceptual and practical consequences. With a certain task or ensemble of tasks given this algorithm allows us to select the message to be sent, namely the one with the biggest p_j. If there are several p_j of the same size it does not matter which message is sent. From the conceptual point of view we may decide, whether by dynamical systems information is annihilated, conserved or generated. To this end we make use of the concept of information in the sense of conventional information theory. But instead of the information content caused by the relative frequency of symbols we use the relative importance within a set of messages, i.e. we introduce the quantities

$$S^{(o)} = - \Sigma p_j \ln p_j \qquad (3.1)$$

$$S^{(1)} = - \Sigma p'_k \ln p'_k . \qquad (3.2)$$

where p_j, p'_k have been introduced in § 2. If $\sum_k p'_k = 1$, as is always assumed here, and $\Sigma p_j < 1$, an information deficiency is present. In the case $\Sigma p_j = 1$ we shall speak of annihilation of information if

$$S^{(1)} < S^{(o)} \qquad (3.3)$$

holds, of conservation of information if

$$S^{(1)} = S^{(o)} \qquad (3.4)$$

holds and of generation of information if

$$S^{(1)} > S^{(o)} \qquad (3.5)$$

holds. The meaning of this definition quickly becomes clear when we treat special cases. If, for instance, two messages lead to the same attractor there is a redundancy in the system and the information content (in the traditional technical sense of the word) becomes smaller. It is reduced from

$$S^{(o)} = - K(\frac{1}{2} \ln (1/2) + \frac{1}{2} \ln (1/2))$$
$$= K \ln 2 \qquad (3.6)$$

to

$$S^{(1)} = - K \cdot 1 \cdot \ln 1 = 0. \tag{3.7}$$

In the case of a one-to-one mapping of p_j onto p'_k we find the transfer of $\{p_j\}$ into the same set $\{p'_k\}$, maybe except for the permutation of indices, i.e. for different numeration of states. In such a case clearly (3.4) holds. In the case (3.5) finally, for instance one $p_j = 1$ and all others $= 0$, are transfered into e.g. $p' = p'' = 1/2$, all others equal to 0. Then $S^{(0)} = - K \cdot 1 \cdot \ln 1 = 0$ is enlarged to

$$S^{(1)} = - K \left(\frac{1}{2} \ln (1/2) + \frac{1}{2} \ln (1/2)\right) = K \ln 2. \tag{3.8}$$

Of course these examples are not meant to prove the definitions (3.3) to (3.5) but rather to illustrate their meaning.

Our approach based on Synergetics [3], [4] has some further pleasent features. Semantics has become a problem of the study of the response (attractors) of the dynamic system. The system may be error-correcting (or supplement lacking information). If the incident message does not set the initial state q on the attractor (i.e. not correctly), it may set the initially state q within the basin of the attractor and the system pulls the state vector into the attractor corresponding to that basin, i.e. into the correct state. An interesting problem will it be to determine the minimum number of bits required to realize a given attractor (or to realize a given value of "relative importance").

Within our present scheme, learning of a system can also be modelled. A system can be "sensitized" or "desensitized" with respect to messages j e.g. by letting more or fewer parameters react on specific messages.

In our above treatment we have assumed that the value of the messages is measured with respect to the same initial state of the receiver. In the next step of our considerations we may assume that messages apply to a receiver in another initial state which has been set for instance by a previous message.

In such a way we obtain an interference of messages and the relative importance of a message depends on messages the receiver has received before. In particular in the general case the relative importance of the message will depend in a non-commuting way on subsequent messages. In this way the receiver is transformed by messages again and again and clearly the relative importance of messages will become a function of time.

Another remark might be useful in particular with respect to synergetic processes. A synergetic system need not only be a dynamical system showing e.g. limit cycle or chaotic behavior but it might also be one in which irreversible processes leading for instance from an organized liquid state into a structured solid state may happen.

The details of my approach will be published elsewhere.

4. A comment on pattern recognition

Pattern recognition can be considered as a processing of incoming messages by a receiver, e.g. the brain or a machine. It is therefore an interesting task to discuss pattern recognition using ideas outlined above. I suggest that pattern recognition, at least in general, is a multistep process in which the receiver takes an active part. In a first step, the pattern is received at a global level where, in general several attractors can be reached. Then, the sensory system is requested to focus its attention on the exploration of additional features so that a finer set of attractors can be selected. To be more explicit: In a first step the global shape of the contour lines of an object are determined e.g. close to a circle, rectangular, etc. Then, in the case of a circle, there are several attractors: Apple, face, wheel, tree. Then the receiver asks back to deliver further details, e.g. color, vertical lines (nose?), etc. In this way the process can be continued.

Note that our interpretation of pattern recognition differs from the "traditional" approach [5]. There the pattern is first decomposed into its "primitives" or "features". Here we start from the global pattern (contour line) and then proceed to more and more details.

This approach offers us an explanation (or at least a hint) why in human pattern recognition even interrupted contour lines are complemented so that a continuous line is "seen".

References

[1] Shannon, C.E., A mathematical theory of communication, Bell System Techn. J. (1948) 27, v. 37o-423, p. 623-656; Prediction and Entropy of printed English, Bell System Techn. J., 1951, v. 3o, p. 5o-64

Shannon, C.E., Weaver, W., The Mathematical Theory of Communication, Urbana: University of Illinois Press, 1949

[2] Brillouin, L., Science and Information Theory, New York, London: Academic Press, 1962

Brillouin, L., Scientific Uncertainty and Information, New York, London: Academic Press, 1964

[3] Haken, H., Synergetics. An Introduction, 2nd ed. New York, Berlin, Heidelberg, Springer 1978 in Russian: Sinergetica, Moskau: Mir 198o

[4] Haken, H., Advanced Synergetics, New York, Berlin, Heidelberg, Springer 1983

[5] cf. for example the contributions by Fu and Kohonen in: Pattern Formation in Dynamic Systems and Pattern Recognition, ed. H. Haken, Springer, 1979

Aspects of Optimization and Adaptation

Y.M. Ermoliev
International Institute for Applied Systems Analysis (IIASA)
A-2361 Laxenburg, Austria

Abstract

This paper provides a brief review of some aspects of optimization in the presence of uncertainty.

1. Uncertainty - a Fundamental Theme

The rapid changes taking place in the world today emphasize the need for methods capable of dealing with the uncertainties inherent in virtually all modern systems. We cannot know, or measure, everything. In addition, systems are now undergoing disturbances which may be unlike anything they have experienced before. To cope with such uncertainty we must develop new adaptive mechanisms, which can respond to changing conditions.

2. Adaptation as a Response to Changing Conditions

According to classical theory, a system should be thoroughly investigated before it can be optimized. However, such an investigation is generally ruled out under conditions of uncertainty due to the possibility of unexpected changes in both system and environment. Adaptation must therefore be a continuous process, a series of steps. During this process, various estimation and prediction procedures are used to update the system parameters, and an optimization technique is employed to update the control law.

3. Long-Term Strategies

The process of adaptation is associated with the short-term (control) actions taken after an observation of the current state of the system has become available. However, in practice it is possible to identify problems which require long-term actions (engineering design, allocation of resources, investment strategies, etc.) as well as those requiring only short-term adjustment (flying an aeroplane, marketing, inventory policy, etc.). Thus, to deal with uncertainty successfully we need to develop approaches (models, computational methods) which integrate a long-term strategy with short-term adaptive actions. In other words, we need approaches which combine the idea of preparing for several possible futures (anticipatory optimization) with that of learning from experience (adaptation).

4. Stochastic Optimization Problems

As uncertainty is such a broad concept, it is possible (and indeed useful) to approach it in many different ways. One rather general approach is to assign a measure of confidence (which can be interpreted as a probabilistic measure) to various unknown parameters. This leads us to a class of stochastic optimization problems with partially known distribution functions and incomplete observations of unknown parameters for which evaluation of control policy and information collection must take place repeatedly and with systematic adjustments.

5. Drawbacks of Existing Methods

The problems outlined above can very rarely be solved using traditional optimization techniques. Most of these techniques require the evaluation of multiple integrals which characterize the random properties of the system as functions of the control variables: attempts to evaluate these directly fail for systems of dimensionality greater than three. In addition, most of the existing methods have been developed for off-line control and adaptation is essentially an on-line process. Thus, new methods are needed to deal with adaptive processes. One approach could be to reduce stochastic problems to deterministic problems through approximation schemes. There are also promising ideas based on the use of direct stochastic procedures employing available random observations. Methods of this type can be regarded as a sort of formalized process of trial and error, and in fact this cannot be avoided when dealing with real uncertainties.

6. Problems

In this section we mention some typical problems and describe possible solution techniques currently being developed at IIASA.

Short-term actions. Nonmonotonic techniques. Consider the simplest case, in which we have to minimize the loss function

$$F(x) = f(x,\alpha) \quad ,$$

where $\alpha \in R^k$ is a vector of unknown parameters and $x \in R^n$ is a vector of control variables. Function $F(x)$ cannot be optimized directly because of the unknown parameters α. Suppose that at each iteration $s = 0,1,\ldots$ an observation h^s is available which has the form of a direct observation of the parameter vector, i.e.,

$$Eh^s = \alpha \quad .$$

By using h^s at iteration s we can obtain a statistical estimate α^s of α such that $\alpha^s \to \alpha$ with probability 1. In this case $F(x)$ must be replaced at iteration s by

$$F^s(x) = f(x,\alpha^s) \quad ,$$

where $F^s(x) \to F(x)$ with probability 1 for $s \to \infty$. The question is: can we use the sequence of functions $F^s(x)$ to find the minimum of $F(x)$? One possibility is to use the procedure

$$x^{s+1} = x^s - \rho_s F_x^s(x^s) \quad , \quad s = 0,1,\ldots \quad , \tag{1}$$

where $F_x^s(x)$ is the gradient of $F^s(x)$ or its analogue (for nondifferentiable functions), and ρ_s is a step-size multiplier. This procedure, together with a procedure for calculating α^s, allows us to carry out the optimization while simultaneously estimating α. The properties of such procedures are described in [1]. It should be emphasized, however, that the behavior of approximations $F^s(x^s)$ is not necessarily monotonic, that is, $F^s(x^s)$ might be greater than $F^{s-1}(x^{s-1})$, whatever the value of ρ_s. Therefore, even the simplest case of on-line optimization with unknown parameters requires the development of nonmonotonic optimization techniques.

Long-term actions. Suppose that we have to choose x before observing α and that the probabilistic measure $dH(\alpha)$ can be assigned to α. The expected loss function is the following:

$$F(x) = E_\alpha f(x,\alpha) = \int f(x,\alpha)dH(\alpha) \quad .$$

The problem now is to minimize $F(x)$ with respect to the feasible decision variables x. The main difficulty here is concerned with the evaluation of $F(x)$ and its derivatives. The stochastic approximation method and its generalizations (see, for instance, [2]) avoids these difficulties, since it provides a means of minimizing $F(x)$ using information on the random functions $f(x,\alpha)$ only. The approximation schemes for such problems are discussed in [3,4].

Partially known distribution functions. If $H(\alpha)$ is only partially known, i.e., $H \in W$, where W is the class of feasible distributions, then the following minimax problem is of interest:

$$\min_x \max_{H \in W} \int f(x,\alpha)dH(\alpha) \quad .$$

Computational methods for such problems are described in [5].

In general, most problems will involve all of the above-mentioned difficulties simultaneously.

References

1. Y. Ermoliev and A. Gaivoronski: *Simultaneous Nonstationary Optimization, Estimation and Approximation Procedures* (International Institute for Applied Systems Analysis, Collaborative Paper CP-82-16, Laxenbrg, Austria)

2. Y. Ermoliev: *Stochastic Quasigradient Methods and their Application in Systems Optimization* (International Institute for Applied Systems Analysis, Working Paper WP-81-2, Laxenburg, Austria)

3. L. Nazareth and R. Wets: *Algorithms for Stochastic Programs: The Case of Non-stochastic Tenders* (International Institute for Applied Systems Analysis, Working Paper WP-83-5, Laxenburg, Austria)

4. R.T. Rockafellar and R. Wets: *A Dual Solution Procedure for Quadratic Stochastic Programs with Simple Recourse* (International Institute for Applied Systems Analysis, Collaborative Paper CP-83-17, Laxenburg, Austria)

5. Y. Ermoliev and C. Nedeva: *Stochastic optimization Problems with Partially Known Distribution Functions* (International Institute for Systems Analysis, Collaborative Paper CP-82-60, Laxenburg, Austria)

Relaxed Markov Processes, Jackson Networks and Polymerisation

P. Whittle
Statistical Laboratory, 16 Mill Lane, Cambridge CB2 1SB, United Kingdom

Abstract

The concept of relaxing a Markov process is introduced; this is the creation of additional transitions between ergodic classes of the process in such a way as to conserve the existing equilibrium distribution within ergodic classes. The "open" version of a "closed" model of migration, polymerisation etc. often has this character. As further examples, generalised versions of Jackson networks and networks with clustering nodes are given.

1. Relaxed Markov Processes

Consider a Markov process in continuous time with state variable x taking values in a discrete state space $\mathcal{X}$. We suppose that the transition $x \to x'$ has probability intensity $\Lambda(x,x')$, so that the balance equations for the equilibrium distribution $P(x)$ are

$$\sum_{x' \in \mathcal{X}} [P(x')\Lambda(x',x) - P(x)\Lambda(x,x')] = 0 \quad . \tag{1}$$

We suppose all states of the process positive recurrent so that an equilibrium distribution indeed exists. We also suppose the process reducible, so that, under $\Lambda(\cdot,\cdot)$, $\mathcal{X}$ decomposes into irreducible sub-spaces $\mathcal{X}(b)$, the ergodic classes of the process. If $b(x)$ is the label b of the ergodic class within which x lies, then $b(x)$ is the maximal invariant of the motion. That is, it is a function of x which cannot change under any possible transition, and any such invariant is a function of $b(x)$. Let $\mathcal{B}$ be the set of values of b.

In this reducible case the full balance equation (1) reduces to a number of partial balance equations

$$\sum_{x' \in \mathcal{X}(b')} [P(x')\Lambda(x',x) - P(x)\Lambda(x,x')] = 0 \quad (b' \in \mathcal{B}) \tag{2}$$

Let $P(x) = \Phi(x)$ be any strictly positive solution of (1) (or (2)). Then the general solution is of the form

$$P(x) \propto f(b(x))\Phi(x) \tag{3}$$

for arbitrary $f(\cdot)$. This can be regarded as a solution corresponding to a general prior distribution over ergodic classes, the form of $f(\cdot)$ being related to this distribution. Alternatively, it can be regarded as the equilibrium distribution induced if suitable additional mechanisms are introduced which permit transition between different values of b , and so open communication between what were formerly the ergodic classes.

Under such a modification b will no longer be an invariant, but it will have a particular role; let us call it the <u>count</u>. The term is appropriate since in many cases $b(x)$ is a measure of the numbers of some kinds of basic units whose abundances were prescribed in the initial version of the process. Correspondingly, the former ergodic classes will be termed <u>count classes</u>. Transitions within and between classes (i.e. which are count-conserving or count-mutating) will be termed <u>internal</u> and <u>external</u> transitions respectively.

We shall say that the effect of introducing such external transitions is to <u>relax</u> the process, and shall say that the resultant Markov process is a <u>relaxed Markov process</u> if its equilibrium distribution is of the form (3). That is, if the previous equilibrium distribution within count-classes is conserved. This requirement places constraints on the pattern of relaxation. However, because the Markov property and the equilibrium distribution are conserved under time-reversal we can state immediately:

<u>Theorem 1</u> <u>The time-reversed version of a relaxed Markov process is a relaxed Markov process</u>.

Also, since the relaxed process has equilibrium distribution of the form (3) one can state

<u>Theorem 2</u> <u>A relaxed Markov process satisfies the partial balance equations (2)</u>.

Let us now deduce sufficient conditions that a modified transition intensity $\Lambda(x,x')$ be consistent with the relaxed Markov property. We assume internal transition intensities unmodified.

Define the quantity

$$\theta(x,x') = \Phi(x)\Lambda(x,x') \tag{4}$$

for the unrelaxed process. The partial balance equations (2) then become a kind of quasi-symmetry condition for θ

$$\sum_{x' \in \mathcal{X}(b')} \theta(x',x) = \sum_{x' \in \mathcal{X}(b')} \theta(x,x') \tag{5}$$

which holds for all x, b', since both components of the sum will be zero if x, x' are in different count-classes.

Let us write (5) as

$$\theta(b',x) = \theta(x,b') \tag{5'}$$

so implicitly defining these quantities as the respective sums in (5).

With this preparation we proceed to

Theorem 3 Suppose external transitions are introduced so that the transition intensity has the form

$$\Lambda(x,x') = \frac{\lambda(b(x),b(x'))\theta(x,x')}{\Phi(x)} \tag{6}$$

where $\theta(x,x')$ is non-negative, satisfies the quasi-symmetry condition (5) for all x, b', and is such that both sums in (5) are independent of b' for $b' \neq b(x)$.

Then the modified process is relaxed Markov with possible equilibrium distribution (3), where $f(\cdot)$ satisfies the balance equations

$$\sum_{b'} [f(b')\lambda(b',b) - f(b)\lambda(b,b')] = 0 \tag{7}$$

Proof Note that, in view of (4), (5), the conditions of the theorem are consistent with the already-specified internal transition intensities. For x, x' in the same count-class the $\theta(x,x')$ of (6) differs from that of (5) only by the immaterial factor $\lambda(b(x),b(x))$.

Since expression (3) satisfies the internal balance equations (2) the internal transitions cancel out in the balance equation (1), leaving only external transitions. This external balance equation can be written

$$\sum_{b' \neq b(x)} [f(b')\lambda(b',b)\theta(b',x) - f(b)\lambda(b,b')\theta(x,b')] = 0 \tag{8}$$

The conditions of the theorem now ensure that (8) is consistent with equation (7) for $f(\cdot)$ □

The extra conditions on $\theta(x,x')$ of the theorem can be written

$$\theta(b',x) = \theta(x,b') = \theta(x) , \quad (b' \neq b(x)) \tag{9}$$

say. Since $\theta(x)$ is non-negative then it is either positive or zero; i.e. any given state x either communicates directly (and in both directions) with all count-classes outside its own that $\lambda(\cdot,\cdot)$ permits or with none. We may assume, as a matter of normalisation, that, for any given b, the quantity $\theta(x)$ is positive for some x in $\mathfrak{X}(b)$, otherwise one might as well assume $\lambda(\cdot,\cdot)$ such as to isolate $\mathfrak{X}(b)$ from the other count-classes. With this normalisation, equation (8) is not only consistent with (7), but implies it.

Let us term the process with state variable b and transition intensity $\lambda(b,b')$ the relaxing process.

Corollary 1 The relaxed process constructed as in Theorem 3 is irreducible if and only if the relaxing process is irreducible.

Proof In virtue of the normalisation of $\theta(\cdot,\cdot)$ the count-classes will communicate under intensity (6) if and only if the relaxing process is irreducible. □

Corollary 2 The time-reversed version of the relaxed process constructed in Theorem 3 has transition intensity

$$\Lambda^*(x,x') = \frac{\lambda^*(b(x),b(x'))\theta(x',x)}{\Phi(x)} \tag{10}$$

where

$$\Lambda^*(b,b') = \frac{f(b') \quad (b',b)}{f(b)}$$

It thus has the same structure as the relaxed process of Theorem 3, and is reversible if $\lambda=\lambda^*$ and $\theta(x',x) = \theta(x,x')$.

Expression (10) follows immediately from the general evaluation

$$\Lambda^*(x,x') = \frac{P(x')\Lambda(x',x)}{P(x)}$$

and (3), (6). □

Corollary 2 does not follow from Theorem 1 directly, as we have not proved that the conditions of Theorem 3 are necessary for the relaxed Markov property.

The general implications of the conditions on $\theta(\cdot,\cdot)$ are not very clear. They can always be satisfied, however, by the following simple prescription.

Theorem 4 Relaxed Markov structure can be achieved for any Markov process by choosing a single state $x(b)$ in each count-class $\mathfrak{X}(b)$ and allowing transitions into and out of $\mathfrak{X}(b)$ only through $x(b)$.

More specifically, term the states $x(b)$ port states. Then the transition intensity is chosen to have the form (6) with $\theta(x,x')$ for x, x' in distinct count classes prescribed as unity if x and x' are both port states, zero otherwise.

<u>Proof</u> It is clear that $\theta(\cdot,\cdot)$ as thus prescribed satisfies the conditions of Theorem 3, with $\theta(x)$ equal to 1 or 0 according as x is a port state or not. □

The function $\Phi(x)$ is arbitrarily scaled to some extent, since it is determined only to within an arbitrary positive factor $h(b(x))$. In statistical-mechanical contexts (the canonical ensemble: a system in a heat bath of temperature T, closed to all exchanges other than energy exchanges) it would have the form:

$$\Phi(x) = \omega(x)e^{-V(x)/kT} .$$

Here $V(x)$ is the potential energy of configuration x, and $\omega(x)$ is a combinatorial term, reflecting the number of configurations adequately summarised by x, in that they are not distinguished on the grounds of energy or other relevant factors; k is Boltzmann's constant. We shall term such a factor a <u>Gibbs factor</u>.

In such cases $b(x)$ will represent something like the total number of atoms in the closed system. The term $V(x)$ is defined for all x, and represents something like the interactive energy of configuration x. To multiply $\Phi(x)$ by $f(b(x))$ is to concede the possibility that the closed system can be opened, and to assign a potential ("chemical potential") to the additional particles in virtue of their mere presence. Since this chemical potential expresses no interaction, $f(b(x))$ is restricted in form. Furthermore, in the statistical-mechanical context one will demand reversibility, which means that only certain "openings" of the system are natural. However, in our more general case of a relaxed Markov process, with no mention of reversibility, there are no such constraints. If there is a relaxing of the process which preserves $\Phi(x)$, then there is a relaxing that realises any $f(b(x))$. See further the remarks at the end of sections 3 and 4.

2. Insensitivity of a relaxed process

By "insensitivity" is meant that an equilibrium distribution is not dependent upon the Markov property: this is a familiar phenomenon in semi-Markov processes, network theory, etc. A relaxed process possesses the following insensitivity property; for a proof, see Whittle (1983).

<u>Theorem 5 Suppose the conditions of Theorem 4 satisfied. Then the equilibrium distribution (3) persists if the Markov relaxing process is replaced by a semi-Markov process with the same transition probab-</u>

ilities and expected sojourn times in count-classes, provided that this sojourn time is worked off at the x-dependent rate $\theta(x)/\Phi(x)$.

3. Generalised Jackson networks

The nature and properties of a Jackson network are very well-known in the literature on networks of queues, having been discovered initially by Jackson (1963), repeatedly rediscovered since (e.g. Whittle, 1967, 1968; Gordon and Newell, 1967; Spitzer, 1971, partially) and generalised, notably by Kelly (1979 and references therein).

We now indicate a further generalisation, which exhibits the network as a relaxed Markov process.

Consider a Markov system with state variable x , which prescribes among other things the numbers of units n_j at m nodes $(j=1,2,\ldots,m)$ Let e_j denote a unit at node j , and suppose that units can enter and leave nodes in only one way, so that the states $x \pm e_j$ are well-defined, and $x-e_j+e_j = x+e_j-e_j = x$. (c.f. the "port state" assumptions of Theorem 4). Let the internal transitions $x \to x'$ be those which do not change $n=(n_1,n_2,\ldots,n_m)$ and let $\Phi(x)$ be a strictly positive solution of the balance equations (1) of the process in which only internal transitions are permitted. Assume the external transition rates

$$\Lambda(x,x+e_j) = \nu_j \tag{11}$$

$$\Lambda(x,x-e_j) = \mu_j \frac{\Phi(x-e_j)}{\Phi(x)} \tag{12}$$

$$\Lambda(x,x-e_j+e_k) = \lambda_{jk} \frac{\Phi(x-e_j)}{\Phi(x)} \tag{13}$$

Theorem 6 The balance equations for the process described above are solved by

$$P(x) \propto \Phi(x) \prod_j w_j^{n_j(x)} \tag{14}$$

where the w_j are the solutions of

$$\nu_j - \mu_j w_j + \sum_k (w_k \lambda_{kj} - w_j \lambda_{jk}) = 0 \qquad (j=1,2,\ldots,m) \tag{15}$$

and so satisfy also

$$\sum_j (\nu_j - \mu_j w_j) = 0 \tag{16}$$

Solution (14) satisfies also the partial balance equations

$$P(x-e_j)\Lambda(x-e_j,x)-P(x)\Lambda(x,x-e_j)$$
$$+\sum_k[P(x-e_j+e_k)\Lambda(x-e_j+e_k,x) - P(x)\Lambda(x,x-e_j+e_k)] = 0$$
$$(j=1,2,\ldots,m) \qquad (17)$$

$$\sum_k[P(x+e_k)\Lambda(x+e_k,x) - P(x)\Lambda(x,x+e_k)] = 0 \qquad (18)$$

$$\sum_{x'\in\mathcal{E}(x)}[P(x')\Lambda(x',x) - P(x)\Lambda(x,x')] = 0 \qquad (19)$$

where $\mathcal{E}(x)$ is the set of states x' for which $n(x') = n(x)$.

Verification is immediate, and the general form of results is familiar. Distribution (14) will be the unique and proper distribution if the process is irreducible and positive recurrent. Kelly (1975 and 1979 p. 199) has given the result in the case when there are no internal transitions and one can make an identification $\Phi(x) = \Phi(n)$. The novelty of the present case is that $\Phi(x)$ is generated (to within an $n(x)$-dependent factor) as the solution of the internal balance equations (19).

The process is obviously a relaxed version of the process in which $n(x)$ is frozen - immigration, emigration and migration are forbidden, and only internal transitions are possible.

The assumptions behind the transition intensities (11)-(13) seem the weakest possible, which lead to a result of form (14): that the effect of opening communication between sites is to modify the "internal" distribution $\Phi(x)$ by the "external" factor $\prod_j w_j^{n_j(x)}$. In a statistical-mechanical context, with $\Phi(x)$ the Gibbs factor for the interactive energy of configuration x , then the quantity $u_j = -kT\log w_j$ is effectively the chemical potential of a unit at site j . The assumptions (11)-(13) seem to be the weakest which ensure that constant (x-independent) unit potentials exist.

An economist would see the w_j in terms of prices, w_j being related to the value of a unit when at node j . Prices need exist no more than potentials need exist, but, when they do, they are the mediating agents which allow systems to interact in equilibrium by adjustment of parameters rather than by fundamental adjustment of internal structure.

Although $u_j = -kT\log w_j$ is effectively an equilibrium potential, it should be emphasised that movement between the nodes is not induced by a potential. The nature of the migration process induced by transitions (11)-(13) is in general irreversible, and incompatible with a potential field; see the remarks at the end of section 4.

4. Networks with clustering nodes

The treatment of section 3 extends easily to the case when there are units of several types. A unit may even change its type while moving along an arc of the network - as Kelly has demonstrated (1979), this is a useful way of "programming" a unit so that it follows a planned route through the net.

To allow a unit to change type at a node is more difficult, although possible in some cases. A more extreme case of such behaviour is to allow the possibility that units are "molecules" made up of several elements, possibly mutable. That is, one allows clustering and dissocation of units.

The author has studied clustering ("molecular") statistics extensively in the time-reversible case (Whittle, 1965, 1972, 1977, 1980, 1981) and Kelly (1982) has studied some special cases of networks with clustering nodes. In this section we sketch the theory for a moderately general case.

Let a unit (molecule) of type r at node j be denoted e_{jr}, and let us suppose, as in section 3, that the mode of entry is unique and is the reverse of the mode of exit, so that $x \pm e_{jr}$ are well-defined. The number of e_{jr} is denoted n_{jr}; the overall state variable x specifies the n_{jr}, among other things.

Suppose that in an internal (within-node) transition the quantities

$$b_{j\alpha} = \sum_r n_{jr} a_{r\alpha} \tag{20}$$

are conserved. These are to be regarded as the abundances of some basic immutable constituents, labelled by α. Thus, if $P(x) \propto \Phi(x)$ solves the internal balance equations (2), so too does

$$\begin{aligned} P(x) &\propto \Phi(x) \prod_{j\alpha}\prod w_{j\alpha}^{b_{j\alpha}} \\ &= \Phi(x) \prod_{jr}\prod \zeta_{jr}^{n_{jr}} \quad \text{where} \end{aligned} \tag{21}$$

$$\zeta_{jr} = \prod_\alpha w_{j\alpha}^{a_{r\alpha}} \tag{22}$$

Specify now the external transition intensities by

$$\Lambda(x, x+e_{jr}) = \nu_{jr} \tag{23}$$

$$\Lambda(x, x-e_{jr}) = \mu_{jr} \frac{\Phi(x-e_{jr})}{\Phi(x)} \tag{24}$$

$$\Lambda(x, x-e_{jr}+e_{ks}) = \lambda_{jk}^{(rs)} \frac{\Phi(x-e_{jr})}{\Phi(x)} \tag{25}$$

The last transition is thus a migration/mutation; a molecule of type r leaves node j and arrives at node k as a molecule of type s.

Theorem 7 Suppose that the equations in the w_j

$$\nu_{jr} - \mu_{jr}\zeta_{jr} + \sum_{ks}\sum (\zeta_{ks}\lambda_{kj}^{(sr)} - \zeta_{jr}\lambda_{jk}^{(rs)}) = 0 \tag{26}$$

with the $\ _{jr}$ given by (22) possess solutions. Then the balance equations for the process described above have solutions (21). This solution satisfies partial balance equations analogous to relations (17)-(19).

Verification is immediate. If there are p invariants in an internal transition and q types of molecule are mobile (i.e. allowed to move in the network) then (26) constitutes mq equations in mp unknowns. If $p>q$ then there are invariants even under external transitions. If $p<q$ then the rates λ, μ, ν will have to obey certain consistency conditions if equations (22), (26) are to have a solution.

There is nothing in the specification of the model which requires it to be reversible, or to show detailed balance in either internal or external transitions. The irreversible case is indeed what Prigogine and his school (see e.g. Prigogine, 1962; Balescu, 1975; Lavenda, 1978; Kreuzer, 1981) refer to as the "irreversible" or "non-equilibrium" case. The term "non-equilibrium" is unfortunate, since a stochastic equilibrium will indeed exist in most cases; what is meant is that it is not a Gibbs equilibrium, and the stochastic dynamics are not induced by the existence of a potential. So, although the model will usually attain a perfectly respectable stochastic equilibrium, there will in general be a net transport of matter through or around the system, inconsistent with reversibility.

We see from the theorem that the irreversible cases of this model break into two classes, depending on whether equations (22), (26) have a solution or not. In the first case the equilibrium distribution (21) is much of Gibbs form, with chemical potential $-kT\log w_{j\alpha}$ attributable to a basic unit of type α at node j; this even though the dynamics of the process are not reversible. In the second case there does not seem to be much one can say about the form of the equilibrium distribution.

The condition that equations (22), (26) have a solution could be regarded as demanding something like balance for the components of the vectors $n_j = (n_{j\alpha})$ which are not conserved in internal transitions.

References

Balescu, R. Equilibrium and nonequilibrium statistical mechanics (Wiley, 1975)

Gordon, W.J. and Newell, G.F.: Operations Research, 15, 254-265 (1967)

Jackson, J.R.: Operations Research, 5, 518-521 (1957)

Jackson, J.R.: Management Science, 10, 131-142 (1963)

Kelly, F.P.: Bull. Int. Inst. Statist. 46, 397-404 (1975)

Kelly, F.P.: Adv. Appl. Prob. 8, 416-432 (1976)

Kelly, F.P. Reversibility and stochastic networks (Wiley 1979)

Kelly, F.P.: "Networks of quasi-reversible nodes", in Applied Probability-Computer Science, the Interface: Proceedings of the ORSA-TIMS Boca Raton Symposium (ed. R. Disney) (Birkhauser Boston, Cambridge, Mass. 1982)

Kreuzer, H.J.: Nonequilibrium thermodynamics and its statistical foundations (Oxford 1981)

Lavenda, B.H.: Thermodynamics of irreversible processes (MacMillan 1978)

Muntz, R.R.: "Poisson departure processes and queueing networks", in IBM Research Report RC 4145 (IBM Thomas J. Watson Research Center, Yorktown Heights, New York 1972)

Prigogine, I.: Nonequilibrium statistical mechanics (Wiley Interscience 1962)

Spitzer, F.: Random fields and interacting particle systems (Mathematical Association of America 1971)

Whittle, P.: Proc. Camb. Phil. Soc. 61, 475-495 (1965a)

Whittle, P.: Proc. Roy. Soc. Ser. A 285, 501-519 (1965b)

Whittle, P.: Bull. Int. Inst. Statist. 42, 642-647 (1967)

Whittle, P.: J. Appl. Prob. 5, 567-571 (1968)

Whittle, P.: "Statistics and critical points of polymerisation processes", in Proceedings of Symposium on Statistical and Probabilistic Problems in Metallurgy (supplement to Adv. Appl. Prob.) 1972 pp. 199-215

Whittle, P.: "Co-operative effects in assemblies of stochastic automata", in Proceedings of Symposium to honour Jerzy Neyman, pp. 335-343 (Polish Scientific Publishers, Warsaw 1977)

Whittle, P.: Adv. Appl. Prob. 12, 94-115 (1980a)
Whittle, P.: Adv. Appl. Prob. 12, 116-134 (1980b)
Whittle, P.: Adv. Appl. Prob. 12, 135-153 (1980c)
Whittle, P.: Teoriya Veryatnostei 26, 350-361 (1981)
Whittle, P.: J. Appl. Prob. (to appear) (1983)

Part II

Chaotic Dynamics-Theory

Noodle-Map Chaos: A Simple Example

O.E. Rössler: Institute for Physical and Theoretical Chemistry, University of Tübingen, D-7400 Tübingen, Fed. Rep. of Germany

J.L. Hudson: Department of Chemical Engineering, Thornton Hall, University of Virginia, Charlottesville, VA 22901, USA

J.D. Farmer: Center of Nonlinear Studies, Los Alamos National Laboratory Los Alamos, NM 87545, USA

Chaos-generating folded 2-dimensional maps can be generalized to higher dimensions in two ways: as folded-towel (or pancake) maps and as bent-walking-stick (or noodle) maps. The noodle case is of mathematical interest because the topologically one-dimensional attractors involved may, despite their thinness, be of the "non-sink" type (that is, stand in a bijective relation to their domain of attraction). Moreover, Shtern recently showed that the well-known Kaplan-Yorke conjecture on the fractal dimensionality of chaotic attractors may fail in the case of noodle maps. We present here an explicit 3-variable noodle map with constant divergence (constant Jacobian determinant). The example is a higher analogue to the Hênon diffeomorphism. A map of similar shape was recently found experimentally by Rob Shaw in a study of the irregularly dripping faucet.

1. Introduction

Chaos is coming en vogue, and along with it is a whole new science, "Nonlinear Science." What, then, does chaos mean? Taken at face value, it refers to the somewhat counterintuitive mixing behavior that is possible when everywhere locally parallel hair lines (trajectories) are bent into a recurrent bundle in 3-space. A taffy puller placed onto a synchronously rotating lazy susan provides the simplest example [1].

When looked at from a somewhat greater distance, chaos can be said to reflect a mathematical discovery made by Peano, Cantor, and Brouwer around the turn of the century: that bijections across dimensions are possible, although only at the price of discontinuity (see [2]). The stepwise weaving in time of such a discontinuous, yet lower-dimensional representation of an originally higher-dimensional set (cross section) is precisely what is being observed in chaos. A deterministic trajectory keeps shuttling back and forth busily until, in the limit, any two originally arbitrarily close points of a transversal square arrive at a completely different place along a transversal line, in the simplest case.

This is tantamount to saying that a new class of attractors exists. These attractors are, like their regular cousins (the point attractor, the closed-line attractor, the toroidal attractor, etc.) of zero measure in state space; for otherwise they would not be attractors. On the other hand, they "care." Unlike their cousins, they are not objects that sit there in splendid isolation for an infinite time while everything around them is crushed into disappearance. Not being invariant sets ("sinks") themselves, these "non-sink" attractors [2] never sever their ties with their higher-dimensional constituency (their domain of attraction). So does only a zero-measure subset (called their skeleton).

All of this is not easy to prove since finitistic (ε,δ) methods break down. A sketch how to make use of a commuting functor was given in ref. [2]. Basically, it was first shown that a certain area-preserving map (which by definition does not produce an attractor) stays bijective in the limit except for a zero-measure subset. Then another map was considered which differed from the former only by the introduction of a little shrinking which turned the full-measure Cantor set, found to apply in the limit of the former map, into an ordinary (zero-measure, and hence attracting) Cantor set. Between this map and the former, a new (arbitrarily defined) bijection was then shown to exist up to the limit. Hence the diagram commuted, meaning that the second map was, in the limit, just as bijective as the first.

What has all this to do with reality? Perhaps nothing since reality may be of the ε,δ type (who knows?). But most recently, a first noodle map was found experimentally. SHAW [3] investigated the dripping faucet [4] experimentally and found not only ordinary chaos (governed by a single-humped map), but also a new type of chaos (cf. [5]) governed by a locally one-dimensional attractor that appeared to cross itself several times in a one-dimensional projection (next-interval plot).

In the following, an explicit example of a simple 3-dimensional diffeomorphic map that may be used to qualitatively represent the new observation will be presented.

2. A Class of Hénon-Type Maps

HENON [6] described the following map:

$$\begin{aligned} x_{n+1} &= f(x_n) - y_n \\ y_{n+1} &= \varepsilon x_n \end{aligned} \tag{1}$$

with $f(x_n) = \alpha x_n(1-x_n)$, the right hand-side of the logistic difference equation, and $\varepsilon < 1$. This map is a global diffeomorphism with constant contraction (the Jacobian determinant equals ε). Taking the right-hand side of (1) literally as a prescription, one sees that a horizontal strip (with vertical stripes) is turned into a bent vertical one (with horizontal stripes):

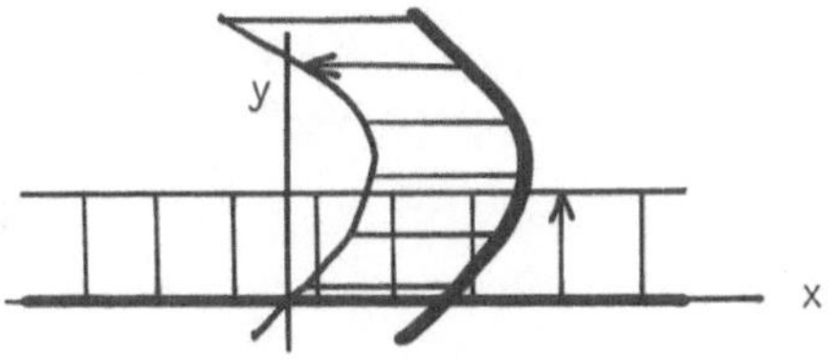

Most noteworthy, there exists for certain ranges of values of α and ε , an attracting rectangular (say) box that contains its own image inside itself:

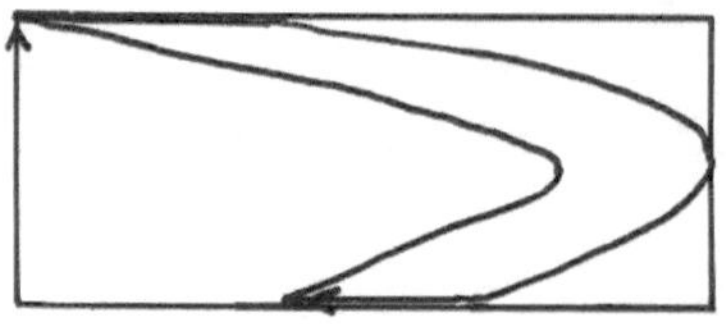

(Take, for example, $\alpha = 3.8$, $\varepsilon = 0.062$ and, for the box, $y = 0.0025$ to 0.0609 and $x = 0.009$ to 0.949.)

This map can be said to describe how the baker rolls out dough (with a loss of volume, due to the squeezing out of some shortening) and puts it then back into the original form (with some new dough added as a fill-in), in order to repeat the whole process over and over, when making fluffy pastry (cf. [7, 1]).

The only unphysical part is the exact horizontality of originally vertical stripes implied in (1) - there is only a slight sliding between adjacent layers allowed, but no bending. This drawback is immediately amended, however, by going to the second iterate. In it, no trace of a preferred parallel orientation is left. The price: the second iterate is governed by an equation that contains two quadratic terms on its right-hand side rather than just one.

Instead of going to the second iterate, one may therefore introduce an additional quadratic term directly into (1) already. This yields, for example,

$$\begin{aligned} x_{n+1} &= \alpha x_n(1 - \alpha_n) - y_n \\ y_{n+1} &= (\varepsilon - \delta y_n)(x_n - 1/2) \ . \end{aligned} \tag{2}$$

This map becomes the former map as δ approaches zero (except for a linear translation by $\varepsilon/2$ along x and $-y$). Conversely, as the parameter ε is made approach zero in (2), another limiting map is obtained. In it, x_{n+1} no longer depends on y_n (so that only the logistic one-dimensional map remains for the first variable), while y_{n+1} still depends on x_n. (This is because y in (2) can be replaced by a new variable, $\varepsilon\tilde{y}$, which obviously disappears from the first line as $\varepsilon \to 0$ while at the same time the second line becomes $\tilde{y}_{n+1} = (1 - \delta\tilde{y}_n)\cdot(x_n - 1/2)$, independent of ε .) The resulting picture is

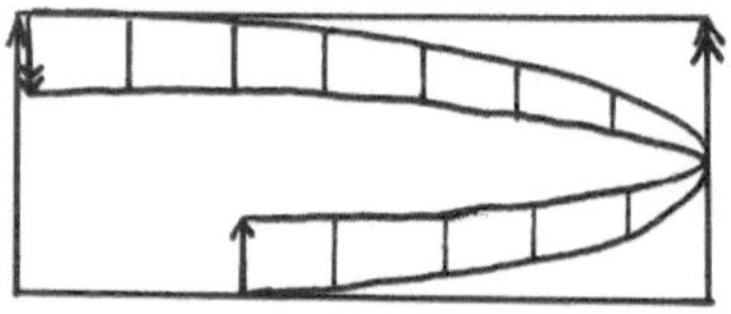

This time, vertical stripes are mapped onto vertical stripes, yet with a change of orientation occurring at $x = 1/2$ where (as the only non-uniqueness) a vertical line is mapped onto a point. (Take, for example, $\alpha = 3.8$, $\delta = 0.8$ and $\varepsilon = 0$; an appropriate box then goes from $\tilde{y} = -0.0294$ to $\tilde{y} = 0.0306$ and from $x = 0.071$ to $x = 0.949$.)

If now a small nonzero value of ε is introduced, a combination of the preceding two "box pictures" applies, yielding

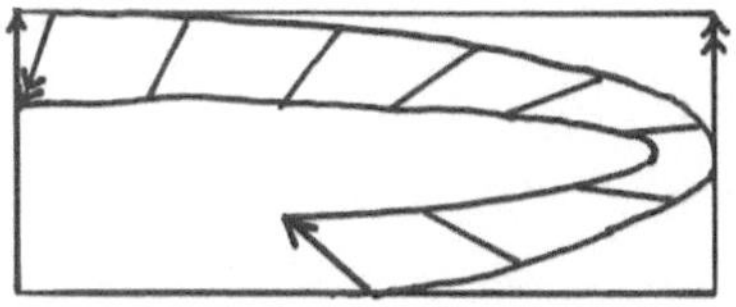

(Take, for example, $\alpha = 3.8$, $\delta = 0.8$, and $\varepsilon = 0.04$; a box is then given by $y = -0.0294$ to $y = 0.0306$ and $x = 0.04$ to $x = 0.98$.)

The disadvantage of map (2) is that it no longer has a constant local shrinking factor (Jacobian determinant) independent of the original position in the dough, and that it no longer is a global diffeomorphism. (When the parameter δ is made sufficiently large, the two legs of the inner lining of the image may cross each other.)

An advantage of (2) is that the underlying principle can more readily be generalized to higher dimensions. A "folded towel" diffeomorphism that makes use of the same folding principle, applied twice, is

$$
\begin{aligned}
x_{n+1} &= \alpha x_n(1 - x_n) + (\varepsilon + \delta y_n)(z_n - 1/2) \\
y_{n+1} &= [\varepsilon' + \delta'(\varepsilon + \delta y_n)(z_n - 1/2)](x_n - 1/1.9) \qquad (3) \\
z_{n+1} &= \alpha' z_n(1 - z_n) + \beta y_n .
\end{aligned}
$$

An appropriate set of parameters is $\alpha = 3.8$, $\beta = 0.2$, $\delta = 0.05$, $\varepsilon = 0.0175$, $\alpha' = 3.78$, $\delta' = \varepsilon' = 0.1$. (See [8] for an appropriate initial box as well as a stereo picture.) Possibly, one of the two δ's contained in (3) can be eliminated again.

In other words, it seems as if a whole hierarchy of "folded hypertowel" diffeomorphisms could be generated on the basis of (2). This is one possible avenue to generalize Hénon's map to higher dimensions - by introducing more and more directions of lateral expansion (each with a positive Lyapunov characteristic exponent of its own).

3. A Noodle Map

Unlike folded-towel maps, folded-noodle maps possess only one direction of lateral expansion. They nonetheless in a special case (volume preservation) can be equivalent (identical) to the former under time reversal [2]. The simplest non-trivial shape (with at least 3 non-complanar "legs") looks like a letter S in one projection and like a letter U in another.

Finding an S-shaped submap is not difficult: take (1) with $f(x_n) = -\alpha x_n(1 - x_n^2)$; that is,

$$\begin{aligned} x_{n+1} &= -\alpha x_n(1 - x_n^2) - y_n \\ y_{n+1} &= \varepsilon x_n \, . \end{aligned} \tag{4}$$

The inset made up of horizontal stripes (see the drawing following (1) above) is now based on a cubic () rather than parabolic () underlying curve.

As a second ingredient, a U-shaped submap (x,z) that shares x with the above map (x,y) has to be found. The following map does yield a U- ( - or rather, if α is not small,  -) shaped, locally diffeomorphic image:

$$\begin{aligned} x_{n+1} &= -\alpha x_n(1 - x_n^2) - \beta x_n z_n \\ z_{n+1} &= (\delta z_n + \varepsilon)(x_n^2 - 1/3) \, . \end{aligned} \tag{5}$$

(Take, for example, $\alpha = 2.55$, $\beta = 3$, $\delta = 0.1$, $\varepsilon = 0.05$; an appropriate box then goes from $z = -0.02$ to 0.042 and from $x = -1.05$ to 1.05 .)

Combining these two maps ((4) and (5)) does not work out, however. Two diffeomorphic projections combined do not necessarily make a diffeomorphism. In a non-planar case like the present one, there must be a folded over (endomorphic) region at the right place somewhere in one of the two maps at least. Therefore, a simpler (in the central region self-overlapping) submap offers itself in place of (5):

$$\begin{aligned} x_{n+1} &= -\alpha x_n(1 - x_n^2) - z_n \\ z_{n+1} &= \varepsilon(x_n^2 - 1/3) \, . \end{aligned} \tag{6}$$

It yields a  -shaped picture. Combining it with (4) also does not work as expected, however: The Jacobian determinant of the combined system is zero for an internal slice ($x = 0$).

Then the following minor modification (addition of the δ-term in the third line) turned out to work fine:

$$x_{n+1} = -\alpha x_n(1 - x_n^2) - y_n - z_n$$
$$y_{n+1} = \varepsilon x_n \qquad (7)$$
$$z_{n+1} = \varepsilon'(x_n^2 - 0.33) + \delta z_n \; .$$

Figure 1 below gives a set of pictures along with the measures of an appropriate box. The first iterate always sticks its "legs" out of any rectangular box when α is not small. For the present set of parameters, this does not matter, however, since a subset of the box stays within itself under iteration (as the upper pictures in Fig.1 are proof of).

Equation (7) is a constant-contraction diffeomorphism for ε and δ sufficiently small. (The Jacobian determinant equals $\varepsilon\delta$.) The equation in this respect closely resembles Hénon's map. Presumably, no simpler nontrivial 3-dimensional noodle map than (7) can be found. Also, it is safe to predict that its only "unnatural" property, from the point of view of dough processing, (namely, that z=constant slices in the pre-image always yield y=constant slices in the image), will disappear under iteration in the same way as this was the case above with an analogous property in (1).

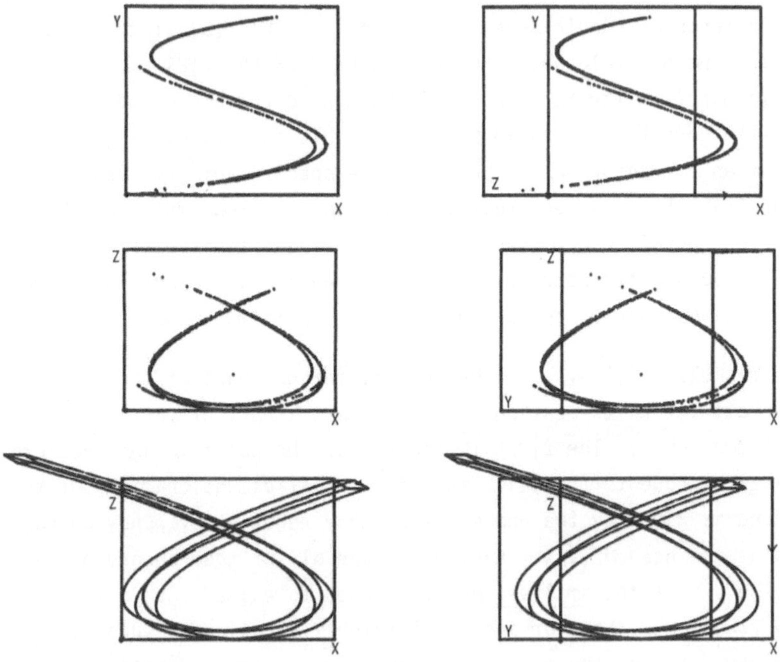

Fig.1 Stereoscopic pictures of the "noodle map" (7). An initial rectangular box (with $x = -1.3$ to 1.2, $y = -0.12$ to 0.12, $z = -0.035$ to 0.119) is shown in 3 stereo pairs (parallel projections). In the first two pairs, an initial point ($x_1 = y_1 = z_1 = 0$) is followed through 1,600 iterations. In the bottom pair, the first iterate of the frame of the box is shown instead. Parameters chosen: $\alpha = 2.726$, $\varepsilon = \varepsilon' = 0.1$, $\delta = 0.05$. Numerical calculation performed on a HP 9845B desk-top computer with peripherals

5. Discussion

So far, no flows (ordinary differential equations) possessing a noodle map as a cross section are known. Still, it should not be difficult to obtain such equations. (One, based upon two single-variable, letter-Z shaped slow manifolds coupled to an expanding two-variable linear subsystem, is in preparation.) As a rule, O.D.E.s producing a particular type of chaos are easier to obtain than difference equations of invertible type doing the same thing. (A folded-towel map, for example, needs only a single quadratic term for being generated in a 4-variable O.D.E. [8].)

It is not known whether noodle and pancake maps (as well as their combinations) actually exhaust the "major" possibilities of doing repetitive operations on dough (in such a way that the result is an orientation-preserving diffeomorphism that can be obtained directly without an intervening cutting and re-pasting process). There might exist another first-magnitude case in 3D maps that has been completely overlooked so far (cf. [1]). Further observation of chaotic systems of natural type (like idling [9] - and non-idling [10] - automobile motors, to mention only one readily accessible class of realistic systems aside from chemical ones [11]) can be counted on settling this question in the long run.

As far as the importance of noodle maps is concerned, SHAW's [3] finding has already been mentioned. Another relevant finding is due to SHTERN [12]. He showed that a certain noodle map (similar in shape to those considered and named in [2]) does not obey the well-known KAPLAN and YORKE [13] formula for calculating the fractal dimensionality of an attractor on the basis of the spectrum of its Lyapunov characteristic exponents. Even though fractal dimensionality (and its generalizations, see [14]) is in general insufficient to characterize a topologically higher than two-dimensional chaotic attractor [2], it is apparently sufficient in the case of noodle maps. This gives noodle maps an added theoretical importance.

The Kaplan-Yorke formula [13] says, concerning the integer part of the fractal dimensionality of an attractor (to mention only this most important part), that it is equal to a sum of two terms. The first is the sum of the positive and the zero Lyapunov characteristic exponents (that is, the topological dimensionality) of the attractor. The second is given by the number of so many negative Lyapunov characteristic exponents as are needed (after having ordered all of them according to size) in order for the sum of the smallest negative ones to exceed the former (positive) sum. SHTERN [12] found that for a special (highly idealized - such as to be factorizable) noodle map - to use the present terminology - , fewer negative terms suffice because one of the smallest negative terms is ineffectual and has to be skipped. It will now be interesting to check (with the methods described in [14] and [15], for example) how more realistic noodle maps like (7) fare in the same respect.

To conclude, the class of simple diffeomorphic maps producing relevant types of chaos is apparently still expanding.

Acknowledgments

O.E.R. would like to thank the Center for Nonlinear Studies, University of California, Los Alamos National Laboratory, Los Alamos, New Mexico, for a two-months visiting appointment. J.L.H. acknowledges support by the Fulbright Commission and the National Science Foundation, grant CPE 80.21950.

References

1. O.E. Rossler: "The Chaotic Hierarchy", Z. Naturforsch. 38 a (1983) in press
2. O.E. Rossler: "Chaos and Bijections Across Dimensions", in New Approaches to Nonlinear Problems in Dynamics, P. Holmes. Ed., pp. 477-486 (SIAM, Philadelphia 1980)
3. R. Shaw: "The Dripping Tap" (Preprint 1983)
4. O.E. Rossler: "Chemical Turbulence - A Synopsis", in Synergetics - A Workshop, H. Haken, Ed., pp. 174-183 (Springer-Verlag, New York 1977)
5. O.E. Rossler: "Continuous Chaos", ibidem, pp. 184-199
6. M. Hénon: "A Two-Dimensional Mapping with a Strange Attractor", Commun. Math. Phys. 50, 69-78 (1976)
7. E. Hopf: Ergodentheorie (Ergodic Theory), p. 42 (Springer, Berlin 1937)
8. O.E. Rossler: "An Equation for Hyperchaos", Phys. Lett. 51 A, 155-157 (1979)
9. O.E. Rossler: "Different Types of Chaos in Two Simple Differential Equations", Z. Naturforsch. 31 a, 1664-1670 (1976)
10. J. Kantor (Personal Communication 1983)
11. J.L. Hudson and O.E. Rossler: "Chaos and Complex Oscillations in Stirred Chemical Reactors", in Dynamics of Nonlinear Systems, V. Hlavacek, Ed. (Gordon and Breach, New York 1983) in press
12. V.I. Shtern: "Arrangement and Dimension of Turbulent-Motion Attractors" (Preprint Novosibirsk 1982)
13. J.L. Kaplan and J.A. Yorke: "Chaotic Behavior of Multidimensional Difference Equations", Springer Lect. Notes Math. 730, 204-227 (1979)
14. J.D. Farmer: "Information Dimension and the Probabilistic Structure of Chaos", Z. Naturforsch. 37 a, 1304-1325 (1982)
15. J.D. Farmer, E. Ott and J.A. Yorke: "The Dimension of Chaotic Attractors", Physica D (1983) in press

A Mechanism for Spurious Solutions of Nonlinear Boundary Value Problems

H.-O. Peitgen *

Forschungsschwerpunkt "Dynamische Systeme", Universität Bremen
D-2800 Bremen 33, Fed Rep. of Germany

Abstract

Typically numerical approximation schemes for many nonlinear boundary value problems generate spurious solutions. By developing a dynamical system approach a mechanism is presented which is able to explain a certain class of such solutions. In essence the spurious solutions here are a consequence of structural changes such as bifurcations in the homoclinic structure of the associated dynamical system.

1. The Phenomenon of Spurious Solutions and Dynamical Systems

In recent years (see [Al, Bo, BD, BL, P1, PS, PSS, SA]) it has been observed that the discretization of nonlinear elliptic boundary value problems sometimes possess solutions (spurious solutions) which do not represent approximations to actual solutions of the original problem as the mesh size decreases toward zero. Of course this does not contradict classical truncation error estimates, but merely reflects the fact that the approximating equations have certain singularities which have no analogue in the original problem. Such phenomena arise also in other areas such as, e.g., fluid dynamics [SS] or differential delay equations [NP]. From what is known already it would be inadequate to look for one and only one mechanism. Therefore our goal here is to present one mechanism for one type of spurious solutions for one particular class of problems. To achieve this goal in this short note we will make several simplifications, which allow for a non-technical exposition, but at the same time allow us to understand what kind of singularity is introduced by the process of discretization. Another more practical problem would be the developement of devices which allow to detect or avoid spurious solutions, such as adaptive mesh refinement strategies, embedding methods or a posteriori error bounds.

A typical problem for which spurious solutions have been found [PS] is

$$\begin{cases} \Delta u + \lambda f(u) = o & \text{in } \Omega \subset \mathbb{R}^N \\ u = o & \text{on } \partial\Omega \end{cases}, \qquad (1)$$

*) Research was supported by "Stiftung Volkswagenwerk"

where $\lambda \in \mathbb{R}$, $\Omega \subset \mathbb{R}^N$ is a bounded domain with sufficiently smooth boundary $\partial\Omega$ and $f: \mathbb{R} \to \mathbb{R}$ is a Lipschitz continuous function having several simple zeros (see (3)). Our first simplification here is to investigate equation (1) for $N = 1$, i.e.

$$\begin{cases} u_{tt} + \lambda f(u) = o \\ u(o) = o = u(1) \end{cases} , \tag{2}$$

where $\Omega = (o,1)$. In [PSS, PS] one finds a first attempt to classify spurious solutions of (2), which arise, if the discretization is by means of finite differences, provided f , the nonlinearity, satisfies the following conditions:

$$\begin{cases} f(s) = m_o s + o(s) , & \text{as } s \to o \\ f(s) = m_\infty s + o(s) , & \text{as } |s| \to \infty \\ \{s : f(s) = o\} \neq \emptyset & \\ m_o > o , \; m_\infty > o & \end{cases} \tag{3}$$

In particular three types of spurious solutions are distinguished, one of which is constituted by a bifurcation of asymmetric spurious solutions from geunine symmetric solutions. In general, e.g. for problem (1) symmetry properties of solutions belong to the more sophisticated problems of elliptic equations (cf. [GNN]). For our simplified model problem (2), however, we have the following elementary observation: Let Φ_λ^t , $t \in \mathbb{R}$, denote the phase flow corresponding to the vectorfield

$$X_\lambda(u,v) = (v, - \lambda f(u)) \tag{4}$$

which establishes the elementary dynamical system in $\mathbb{R}^2$:

$$\begin{cases} u_t = v \\ v_t = - \lambda f(u) . \end{cases} \tag{5}$$

Furthermore, let $V = \{(u,v) : u = o\}$. Then, solving (2) is equivalent to

$$\begin{cases} \Phi_\lambda^t(V) \cap V \neq \emptyset \\ t = 1 . \end{cases} \tag{6}$$

In other words, if $V_\lambda^t := \Phi_\lambda^t(V)$, then solving (2) is a problem of understanding intersection properties of smooth 1-manifolds (Φ^t is a diffeomorphism). If we rescale the time in (1) by $s = \sqrt{\lambda}t$ and set $w(s) = u(t)$, then (1) is equivalent to

$$\begin{cases} w_{ss} + f(w) = o \\ w(o) = o = w(\sqrt{\lambda}) . \end{cases} \tag{7}$$

Hence, we obtain immediately a qualitative insight into (6) if we analize the phase portrait of the rescaled vectorfield and look for trajectories, which are governed by $w(o) = o = w(\sqrt{\lambda})$ for large λ. These are complitely understood by the existence of homoclinic orbits (see figure 1):

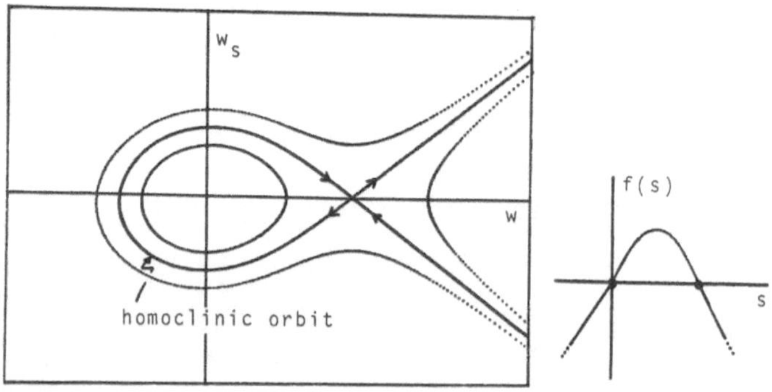

Figure 1.

Furthermore, we let $\Sigma(u,v) = (u,-v)$ be the reflection in the u-axis, then we have that

$$\Sigma \circ X_\lambda \circ \Sigma = - X_\lambda \tag{8}$$

and, hence,

$$\Sigma \circ \Phi_\lambda^t \circ \Sigma = \Phi_\lambda^{-t} \Leftrightarrow (\Sigma \circ \Phi_\lambda^t)^2 = \mathrm{Id} . \tag{9}$$

Thus, we make the elementary but for our further discussion important observation that

$$\Phi_\lambda^t = R_\lambda^t \circ \Sigma \quad , \quad \text{where } R_\lambda^t = \Phi_\lambda^t \circ \Sigma \tag{10}$$

i.e. our phase flow is the product of two involutions. Another immediate consequence is the following symmetry of solutions of (2): Let u be a solution of (2) and let $u(a) = o = u(b)$, where $o < a < b < 1$ and either $u(t) > o$ for $t \in (a,b)$ or $u(t) < o$ for $t \in (a,b)$. Then

$$\left\{ \begin{array}{l} u(m-t) = u(m+t) \\ \text{for } t \in [o,m-a] \ , \ m = (a+b)/2 . \end{array} \right. \tag{11}$$

Thus, in solving (2) we may equivalently analyze (6) and take advantage from (11). We have explained this elementary analysis in such a detail because we will proceed in the same spirit for the discretized problem. In particular we will see, that spurious solutions are due to a singular behaviour of the homoclinic structure as the meshsize varies. Before we proceed to discuss the discretization we introduce

$$\begin{array}{l} N^k = \{f: \mathbb{R} \to \mathbb{R} : f \in C^k , f(o) = o, f(s_o) = o \text{ for some } s_o > o , \\ \qquad f'(o) > o , f'(s_o) < o , f(s) \neq o \text{ for all } s \in (o,s_o)\} , \end{array} \tag{12}$$

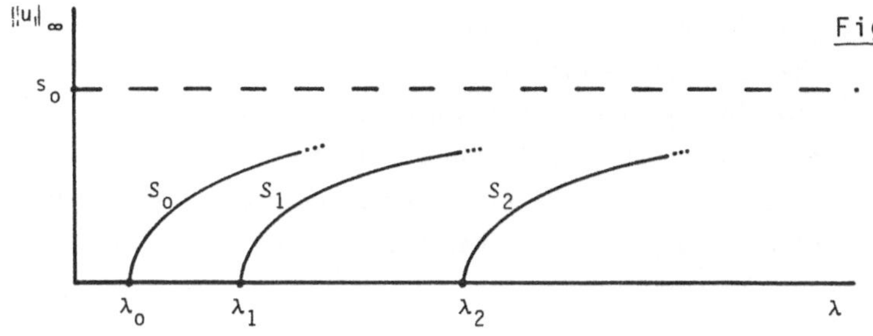

Figure 2.

where $k > 1$. If $f \in N^k$ then the following bifurcation diagram (figure 2) is well known from classical bifurcation theory (see figure 1, cf. [PS]) .
Here $\lambda_n = (n+1)^2\pi^2$, $n = o, 1, 2, \ldots,$ and each branch emanating from (o,λ_i) is constituted by solutions having precisely i internal zeros in $(o,1)$ and being symmetric in the sense of (11).

We will now pass to a simplified discretization and show how the problem of spurious solutions can be embedded into a dynamical systems frame work. For this we choose the central difference approximation

$$\begin{cases} u_{tt}(t_i) = (u_{i+1} - 2u_i + u_{i-1})/h^2 + O(h^2) \\ t_i = i\cdot h \ , \quad h = {}^1/_{n+1} \ , \quad u_i = u(t_i) \ , \end{cases} \tag{13}$$

where the above equality holds, if u is sufficiently smooth. Solving for u_{i+1} and introducing $v_{i+1} = u_i$ constitutes the discrete time dynamical system (compare with Φ_λ^t)

$$T_\mu(u,v) = (2u-v-\mu f(u),u) \tag{14}$$

associated with the second order differential equation in (2), where $\mu = \lambda h^2$.Obviously, any solution $x = (u_1,\ldots,u_n)^T$ of the n-dimensional system

$$Ax - \mu F(x) = 0 \qquad \text{(discrete boundary value problem)} \tag{15}$$

is a particular orbit of the free dynamical system $T_\mu : \mathbb{R}^2 \to \mathbb{R}^2$, where

$$A = \begin{pmatrix} 2 & -1 & & \bigcirc \\ -1 & \ddots & \ddots & \\ & \ddots & \ddots & -1 \\ \bigcirc & & -1 & 2 \end{pmatrix} \qquad \text{and} \quad F(x) = (f(u_1),\ldots, f(u_n))^T.$$

In strict analogy to (6) we can rewrite (15) by

$$\begin{cases} T_\mu^k(V) \cap V \neq \emptyset \qquad (T_\mu^k = \underbrace{T_\mu \circ \ldots \circ T_\mu}_{k\text{-times}}) \ , \\ k = n + 1 \end{cases} \tag{16}$$

which distinguishes the particular orbits mentioned above. More precisely, if $(o,v) \in V$ is a point such that $T_\mu^{n+1}(o,v) \in V$, then setting

$$(u_o,v_o) = (o,v) \quad \text{and} \qquad (u_i,v_i) = T_\mu^i(u_o,v_o) \qquad i=1,\ldots,n+1$$

we have that $x = (u_1, \dots, u_n)^T$ solves (15) (and vice versa). It is well known (see [Be, M, P2]), that area preserving dynamical systems like (14) have all the typical features of complexity (i.e. coexistence of chaos and order on each scale). In view of this the phenomenon of spurious solutions of (15) is not surprising even under the rigid restrictions of (16). Before we proceed to explain our mechanism we examine (14) with respect to the basic properties (9) and (10): Let $S(u,v) = (v,u)$, then it follows that

$$(T_\mu^n \circ S)^2 = Id \ , \quad n \in Z \ , \quad \text{and} \tag{17}$$

$$T_\mu^n = R_n^\mu \circ S \ , \quad \text{where } R_n^\mu := T_\mu^n \circ S \ . \quad \text{Moreover, we let} \tag{18}$$

$$\mathcal{R}_n^\mu := Fix(R_n^\mu) = \{(u,v) : R_n^\mu(u,v) = (u,v)\} \tag{19}$$

and have

$$\begin{cases} \mathcal{R}_{2n}^\mu = T_\mu^n(\mathcal{R}_o^\mu) \ , \quad \mathcal{R}_{2n+1}^\mu = T_\mu^n(\mathcal{R}_1^\mu) \\ P \in \mathcal{R}_n^\mu \cap \mathcal{R}_m^\mu \Rightarrow T_\mu^{n-m}(P) = P \ . \end{cases} \tag{20}$$

These basic properties were already exploited by Birkhoff in his studies of periodic points [Bi] and later on by many others (see[G,V,U]). It is noteworthy to remark that by (20) the study of periodic points of T_μ and analyzing the solutions of (15) has been reduced practically to the same general settings (16), (20), which are intersection properties of smooth 1-manifolds (the $\mathcal{R}_n^\mu$ are easily seen to be smooth 1-manifolds provided the generating function f is smooth). In the next paragraph we will explain that for $f \in N^k$ the dynamical system (14) will have a homoclinic structure which interacts with the manifolds $\mathcal{R}_n^\mu$ and $T_\mu^k(V)$. This interaction is then analyzed for varying $\mu = \lambda h^2$ by means of the λ-lemma (see [Pa]), explaining eventually one class of spurious solutions. It remains to identify this class. In (11) we have noted that all solutions of (2) have certain symmetries which were described in terms of the flow Φ^t with respect to the u-axis. For discretized solutions, i.e. for solutions of (15), it turns out that the notion of symmetry depends on the parity of interior meshpoints. Problem (2) has essentially two classes of solutions and these are solutions with

- an even number of internal zeros (S_{2k}), and
- an odd number of internal zeros (S_{2k+1}).

Both classes have the symmetries (11). However, those of the first kind have the additional property that

$$u(m-t) = u(m+t) \ , \quad t \in [o,m] \ , \quad m = 1/2 \tag{21}$$

The breaking of this symmetry will be the distinguishing feature between genuine versus spurious solutions of (15). All other classes given in [PS] and solutions of the second kind will not be discussed here. In the following we restrict our attention to solutions of (2) with an even number of internal zeros and formulate the symmetry (21) for discrete solutions of (15) in terms of the dynamical system T_μ and its involution structure. Let $V_{+(-)} = \{(u,v) : u = o \text{ and } v > o \ (v<o)\}$. Then a discrete solution of (15) corresponding to a solution with an even number of internal zeros can be described in the following way:

$$\begin{cases} T_\mu^k(V_{+(-)}) \cap V_{-(+)} \neq \emptyset \\ k = n+1 \end{cases} \tag{22}$$

Now let $(o,v) \in V_{+(-)}$ and assume that $T_\mu^{n+1}(o,v) \in V_{-(+)}$. Set $x = (u_1,\dots,u_n)$, where $(u_i,v_i) = T_\mu^i(o,v)$. Then x is called *symmetric*, provided

$$\begin{cases} u_{m-t} = u_{m+t}, & \text{for } t = 1,\dots,m, \quad \text{if } n+1 = 2m \\ u_{m-t-1} = u_{m+t}, & \text{for } t = o,\dots,m-1, \quad \text{if } n+1 = 2m-1. \end{cases} \tag{23}$$

This notion of symmetry is equivalent to

$$\begin{cases} T_\mu^{m+1}(o,v) \in R_1^\mu, & \text{if } n+1 = 2m \\ T_\mu^m(o,v) \in R_o^\mu, & \text{if } n+1 = 2m-1, \end{cases} \tag{24}$$

as follows essentially from the definitions, and this shows the importance of the manifolds $R_{o(1)}^\mu$ for the discussion of symmetric solutions of (15).

To explain spurious solutions in this context we will present a mechanism which yields bifurcation of symmetric solutions into non-symmetric ones. This is the content of the next paragraph.

2. Homoclinic Bifurcation

In [P2] we discussed the phenomenon of homoclinic bifurcation for the dynamical system (14). The discussion was continued in [NP] and [PR]. We will summarize one of the main results from [PR], because it will essentially solve our problem here. Rather than describing notions and results in their greatest generality we prefer here to stay as close as possible to our concrete dynamical system T_μ. We assume throughout this paragraph that $f \in N^k$. Let $E=(o,o)$ and $H = (s_o,s_o)$. Then H is a hyperbolic fixed point of T_μ and we let ($\mu > o$ fixed)

$$\begin{aligned} W^s(\mu) &= \{P \in \mathbb{R}^2 : T_\mu^k(P) \to H, \text{ as } k \to \infty\} \\ W^u(\mu) &= \{P \in \mathbb{R}^2 : T_\mu^{-k}(P) \to H, \text{ as } k \to \infty\} \end{aligned} \tag{25}$$

denote the stable (respectively unstable) manifolds of H. An immediate consequence of (17) is the useful fact that

$$R_n^\mu(W^s(\mu)) = W^u(\mu) \quad \text{and} \quad R_n^\mu(W^u(\mu)) = W^s(\mu) \ , \quad n \in \mathbb{Z} \tag{26}$$

i.e., if $P \in R_n^\mu \cap W^s(\mu)\setminus\{H\}$, then P is a *homoclinic point*, i.e. $P \in W^s(\mu) \cap W^u(\mu)\setminus\{H\}$.

Taking into account (20) it then makes sense to study distinguished homoclinic points as the parameter μ is varied:

$$R_o^\mu \cap W^s_{loc}(\mu) \ , \quad R_1^\mu \cap W^s_{loc}(\mu) \tag{27}$$

More precisely, it will be typical that T_μ has a rather complex homoclinic structure. However, by means of the selection (27) we are able to set up a 1-parameter family of homoclinic points ($W^s_{loc}(\mu)$ denotes the local stable manifold).

We now want to explain the notion of *homoclinic bifurcation of odd type*: Assume that we have a smooth 1-parameter family of

$$P(\mu) \in R^\mu_{o(1)} \cap W^s_{loc}(\mu) \ , \quad \mu \in (\mu_*-\varepsilon, \mu_*+\varepsilon) \ , \quad \mu_* > o \ \text{ fixed,}$$

homoclinic points. Then μ_* is said to be a point of *homoclinic bifurcation of odd type for* $P(\mu)$ *at* R_o^μ *(resp.* R_1^μ*)*, provided there exists a smooth 1-parameter family of diffeomorphisms

$$h_\mu : U(\mu) \to \mathbb{R}^2 \ ,$$

where $U(\mu)$ is a family of open neighborhoods of $P(\mu)$ and $h_\mu(P(\mu)) = (o,o)$, such that

$$\begin{cases} h_\mu : W^s_{loc}(\mu) \cap U(\mu) \cong \{(x,y) \in \mathbb{R}^2 : y = o\} \ , \\ h_\mu : W^u_{loc}(\mu) \cap U(\mu) \cong \{(x,y) \in \mathbb{R}^2 : y = x^3-\alpha(\mu)x\} \ , \\ \text{where } \alpha(\mu) \text{ is smooth with respect to } \mu \text{, and} \\ \alpha(\mu_*) = o \text{, and } \alpha'(\mu_*) \neq o \end{cases} \tag{28}$$

Figure 3 is an illustration of bifurcation at R_o^μ. Note that in the example shown there the emanating homoclinic points are <u>not</u> on $R_o^\mu \cup R_1^\mu$. However, they are related, i.e. if $Q_{1,2}(\mu)$ is a pair of bifurcating homoclinic points then $S(Q_1(\mu) = Q_2(\mu)$. Similarily, if the bifurcation occurs at R_1^μ, then $R_1^\mu(Q_1(\mu)) = Q_2(\mu)$. We remark in passing that this fact is the deeper reason for the occurance of an odd type bifurcation. In other words it is the symmetry (invo-

Figure 3.

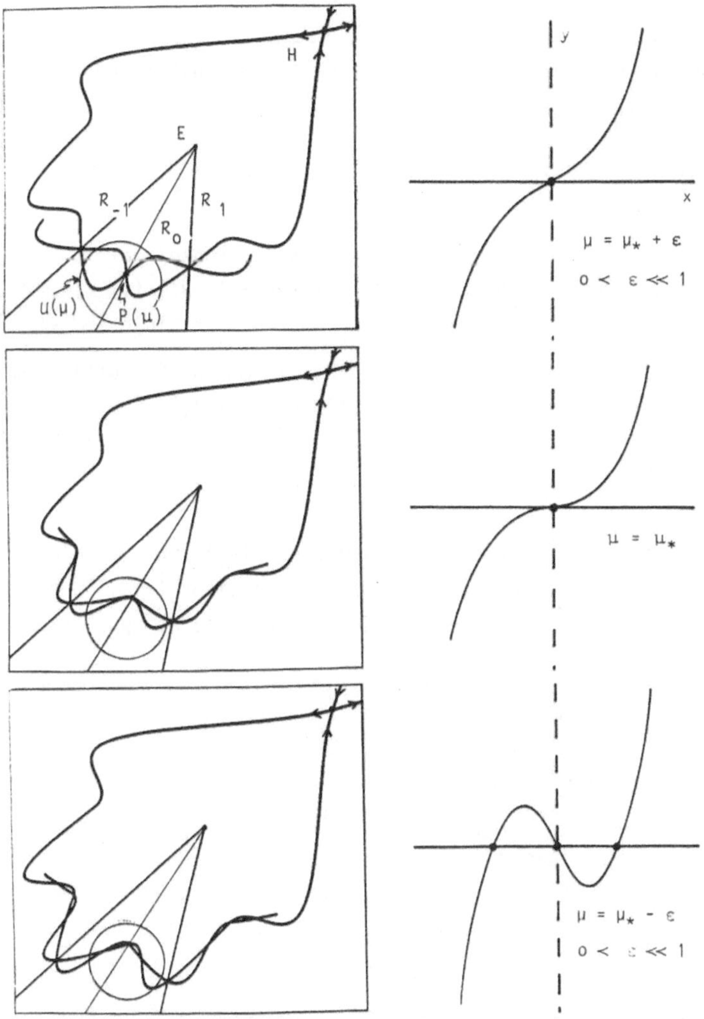

lution structure) in $T\mu$, which yields the odd type bifurcation. We are now in a position to explain a mechanism for spurious solutions. The idea is that homoclinic bifurcation of odd type initiates a bifurcation of discrete symmetric solutions of (15) in a symmetry-breaking way. The essential ingredient to pass from homoclinic bifurcation to bifurcation of solutions of (15) is the λ-lemma [Pa]. In essence the λ-lemma in $\mathbb{R}^2$ is the following: Given a diffeomorphism $T: \mathbb{R}^2 \to \mathbb{R}^2$ with a hyperbolic fixed point H and stable (resp. unstable) manifolds $W^s(H)$ (resp. $W^u(H)$). Let I be a smooth 1-cell which intersects $W^s(H)$ transversely. Then for any $\delta > 0$ there exists $k \in \mathbb{N}$ and a smooth 1-cell $I^* \subset I$ such that the C^1-distance between $T^k(I^*)$ and $W^u_{loc}(H)$ is at most δ. In other words, the 1-manifolds $T^k(I)$ approximate $W^u_{loc}(H)$ in a C^1-sense, as $k \to \infty$.

Let $f \in N^k$ and T_μ as in (14). Let μ_* be fixed and assume that there exists a 1-parameter family of homoclinic points

$$P(\mu) \in R^{\mu}_{o(1)} \cap W^s_{loc}(\mu) ,$$

which undergoes a homoclinic bifurcation of odd type in the interval $(\mu_*-\varepsilon , \mu_*+\varepsilon)$, $o < \varepsilon \ll 1$. Furthermore, assume that for all $\mu \in (\mu_*-\varepsilon,\mu_*+\varepsilon)$ $W^s_{loc}(\mu)$ intersects $R^{\mu}_{o(1)}$ transversely. Finally, assume that V_- intersects $W^s_{loc}(\mu)$ transversely for all $\mu \in (\mu_*-\varepsilon,\mu_*+\varepsilon)$. Then, if $h > o$ is sufficiently small, there will be a 1-parameter family of solutions of (15),

$$(\mu_*-\varepsilon,\mu_*+\varepsilon) \ni \mu \to x_\mu \in \mathbb{R}^n$$

which satisfy (22) and which are symmetric in the sense of (23) (or equivalently (24)). Furthermore, this 1-parameter family of symmetric solutions (see figure 4) bifurcates into non-symmetric solutions (i.e. spurious solutions) as μ varies in $(\mu_*-\varepsilon, \mu_*+\varepsilon)$.

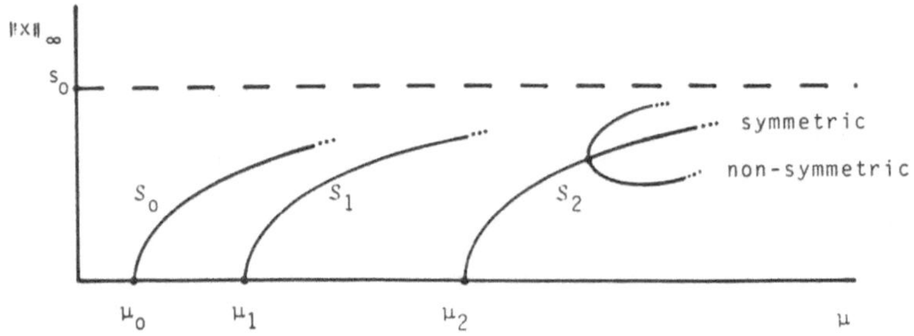

Figure 4. (Secondary bifurcation of spurious solutions)

To give a quick argument for the above conclusions we apply the λ-lemma. We discuss the situation, for a homoclinic bifurcation at R^{μ}_{o} . The other case follows analogously. Let V_s be a sufficiently small 1-cell on V_- , which intersects $W^s_{loc}(\mu)$ transversely in a unique point. Then $T^r_\mu(V_s)$ approximates $W^u_{loc}(\mu)$ in a C^1-sense, as $r \gg 1$. Since $W^u_{loc}(\mu) = S(W^s_{loc}(\mu))$, we may conclude that $ST^r_\mu(V_s)$ approximates $W^s_{loc}(\mu)$ in a C^1-sense, as $r \gg 1$. Now $W^s_{loc}(\mu)$ and $W^u_{loc}(\mu)$ undergo an odd type bifurcation as μ varies in $(\mu_*-\varepsilon,\mu_*+\varepsilon)$ and therefore also the 1-manifolds $T^r_\mu(V_s)$ and $ST^r_\mu(V_s)$ undergo an odd type bifurcation, as $r \gg 1$. However, $ST^r_\mu(V_s) = ST^r_\mu SS(V_s) = T^{-r+1}_\mu(V_u)$, where $V_u := T^{-1}_\mu(S(V_s)) \subset V_+$. Thus, (see (22)) in particular $T^{2r-1}_\mu(V_s) \cap V_u \neq \emptyset$, i.e. we are dealing with solutions, which correspond to solutions of (2) with an even number of internal zeros. Moreover, since $W^s_{loc}(\mu)$ intersects R^μ_o transversely, we must have that $T^r_\mu(V_s) \cap R^\mu_o \neq \emptyset$, i.e.(see (24)) if $(o,v_\mu) \in V_s$ and $T^r_\mu(o,v_\mu) \in R^\mu_o$, then the solution $x_\mu = (u_1,\dots,u_n)$ with

$$(u_i,v_i) = T^i_\mu(o,v_\mu), \quad i = 1,\dots,n \ ; \quad n = 2r-2$$

is symmetric in the sense of (23). Moreover, if we choose $h = {}^1/_{(2r-1)}$ and λ_* , $\bar{\varepsilon}$ so that for $\lambda \in (\lambda_*-\bar{\varepsilon}\ ,\ \lambda_*+\bar{\varepsilon})$ $\lambda h^2 \in (\mu_*-\varepsilon,\mu_*+\varepsilon)$, then we can interpret the familiy x_μ as a 1-parameter family of solutions of (15) for fixed meshsize h . Finally, the bifurcation in $T^{2r-1}_\mu(V_s) \cap V_u$ is into non-symmetric solutions. This is seen as follows: The analysis of $T^{2r-1}_\mu(V_s) \cap V_u$ is equivalent to the analysis of $T^r_\mu(V_s) \cap ST^r_\mu(V_s)$. As we have seen already the latter pair of 1-manifolds approximates $W^s_{loc}(\mu) \cap W^u_{loc}(\mu)$ in a C^1-sense. Since $W^s_{loc}(\mu)$ intersects R^μ_o transversely we must have that the emanating homoclinic points are not on $R^\mu_o \cup R^\mu_1$. Hence, also the emanating points in $T^r_\mu(V_s) \cap ST^r_\mu(V_s)$ will not be on $R^\mu_o \cup R^\mu_1$, i.e. they give rise to non-symmetric solutions of (15).

Finally we discuss one of the main results from [PR] which will provide numerous homoclinic bifurcations for a particular model $f \in N^k$. Let

$$f_\delta(s) = \begin{cases} s & , \quad s < o \\ 2-s & , \quad s > \delta \\ \frac{2-\delta}{\delta}\, s & , \quad o < s < \delta \ . \end{cases} \tag{29}$$

If we use f_δ in (14) as a generating function we obtain a 1-parameter family of piecewise-linear homeomorphisms rather than diffeomorphisms. For $H = (2,2)$ one obtains that H is a hyperbolic fixed point and the corresponding invariant manifolds $W^s(\mu)$ and $W^u(\mu)$ are obtained as 1-dimensional piecewise-linear manifolds, which, as a consequence of the particular model, can be globally constructed by a careful exploitation of the involution structure (see [PR]). Moreover, the notion of homoclinic bifurcation of odd type at $R^\mu_{o(1)}$ can be extended to this case. Now, varying μ and δ simultaneously we obtain a 2-parameter family of piecewise-linear homeomorphisms $T_{\mu,\delta}$ and we can ask for homoclinic bifurcation of odd type at $R^{\mu,\delta}_{o(1)}$. The main result from [PR] with respect to this question is illustrated in figure 5, which we discuss now. Figure 5 is the (μ,δ)-phase diagram of homoclinic bifurcation for $T_{\mu,\delta}$ in a certain domain of the parameter space. The essential information for our discussion here is contained in the line structure and the shaded areas. Each line in figure 5 represents a certain degeneracy in the homoclinic structure of $T_{\mu,\delta}$ (see [PR]), which may or may not lead to a homoclinic bifurcation. In fact, those points on the lines of degeneracy which bound shaded <u>and</u> unshaded areas lead to homoclinic bifurcation of odd type. More precisely, let

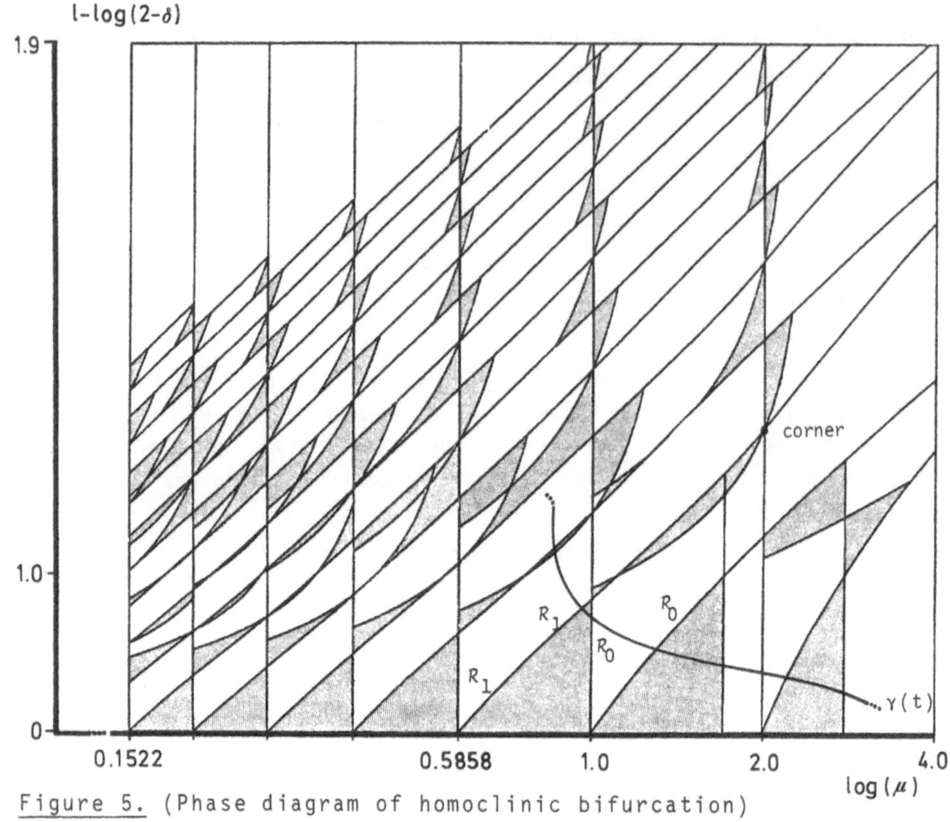

Figure 5. (Phase diagram of homoclinic bifurcation)

$$\gamma : (-1,+1) \to \{(\mu,\delta) : \mu > o \ , \quad o < \delta < 2\}$$

be a smooth path, which intersects the lines of degeneracy in figure 5 transversely away from the corners. Then the 1-parameter family $s \mapsto T_{\gamma(s)}$ undergoes a homoclinic bifurcation of odd type at $R_{o(1)}$ for any s_* at which $\gamma(s)$ enters (resp. leaves) a shaded area. The main ingredients for the proof of figure 5 are the following:

-) A topological index for homoclinic points: Let $P \in W^s \cap W^u$ be a locally unique transversal intersection. Then one defines

$$\operatorname{ind}(W^s,W^u;P) := \text{local intersection number of } W^s \text{ and } W^u .$$

This index is a topological invariant, i.e. it is constant with respect to small C^o-perturbations of W^s and W^u (see any textbook in topology).

-) An explicit construction of the invariant manifolds by means of the involution structure and the special piecewise-linear nature of the model $T_{\mu,\delta}$.

-) If γ is a path in the phase diagram as above, then

$$\mathrm{ind}(W^s(\gamma(t)),W^u(\gamma(t));P(\gamma(t))) = \pm 1$$

for any t such that $\gamma(t)$ is not on a line of degeneracy. If $\gamma(t_*)$ is such that $\gamma(t_*-\varepsilon)$ is in a shaded region and $\gamma(t_*+\varepsilon)$ is not in a shaded region for all $0 < \varepsilon \ll 1$, then

$$\mathrm{ind}(W^s(\gamma(t)),W^u(\gamma(t));P(\gamma(t))) = \begin{cases} \pm 1, & t = t_* - \varepsilon \\ \mp 1, & t = t_* + \varepsilon \end{cases}$$

Thus, a change of index indicates a homoclinic bifurcation of odd type (see figure 6).

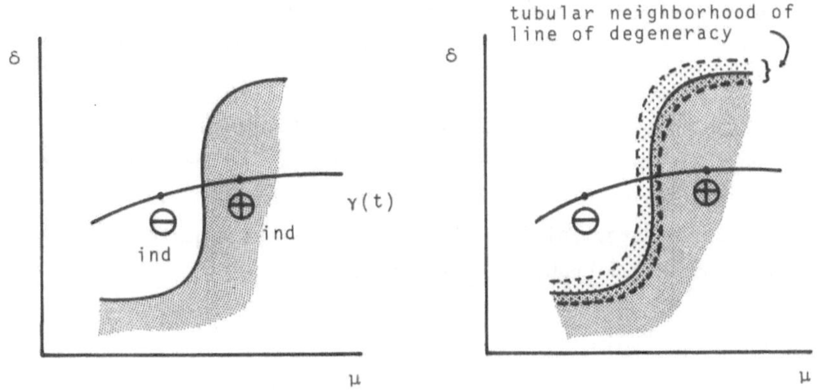

Figure 6. ((a) piecewise-linear, (b) smooth)

We add in passing that the tool of an index for homoclinic points can be used to give an alternative argument for the initiation of bifurcation of spurious solutions as a consequence of homoclinic bifurcation. This argument would be equally valid for the smooth and piecewise-linear case and substitute the application of the λ-lemma. Essentially the argument is as follows: The index is the intersection number of a pair of 1-dim. manifolds. Thus, any pair of 1-dim. manifolds, which is sufficiently C^0-close, will have the same intersection number (up to a factor of ± 1). This fact could be used to obtain, for example, a change in the local intersection number of $T^r_\mu(V_s)$ with $ST^r_\mu(V_s)$ for $r \gg 1$, as μ varies, from a change of the local intersection number of $W^s_{loc}(\mu)$ with $W^u_{loc}(\mu)$. Such a change of index would then imply bifurcation in $T^r_\mu(V_s) \cap ST^r_\mu(V_s)$ at R_0.

If we now pass to a smooth generating function $f^\infty_\delta \in N^k$ such that

$$\begin{cases} \| f^\infty_\delta - f_\delta \| \ll 1 \\ f^\infty_\delta(s) = f_\delta(s) \quad \text{for all } s \in (-\infty,0] \cup [\varepsilon,\delta-\varepsilon] \cup [\delta,\infty) \end{cases} \tag{30}$$

for $0 < \varepsilon \ll 1$, we may obtain a (μ,δ)-phase diagram of homoclinic bifurcation by simply including small tubular neighborhoods for the lines of degeneracy in figure 5 (see figure 6(b) for a symbolic detail) with exactly the same interpretation as for the piecewise-linear model. More precisely, if $\gamma(t)$ is a smooth path as in figure 6(b), then there is a change of the index along the path and that implies homoclinic bifurcation of odd type at $R_{o(1)}$. Here we have used that the index of homoclinic points is a topological invariant together with the very special nature of the smooth approximation f_δ^∞ with respect to f_δ. Another consequence of this special approximation f_δ^∞ is the fact that $W^s_{loc}(\mu,\delta)$ intersects $R_{o(1)}$ transversely (see [PR]).

Thus, in summary we may view figure 5 as a phase diagram of homoclinic bifurcation of odd type for a 2-parameter family of diffeomorphisms and together with the discussion at the beginning of this section we may even interpret figure 5 as a phase diagram for bifurcation of spurious solutions for the discretized boundary value problem (15). As such it reveals a surprising complexity for the phenomenon of spurious solutions.

In view of an interpretation for practical computational purposes one is tempted to neglect the importance of figure 5 for problem (2). Indeed, one could easily choose a discretization with a built in symmetry which would not show the spurious solutions discussed here. This certainly would be the adequate approach for problem (2). For more general problems, e.g. problems where

$$f = f(u,t) \quad \text{or} \quad f = f(u,\dot{u},t) ,$$

this simple idea would be useless, because solutions would not have any (resp. not any obvious) symmetries. However, even in such a situation figure 5 can be accepted as a hint for the kind of complexity one has to be prepared to obtain when solving nonlinear boundary value problems. Indeed, if e.g. $f = g(u(t)) + \varepsilon h(t)$ and $|\varepsilon|$ is sufficiently small one will have a complexity of numerical solutions which is adequately described by figure 5.

Details and a discussion of the other types of spurious solutions given in [PS] will appear elsewhere.

REFERENCES

[Al] ALLGOWER, E. L.: On a discretization of $y'' + \lambda y^k = o$. Proc. Conf. Roy. Irish Acad., J. J. H. Miller (ed.), New York, Academic Press 1975, 1-15.

[Be] BERRY, M. V.: Regular and irregular motion, in: Topics in Nonlinear Dynamics, S. Jorna (ed.), Amer. Inst. Phys. Conf. Proc., New York, 46, 16-120 (1978).

[BD] BEYN, W.-J. and DOEDEL, E.: Stability and multiplicity of solutions to discretizations of ordinary differential equations, SIAM J. Sci. Stat. Comput. 2, 107-120 (1981).

[BL] BEYN, W.-J. and LORENZ, J.: Spurious solutions for discrete superlinear boundary value problems, Computing 28, 43-51 (1982).

[Bi] BIRKHOFF, G.D.: The restricted problem of three bodies, Rend. Circ. Mat. Palermo 39, 265-334 (1915).

[Bo] BOHL, E.: On the bifurcation diagram of discrete analogues for ordinary bifurcation problems, Math. Meth. Appl. in the Sci. 1, 566-571 (1979).

[GNN] GIDAS, B., NI, W. M. and NIRENBERG, L.: Symmetry and related properties via the maximum principle, Commun. Math. Phys. 68, 209-243 (1979).

[G] GREENE, J. M.: A method for determining a stochastic transition, J. Math. Phys. 20, 1183-1201 (1979).

[M] MOSER, J.: Stable and Random Motions in Dynamical Systems, Ann. of Math. Studies 77, Princeton Univ. Press, 1973.

[NP] NUSSBAUM, R. D. and PEITGEN, H. O.: Special and spurious solution of $\dot{x}(t) = - \alpha f(x(t-1))$, to appear

[Pa] PALIS, J.: On Morse - Smale dynamical systems, Topology 8, 365-404 (1968).

[P1] PEITGEN, H. O.: Topologische Perturbationen beim globalen numerischen Studium nichtlinearer Eigenwert- und Verzweigungsprobleme, Jber. d. Dt. Math.-Verein. 84, 107-162 (1982).

[P2] PEITGEN, H. O.: Phase transitions in the homoclinic regime of area preserving diffeomorphisms, Proc. Intern. Symp. on Synergetics, H. Haken (ed.), Springer Series in Synergetics. 17, 197-214 (1982).

[PR] PEITGEN, H. O. and RICHTER, P. H.: Homoclinic bifurcation and the fate of periodic points, to appear.

[PSS] PEITGEN, H. O., SAUPE, D. and SCHMITT, K.: Nonlinear elliptic boundary value problems versus their finite difference approximations: numerically irrelevant solutions, J. reine angew. Math. 322, 74-117 (1981).

[PS] PEITGEN, H. O. and SCHMITT, K.: Positive and spurious solutions of nonlinear eigenvalue problems, in: Numerical solution of Nonlinear Equations, Allgower, E. L., Glashoff, K. and Peitgen, H. O. (eds.), Berlin-Heidelberg-New York, Springer Lecture Notes in Mathematics 878, 275-324 (1981).

[SA] SPREUER, H. and ADAMS, E.: Pathologische Beispiele von Differenzenverfahren bei nichtlinearen gewöhnlichen Randwertaufgaben, ZAMM 57, T 304-T305 (1977).

[SS] STEPHENS, A. B. and SHUBIN, G. R.: Multiple solutions and bifurcation of finite difference approximations to some steady state problems of fluid dynamics, to appear.

[U] USHIKI, S.: Unstable manifolds of analytic dynamical systems, J. Math. Kyoto Univ. 21, 763-785 (1981).

[V] VOGELAERE de, R.: On the structure of symmetric periodic solutions of conservative systems, with applications, in Contributions to the Theory of Nonlinear Oscillations, vol. 4, S. Lefschetz (ed.), Princeton University Press, 1968.

[YU] YAMAGUTI, M. and USHIKI, S.: Chaos in numerical analysis of ordinary differential equations, Physica 3D, 618-626 (1981).

Approach to Equilibrium: Kuzmin's Theorem for Dissipative and Expanding Maps

D. Mayer [1]

Institut für Angewandte Mathematik, Universität Heidelberg
Im Neuenheimer Feld 294, D-6900 Heidelberg, Fed. Rep. of Germany

A method of classical statistical mechanics is used to study existence of asymptotic measures and their approach for higher dimensional nonlinear systems.

1. Introduction

One of the most important aims of the theory of nonlinear dynamical systems is the understanding and if possible even prediction of the long-time behaviour of such systems. We know today that in general only predictions in a statistical sense are possible because many of these deterministic systems show very sensitive dependence on initial conditions in the sense that even microscopically small uncertainties grow exponentially fast in time. It is therefore impossible to predict in such a case the time evolution over longer time scales. The only hope is then that one can at least predict the mean behaviour of an observable along the orbits of the system. Computer calculations show indeed that such averages exist in general and do not depend on the initial state of the system. This experience could be understood if there would be valid some kind of Ergodic Theorem for such systems: in other words there seem to exist on many of these systems very special measures which describe their long time behaviour in a statistical sense. They are therefore also called **asymptotic measures**.

The problem of showing the existence and finding explicit expressions for these asymptotic measures is quite difficult and not much is known for general systems. On the other hand it would be very important to get one's hand on these measures to do finally practical calculations for such non linear systems. There is infact a whole class of nontrivial systems where these measures are well known, namely conservative systems. The relevant measures for those are just the Gibbs ensembles and they work perfectly well as statistical mechanics and thermostatistics tell us, even if we do not know in general why this is so. The situation in

---

1) Heisenberg Fellow on leave of absence from RWTH Aachen

nonconservative systems seems therefore to be more or less hopeless where neither existence is clear nor explicit expressions are known. Nevertheless there exists a continuously growing number of classes of nonconservative nonlinear systems whose long time behaviour is pritty well understood. In many cases one understands even how the system approaches its equlibrium determined by the asymptotic measure. This is also for conservative systems a quite difficult and therefore mostly unsolved problem.

It is amazing that there exist several number theoretic discrete dynamical systems where questions of the above type have been answered for a long time: the most famous example being the continued fraction transformation of the unit interval $Tx = 1/x - [1/x]$, where $[\ ]$ denotes the integer part of the number in the bracket. It was Gauss [1] who found the asymptotic measure for this transformation which turned out to be a smooth, absolutely continuous measure with density $f(x) = 1/(1+x)$ upto normalization constant. He also showed that any other smooth enough measure converges under the iteration of T to this asymptotic measure and asked therefore the famous question how fast this convergence is in fact. The problem was finally settled by KUZMIN [2] who proved the convergence being like $\exp-\gamma\sqrt{n}$ for $n \to \infty$. His result was later improved by LEVI [3] and others who showed that it is even like $\exp-\gamma n$. Exponentially fast convergence of measures to equilibrium states are therefore often called Kuzmin-type Theorems in the mathematical literature.

After giving a very brief review of nonconservative systems with asymptotic measures for which Kuzmin-type Theorems are valid we will discuss a class of dissipative systems where this asymptotic measure can be determined explicitly. Then we will present another class of locally expanding maps in higher dimensions for which we can prove the existence of an asymptotic measure and also a Kuzmin-type Theorem. These results extend well known similar properties of piecewise expanding maps of the unit interval to any dimension n. We show that many of the higherdimensional number theoretic dynamical systems fall into this class. Our results are then a little bit stronger as the ones in the literature giving namely stronger properties of the density functions.

The method we apply is an adaption of the Ruelle-Araki transfer operator method of classical lattice spin systems.

2. Asymptotic measures for discrete dynamical systems

We restrict our discussion to discrete dynamical systems which we denote by (M,T) where M is the phase space and $T: M \to M$ a mapping respecting the given structure of the space M which we do not specify anyfurther.

Def. A T invariant probability measure μ on M is called an asymptotic measure if for almost all $x \in M$ with respect to some smooth measure ν on M and all continuous observables $f: M \to R$ the following is true:

$$\lim_{N} 1/N \sum_{k=0}^{N-1} f(T^k x) = \int_M f(y)\, d\mu(y) \quad .$$

It is possible that a dynamical system (M,T) has more than one asymptotic measure but besides some pathological cases there should be only finitely many of them.This picture is at least true for a whole class of dynamical systems,the so called Axiom-A systems of SMALE [4] . It was shown by RUELLE [5] in the discrete case and by BOWEN and RUELLE [6] in the continuous case that there exist on the attractors of these systems in whose basins of attractions the whole phase space decomposes asymptotic measures, at least for smooth enough T. These measures are unique and any other smooth enough measure converges to them weakly. The rate of convergence however is not known in general.

The Axiom-A systems being invertible can serve as models of reversible dynamics. SHUB [7] introduced a class of noninvertible systems, the so called expanding systems. They expand distances contrary to Axiom-A systems in all directions and one expects therefore smoothing in the whole phase space. In fact, KRZYZEWSKI and SZLENK [8] proved existence of a smooth asymptotic measure for such systems whose support is the whole phase space M and not as in the Axiom-A case a complicated, in general even cantorlike set which are also called strange attractors. Also in the expanding case any smooth enough probability measure converges to the asymptotic measure without knowing however in general its rate.

The results of KRZYZEWSKI and SZLENK have been generalized by LASOTA and YORKE [9] in dimension one. They considered piecewise expanding maps of the unit interval.These are transformations $T:I \to I$ for which there exists a partition $I = \bigcup \bar{I}_n$ into nonintersecting open intervals such that $T:I_n \to TI_n$ is bijective and $|T'(x)| > 1$ for $x \in I_n$. Under certain smoothness conditions they proved the existence of an absolutely continuous invariant measure μ on I. Among others it was BOWEN [10] who studies the conditions under which this μ is an asymptotic measure for T. The most general and complete analysis of such piecewise expanding maps was given by HOFBAUER and KELLER [11]. It follows from their results that also a Kuzmin-type Theorem is valid if T is forinstance weakly mixing and that correlation functions decay then exponentially fast. This in fact is also true for discrete Axiom-A systems [5].

Having established the existence of an asymptotic measure in all of the above cases there remains the problem of giving explicit expressions for them. This task seems not to be without anv hope at least for the piecewise expanding maps where Gauss measure for the continued fraction transformation is a good example. In this case the problem can be rewritten as a spectral problem for the Perron-Frobenius operator whose maximal eigenfunction gives just the wanted density. For Axiom-A or expanding maps there are besides some trivial cases practically no examples known where the asymptotic measure can be determined explicitly. The main reason for this is certainly the fact that these measures at least in the Axiom-A case have their support very often on these strange attractors [12] which themselves can be hardly described analytically. It is therefore quite interesting that there exists even a class of non-Axiom-A systems which can be solved completely by analytical tools and which show a highly nontrivial behaviour in time.

For more general systems besides the ones mentioned above not much is known and a good theory is still missing.

3. A class of dissipative systems with asymptotic measures

We want to discuss now dissipative systems some of which can be solved more or less completely. They are in a certain sense derived from expanding respectively piecewise expanding maps of the interval (see also [13]) a fact which makes them so easily tractable.

Let $M = I \times \mathbb{R}^k$ with I the unit interval and $\mathbb{R}^k$ k-dimensional real space. The mapping $\tilde{T} : M \rightarrow M$ is then defined as

$$\tilde{T}\ (x,y) = (\ Tx,\ V_x(y)\ ), \tag{1}$$

where $T: I \rightarrow I$ is piecewise expanding and $V_x : \mathbb{R}^k \rightarrow \mathbb{R}^k$ is uniformly λ contractive: $\| V_x(y_1) - V_x(y_2) \| \leq \lambda \| y_1 - y_2 \|$ with some $0 < \lambda < 1$.

In the special case k=1 and

$$Tx = 2x \bmod 1, \quad V_x(y) = \lambda y + \cos 2\pi x, \tag{2}$$

the phase space M can be taken to be $S_1 \times \mathbb{R}$ and $\tilde{T}$ becomes a dissipative differentiable noninvertible map of the cylinder. This model was introduced by KAPLAN and YORKE in [14] and treated on a computer. Later JENSEN and OBERMAN [15] investigated its long time behaviour using path integral methods. In [16] a complete solution of this model was found and an explicit expression for its asymptotic measure was given. It was remarked by KELLER [17] that some of these results can indeed be extended to the whole class of models defined in (1). For doing this he made a connection with similar properties of the mapping $T: I \rightarrow I$ established in [10]. Model (2) can be generalized also to a stochstic model and again be solved as a stochastic dynamical system [18], [17]. Let us now briefly recall the solution of model (2) without going too much into the details which can be found in [16].

It is easy to see that any point $(x,y) \in M$ moves under iteration by T after finite many stepsinto the finite cylinders $M_{\lambda,\varepsilon} = \{(x,y) \in M : |y| \leq 1/(1-\lambda) + \varepsilon\}$ for any $\varepsilon > 0$. This tells us that the system is indeed dissipative and that its long time behaviour takes place in the finite cylinder $M_\lambda = M_{\lambda,0}$. Because $M_\lambda \supset \tilde{T}M_\lambda \supset \ldots$ there is an attractor A given by $A = \bigcap_{n \geq 0} \tilde{T}^n M_\lambda$. It is also straightforward to calculate the Liapunov exponents χ_1 and χ_2. They turn out to be ln2 and $\ln\lambda$. This tells us that the system shows sensitive dependence on initial conditions in the whole phase space coming naturally from the expansive character of the mapping T. To give now an explicit analytic description of the attractor A which according to the computer results of KAPLAN and YORKE in [14] should be strange at least for $0 < \lambda < 1/2$ we make use of an almost trivial observation: A is the projection of the attractor of another dynamical system which has been investigated quite a lot! It is just SMALE's solenoid [19]. To make this more explicit let $M_{\mathbb{C}} = S_1 \times \mathbb{C}$ and $\tilde{T}_{\mathbb{C}}: M_{\mathbb{C}} \to M_{\mathbb{C}}$ be defined by $\tilde{T}_{\mathbb{C}}(x,z) = (2x \bmod 1, \lambda z + \exp 2\pi i x)$. Then the finite cylinder $M_{\mathbb{C},\lambda} = S_1 \times D_\lambda$ with $D_\lambda = \{z \in \mathbb{C}: |z| \leq 1/(1-\lambda)\}$ is just the full 2-torus in $\mathbb{R}^3$. In analogy to the real case one has $M_{\mathbb{C},\lambda} \supset \tilde{T}_{\mathbb{C}} M_{\mathbb{C},\lambda} \supset \tilde{T}^2_{\mathbb{C}} M_{\mathbb{C},\lambda} \ldots$ and the set $A_{\mathbb{C}} = \bigcap_{n \geq 0} \tilde{T}^n_{\mathbb{C}} M_{\mathbb{C},\lambda}$ is an attractor for the system $(M_{\mathbb{C}}, \tilde{T}_{\mathbb{C}})$. The projection of the complex system onto the real one is then given by

$$q: M_{\mathbb{C}} \to M \quad \text{with } q(x,z) = (x, \mathrm{Re}z), \tag{3}$$

where Rez denotes the real part of z. The mappings $\tilde{T}_{\mathbb{C}}$ and $\tilde{T}$ are then related in the following way:

$$\tilde{T} \circ q = q \circ \tilde{T}_{\mathbb{C}} . \tag{4}$$

This shows also that A is just the image of $A_{\mathbb{C}}$ under this projection:

$$A = q\, A_{\mathbb{C}} \tag{5}$$

To find an explicit description for A is therefore equivalent to find one for $A_{\mathbb{C}}$. As already mentioned, the attractor $A_{\mathbb{C}}$ has been studied quite a lot at least for $0 < \lambda < 1/2$ where it is called Smale's solenoid. We showed in [16] that $A_{\mathbb{C}}$ is for all $0 < \lambda < 1$ continuous image of a certain compact abelian group S which is known under the name van DANTZIG's dyadic solenoid [20]. It can be most easily described as a subgroup of the infinite product group of U(1):

$$S = \left\{ \underline{x} \in \prod_{i=0}^{\infty} S_1 : \; x_i = 2x_{i+1} \bmod 1 \text{ for all } i \right\} . \tag{6}$$

S is a connected, locally disconnected topological space which is locally the product of the real line and a Cantor set. The explicit form of the continuous mapping $\tilde{h} : S \rightarrow A_{\mathbb{C}}$ mentioned above is

$$\tilde{h}(\underline{x}) = (x_0, h(\underline{x})), \quad h(\underline{x}) = \sum_{k=0}^{\infty} \lambda^k \exp 2\pi i x_{k+1} . \tag{7}$$

It turns out that this mapping is even a homeomorphism as long as $0<\lambda<1/2$. In this case the attractor $A_{\mathbb{C}}$ has therefore the typical Cantor-like structure. The attractor A then is the projection of $A_{\mathbb{C}}$ and will therefore also have such a complicated structure in general. The mapping $\tilde{h}$ in (7) induces an action τ of $T_{\mathbb{C}}$ on S as one easily checks, which turns out to be invertible and whose periodic points are dense in S. Hence the same is true for $T_{\mathbb{C}}$ on $A_{\mathbb{C}}$ and the latter becomes in fact an Axiom-A strange attractor for $0<\lambda<1/2$. The mapping $\tilde{h}$ now gives an explicit analytic description of $A_{\mathbb{C}}$ and therefore also of A:

$$A = \left\{(x_0, y) \in M: \ y = \sum_{k=0}^{\infty} \lambda^k \cos 2\pi x_{k+1} : \ \underline{x} = (x_0, \ldots) \in S \right\} .$$

What now is the asymptotic measure for our system? Because S is a compact group there exists on S a natural measure μ_H, so called Haar measure which is invariant under the group operations. It turns out in our case that this measure is also invariant under the action of the mapping τ. It projects down to $A_{\mathbb{C}}$ via $\tilde{h}$ to an $\tilde{T}_{\mathbb{C}}$ invariant measure on the latter space. As already mentioned, the system $(M_{\mathbb{C}}, \tilde{T}_{\mathbb{C}})$ is part of an Axiom-A system and the Ruelle-Bowen Theorem applies. It tells us that there exists indeed a unique asymptotic measure for this system whose support is just $A_{\mathbb{C}}$. It is then straightforward to see that the above projected Haar measure is identical to this measure .

Having found this way the asymptotic measure for $(M_{\mathbb{C}}, \tilde{T}_{\mathbb{C}})$ we only have to project it once more down to A via the mapping q in (3). The measure μ defined this way turns out to be necessarily the asymptotic measure for the system $(M, \tilde{T})$. Because all the operations described for arriving at μ can be explicitly done, there is also an explicit expression available for this measure: let f namely be a continuous observable of the system $(M, \tilde{T})$ then one has

$$\mu(f) = \lim_{T \to \infty} 1/T \int_0^T f(t \bmod 1, h(t)) \, dt$$

where $$h(t) = \sum_{k=0}^{\infty} \lambda^k \cos \pi t / 2^k . \tag{8}$$

This expression was used in [16] to calculate expectation values and correlation functions for several observables.

The above analysis did not allow us to conclude that the above expression is also the asymptotic measure for values of λ between 1/2 and 1, nor did it allow us to say something about a possible Kuzmin-type Theorem. All this follows however from the more recent work of KELLER [17], who showed that all systems of class (1) have an asymptotic measure if T is for instance weak mixing. Furthermore a Kuzmin-type Theorem is valid in that case: all sufficiently smooth measures converge exponentially fast to the asymptotic measure. Unfortunately his method does not allow an explicit determination of these asymptotic measures nor does it say something about the topological structure of the attractors showing up in all these systems.

4. Locally expanding maps in R^n

We want to extend now the theory of Lasota and Yorke to higher dimensions. This will give us a still more general class of systems with asymptotic measures and nice convergence properties into equilibrium by applying the same technique as in [17]. There exist already in the literature the first steps to such a generalization to dimension two [21] but the method used there seems to meet serious difficulties in higher dimensions. It is an advantage of the method we will now explain that it works in any dimension. Certainly, the systems which allow for such a treatment will belong to a more restricted class. The main ingredient of our method is also the Perron-Frobenius operator but we combine it with an idea which was applied with some succes in the theory of classical spin lattice systems. The Perron-Frobenius operator there is nothing else but the Ruelle-Araki operator [22] acting on the space of observables of the spin system. The discussion of the Perron-Frobenius operator was done in the case of piecewise expanding maps of the unit interval [9], [10] and also in [21] in the space of functions with bounded variation which cannot extended without problems to higher dimensions. One knows on the other hand that the Ruelle-Araki operator has very strong spectral properties when it can be defined on a space of holomorphic functions. Our idea therefore will be to do something similar also for the Perron-Frobenius operator at least for certain of these locally expanding systems. This procedure of using onedimensional spin lattice systems to get results in the ergodic theory of dynamical systems is well known and was one of the cornerstones of the work of Bowen and Ruelle on Axiom-A systems [23]. A detailed discussion of the Ruelle-Araki operator in statistical mechanics can be found in [24] from which we make frequently use in the following.

Instead of stating and proving immediately our main Theorem we prefer to explain our method first on a simple one dimensional system, the so called β-transformation, where all the main ideas already show up.

It is clear that this system can be treated also by the method explained for instance in [10] and the results we get are all well known.

4.1. The β-transformation of Renyi

Let $\beta > 1$ and I be the unit interval. Then consider the transformation $T: I \to I$ defined as $Tx = \beta x \bmod 1$. This transformation was studied by RENYI [25] who showed that it has an absolutely continuous asymptotic measure whose explicit form was found by PARRY [26]. Further stochastic properties of this transformation have been investigated in [27] and [28]. The transformation T is closely related to the always existing β-expansion of any real number in I:

$$x = \sum_{i=1}^{\infty} a_i \beta^{-i} \tag{9}$$

with integers a_i taking values in the set $\{0,1,\ldots,[\beta]\}$ and obeying the relations

$$a_i = [\beta T^{i-1} x] . \tag{10}$$

If therefore x has the β-expansion $x = \sum_{i=1}^{\infty} a_i \beta^{-i}$ then Tx has the expansion $Tx = \sum_{i=1} a_{i+1} \beta^{-i}$.

T is certainly a piecewise expanding map of the unit interval: let I_i denote the open intervals $I_i = \{x \in I: i/\beta < x < (i+1)/\beta$ if $0 \leq i \leq [\beta]-1$, and $I_{[\beta]} = \{x \in I: [\beta]/\beta < x < 1\}$. The last interval appears only if β is not an integer which we will assume from now on. The open intervals define a partition $I = \bigcup \bar{I}_i$ and $T_{|I_i}: I_i \to TI_i$ is bijective with inverse mappings $\psi_i: TI_i \to I_i$ given by

$$\psi_i(x) = (i + x)/\beta . \tag{11}$$

Whereas $\overline{TI_i} = I$ for $0 \leq i \leq [\beta]-1$, $\overline{TI}_{[\beta]}$ is a strict subset of I, in other words whereas ψ_i maps I strictly inside itself for $0 \leq i \leq [\beta]-1$ the map $\psi_{[\beta]}$ maps part of I outside itself. In terms of the expansion (9) this can be reformulated as follows: let the sequence $(a_1, a_2, ..)$ correspond to the β-expansion of the point $x \in I$. Then the sequence $(i, a_1, a_2, ..)$ determines the β-expansion of the point $\psi_i(x)$ for $0 \leq i \leq [\beta]-1$. The sequence $([\beta], a_1, a_2, ..)$ however does not in general belong to the β-expansion of a point in I, because $\psi_{[\beta]}(x)$ can be outside I. It makes therefore sense to speak about allowed and not allowed sequences $(a_1, a_2, ..)$ with a_i in the set $\{0,1,.. ,[\beta]\}$.

Let us briefly recall the definition of the Perron-Frobenius operator $L: L_1(I) \to L_1(I)$ on the space of L_1 functions:

$$\int_I g(Tx)\, f(x)\, dx = \int_I g(x)\, Lf(x)\, dx \tag{12}$$

where g,f are from the space $L_1(I)$ and dx is ordinary Lebesgue measure on I. The following properties follow immediately from this definition:

a) $\int_I Lf(x)\, dx = \int_I f(x)\, dx$

b) $d\nu(x) = f(x)\, dx$ is T invariant iff $Lf(x) = f(x)$.

Remark: All can be done also for more general dynamical **systems**. Property b) above is the main reason for our interest in the operator L. In the case of a piecewise expanding transformation L then can be written also in the following form:

$$Lf(x) = \sum_{i=} |\psi_i'(x)|\ f(\psi_i(x))\ \chi_{TI_i}(x) \tag{13}$$

where the summation is over the partition, in our case just $\{0,\dots,[\beta]\}$ and χ_{TI_i} denotes the characteristic function of the set TI_i. This term is in fact only present if TI_i is strictly contained in I for instance. It is straightforward how to generalize the expression (13) to higher dimensional systems, the absolute value of the derivative of $_i$ has to be replaced by the absolute value of the determinant of the derivative of the mapping ψ_i.

The operator L in (13) looks much like the Ruelle-Araki operator of a onedimensional classical lattice spin system in which the spin variable can take values in the set $\{0,1,\dots,[\beta]$ and where there exist certain restrictions on the allowed configurations. If these configurations would be given by a simple neighbour exclusion principle then the Ruelle -Araki operator has the following form [24]:

$$L_{RA}f(\ _1;\ \underline{\xi}) = \sum_{i=0} A_{i,\xi_1}\quad f(i;(i,\ \underline{\xi})\quad \exp W(i|\underline{\xi}) \tag{14}$$

where the transition matrix A has as entries only 1 and 0 depending if the pair (i,ξ_1) is allowed or not. The function f is an observable of the spin system and depends on the spin variable ξ_1 and the configuration $\underline{\xi}$ of the infinite extended system. The function W finally describes the interaction energy between the spin i and the configuration $\underline{\xi}$. It is therefore natural to try to interpret the operator L in (13) as the Ruelle-Araki operator of such a spin system. The function χ_{TI_i} must then correspond to the transition matrix A in (14).

That this interpretation is indeed possible we will show next. For doing this we have to restrict the allowed numbers β. It must have an eventually periodic β-expansion, that means there exist integers k,N such that the coefficients a_i in the β-expansion of β have the property $a_{i+k} = a_i$ for all $i \geq N$. Then it follows immediately that the orbit of the point 1 in I given by $\{T^n 1\}$ consists of only finitely many points. These points on the other hand determine a new partition of I into disjoint open intervals I_α, $\alpha \in J$. This partition has a very nice property as one checks easily: for any $i \in \{0,1,..,[\beta]\}$ and any $\alpha \in J$ there exists either an unique $\beta \in J$ with $\psi_i(I_\alpha) \subset I_\beta$ or $\psi_i(I_\alpha) \subset \mathbb{R}\setminus I$. This property now allows us to define for any $i \in \{0,..,[\beta]\}$ a transition matrix $A^{(i)}$ indexed by the set J whose entries are only o and 1:

$$A^{(i)}_{\beta,\alpha} = 1 \text{ if } \psi_i(I_\alpha) \subset I_\beta, \; A^{(i)}_{\beta,\alpha} = 0 \text{ if } \psi_i(I_\alpha) \subset \mathbb{R}\setminus I_\beta.$$

Furthermore we define mappings $\varphi_i : J \to J$ as

$$\varphi_i(\alpha) = \beta \text{ if } A^{(i)}_{\beta,\alpha} = 1 \text{ and } \varphi_i(\alpha) = \alpha \text{ otherwise.}$$

There is an obvious geometrical interpretation of the transition matrix $A^{(i)}$: it tells us what intervals can be connected via the transformation ψ_i and therefore also by the transformation T. It is easy to see that in the present case of the β-transformation any two intervals I_α and I_β can be connected by a finite chain of mappings ψ_i in the sense that there exist indices $i_1,...,i_l$ such that $\psi_{i_1} \circ \dots \circ \psi_{i_l}(I_\alpha)$ is contained in I_β and that $A^{(i_1)}_{\beta,\delta} \cdot \ldots \cdot A^{(i_l)}_{\gamma,\alpha} = 1$.

Consider now the operator $\tilde{L}$

$$\tilde{L}\tilde{f}(\alpha,x) = \sum_{i=0}^{[\beta]} A^{(i)}_{\varphi_i(\alpha),\alpha} \, |\psi_i'(x)| \, \tilde{f}(\varphi_i(\alpha), \psi_i(x)) \tag{15}$$

acting on functions depending on both the parameter $\alpha \in J$ and the variable $x \in I$. This operator can now be interpreted as the Ruelle-Araki operator with more complicate restrictions on the allowed configurations.The relation of (15) to our original Perron-Frobenius operator L in (13) is very easy to see: Let

$$(\tau \tilde{f})(x) := \tilde{f}(\alpha, x) \text{ if } x \in I_\alpha, \tag{16}$$

then one gets

$$\tau(\tilde{L}\tilde{f}) = L(\tau \tilde{f}) . \tag{17}$$

What have we achieved by all these manipulations? we can apply to the operator $\tilde{L}$ the theory developped in [24]! It can namely be defined on a space of holomorphic functions. A look at the mappings ψ_i in (11) tells us that there exists a bounded complex domain $\Omega \subset \mathbb{C}$ with $I \subset \Omega$ and the maps ψ_i can be continued to holomorphic mappings $\psi_i : \Omega \to \Omega$ with $\psi_i(\bar{\Omega}) \subset \Omega$. Denote then by $B(\Omega, J)$ the Banach space of all bounded holomorphic functions on Ω with values in $\mathbb{C}\times\cdots\times\mathbb{C}$ together with the sup norm. Using the results of [24] one shows easily

Lemma The operator $\tilde{L}: B(\Omega, J) \to B(\Omega, J)$ is nuclear and strictly positive. It has a positive eigenvalue ρ_0 strictly larger than all other eigenvalues in absolute value and its eigenvector $\tilde{f}_0$ is strictly positive.

Postivity is defined with respect to the cone of functions taking positivevalues on $\Omega_R = \Omega \cap \mathbb{R}$.

Because of (17) the function $\tau \tilde{f}_0$ is a positive eigenfunction of L with eigenvalue ρ_0 which by property a) of L in (12) has to be 1.

Corollary [26], [10] The β-transformation has a absolutely continuous asymptotic measure with a locally holomorphic density if β has an eventually periodic β-expansion. Any other absolutely continuous measure with locally holomorphic density converges exponentially fast to this asymptotic measure. The rate of convergence of the iterations is given by the second highest eigenvalue of the Ruelle-Araki operator $\tilde{L}$ in the space $B(\Omega, J)$ of holomorphic functions.

Let us mention also that exactly the same procedure can be applied to the continued fraction transformation and gives a new proof of Kuzmiń́s original Theorem. The Ruelle-Araki operator for this system was introduced in discussed in another connection in [29].

4.2. Maps in $\mathbb{R}^n$

It should be by now obvious how the foregoing discussion has to be generalized to any dimension n of the phase space. Let M for simplicity be the unit cube in $\mathbb{R}^n$ and let $T: I^n \to I^n$ be a map which fulfills the follwing conditions:

A1) "Local Expansivness"

There exists a countable partition $I^n = \bigcup_{i\in I} \bar{O}_i$ into open sets O_i such that $T: O_i \to TO_i$ is bijective. The inverse mappings $\psi_i: TO_i \to O_i$ can be continued to holomorphic mappings of some bounded domain Ω in $\mathbb{C}^n$ with $I^n \subset \Omega$ and $\psi_i(\bar{\Omega}) \subset \Omega$, $\det \psi_i'(z) \neq 0$, such that $\sum_{i\in I} |\det \psi_i'(z)| < \infty$ uniformly in $\bar{\Omega}$.

A2) "Markov Partition"

There exists a countable partition of $I^n = \bigcup_{\alpha\in J} O_\alpha$ by open sets O_α with the following property: to any $i \in I$ and any $\alpha \in J$ there exists either an unique $\beta \in J$ with $\psi_i(O_\alpha) \subset O_\beta$ or else $\psi_i(O_\alpha) \subset \mathbb{R}^n \setminus I^n$.

A3) "Irreducibility"
To any $\alpha, \beta \in J$ there exists a finite chain $\psi_{i_1}, \ldots, \psi_{i_k}$ depending on α and β with $\psi_{i_1} \circ \ldots \circ \psi_{i_l}(O_\alpha) \subset O_{\gamma_l}$ for some $\gamma_l \in J$ and all l and $\gamma_k = \beta$. We call such a chain an allowed chain.
If $|J| = \infty$ then there exists an integer $M > 0$ such that for all chains of length $r \geq M$ with

$$\psi_{i_1} \circ \ldots \circ \psi_{i_r}(O_\alpha) \subset O_\beta \text{ and } \psi_{i_1} \circ \ldots \circ \psi_{i_r}(O_{\alpha'}) \subset O_{\beta'} \text{ it follows}$$

that $\beta = \beta'$.

A4) "Density of Periodic Points"
The periodic points of T are dense in the phase space I^n.

Remark: It is possible that condition A4) follows already from the other three conditions. A4) is forinstance automatically true if T is a C^1 expanding map [7].

Then we can prove

Theorem If the map $T: I^n \longrightarrow I^n$ fulfills the conditions A1) to A4) then there exists an absolutely continuous asymptotic measure μ on I^n whose density is locally holomorphic. Furthermore a Kuzmin-type Theorem holds for any absolutely continuous measure with locally holomorphic density. The rate of convergence into equilibrium for such measures is determined by the second highest eigenvalue of the Ruelle-Araki operator for the system. The system (I^n, T, μ) has strong mixing properties.

The proof of the Theorem goes along the lines of our discussion for the β - transformation. If $|J| < \infty$ then the proof is exactly the same. If $|J| = \infty$ then we need the second part of condition A3) to show that the operator $\tilde{L}^M$ is a compact operator in the Banach space of all holomorphic mappings of Ω into the complex Banach space c_J of all infinite sequences indexed by J. To get strict positivity of the operator $\tilde{L}$ we use the density of the periodic points which certainly would follow also with less strong condition. The details of the proof will be published elsewhere.

5. An Example

As a nontrivial example where the above Theorem can be applied and gives in fact a slight improvement of otherwise well known results we consider the Jacobi-Perron algorithm as formulated by SCHWEIGER [30]. Let I^k be again the unit cube in $\mathbb{R}^k$ and define the mapping $T: I^k \longrightarrow I^k$ as

$T(\underline{x}) = (x_2/x_1-[x_2/x_1], \ldots, x_k/x_1 -[x_k/x_1], 1/x_1-[1/x_1])$. We claim that conditions A1) to A4) are fulfilled. Let $\underline{n} = (n_1,\ldots,n_k)$ be a vector of integers with $n_k \geq 1$ and $0 \leq n_i \leq n_k$ for all other i. We then define the open sets $O_{\underline{n}}$ as follows:

$$O_{\underline{n}}=\{\underline{x}\in I^k : 1/(1+n_k)<x_1 < 1/n_k, \ldots, n_i x_1 < x_{i+1} < (n_i+1)x_1, \ldots\}$$

if $n_k > n_i$ for all $i=1,\ldots,k-1$, and as

$$O_{\underline{n}}= \{\underline{x}\in I^k : 1/(1+n_k)<x_1 < 1/n_k, n_i x_1 < x_{i+1} < (n_i+1)x_1, \ldots, n_j x_1 < x_{j+1} < 1 \}$$

if some n_j's are equal to n_k. It is then obvious that $I^k = \bigcup_{\underline{n}} \bar{O}_{\underline{n}}$ and $T|_{O_{\underline{n}}} : O_{\underline{n}} \to TO_{\underline{n}}$ is bijective with inverse mapping $\psi_{\underline{n}}$

$$\psi_{\underline{n}}(\underline{x}) = (1/(x_k+n_k), (x_1+n_1)/(x_k+n_k), \ldots, (x_{k-1}+n_{k-1})/(x_k+n_k)).$$

There remains only to find the complex region Ω which is mapped by all the $\psi_{\underline{n}}$'s holomorphically inside itself. In the case k=2 this region can be given analytically, in higher dimensions we did not find up to now a simple proof for its existence.
The existence of what we called a Markov partition follows from a result by Perron cited in [30]. We call an infinite sequence $(\underline{n}_1, \underline{n}_2, \ldots)$ of integer vectors $\underline{n}_i$ as defined before an allowed sequence if there exists a $\underline{x} \in I^k$ with $T^{l-1}\underline{x} \in O_{\underline{n}_l}$ for all l. Perron now showed that given an allowed sequence $(\underline{n}_1,\ldots)$ it is possible to decide if the new sequence $(\underline{n}, \underline{n}_1, \ldots)$ is allowed by looking at a finite number r of elements of the original sequence. If we therefore define $\underline{\alpha} = (\underline{n}_1,\ldots,\underline{n}_r)$ and $O_{\underline{\alpha}}$ by

$$O_{\underline{\alpha}} = \bigcap_{i=1}^{r} T^{-(i-1)} O_{\underline{n}_i}$$ then $\psi_{\underline{n}}(O_{\underline{\alpha}})$ is either contained in $O_{\underline{\alpha}'}$

with $\underline{\alpha}' = (\underline{n}, \underline{n}_1, \ldots, \underline{n}_{r-1})$ or in $\mathbb{R}^k \setminus I^k$. Conditions A3) and A4) are then straightforward.
We can therefore apply our Theorem and get the known results of SCHWEIGER [30] and GORDIN [31] with the slight improvement that the density of the invariant asymptotic measure is even locally holomorphic.
Let us also add the remark that our Theorem can be applied to a whole series of other higherdimensional number theoretic dynamical systems as those discussed in [32] to [34] for instance and gives always the above mentioned improvement.

6. An open Problem

Let us finish with an open problem for which we want to formulate a conjecture. Our Theorem tells us that the approach to equilibrium in the cases covered by the theorem is exponentially fast. But the Theorem makes no statement concerning this rate γ besides that it is given by the second highest eigenvalue of the Ruelle-Araki operator in the space of certain holomorphic functione. It is clear that γ should be related to other dynamical characteristics of the system. In case of the β-transformation γ turns out to be just the Sinai-Kolmogorov entropy of the transformation T. Unfortunately this is not generally true as show already other piecewise linear maps with non constant slopes. There might be however a connection between γ and a certain function which was studied recently by PARRY and TUNCEL in a different problem: Consider the following operator $\tilde{L}_r$ on the space $B(\Omega,J)$:

$$\tilde{L}_r\tilde{f}(\alpha,z) = \sum_{i\in I} A^{(i)}_{\varphi_i(\alpha),\alpha} \left(\det\psi_i'(z)\right)^{1+r} \tilde{f}(\varphi_i(\alpha),\psi_i(z)).$$

There exists again a highest positive eigenvalue $\chi(r)$ at least for $\mathrm{Re}\, r \geqslant 0$ which is even holomorphic in r. Apparently $\chi(0) = 1$. It turns that for very simple one dimensional expanding mappings the derivative of this function at r=0 is given by $-h(T)$, the K-S entropy of the systems [35]. One can therefore conjecture that this is true for all systems governed by our Theorem and that also the convergence rate γ is determined by this function $\chi(r)$.

Acknowledgements: Most of the present work has been done during a stay at the IHES in Bures sur Yvette. I thank its director Prof.N.Kuiper for his kind hospitality and the french foreign ministry for a fellowship which only made this visit possible. I acknowledge also very useful conversations with Dr. Keller at Heidelberg.

Literature:

1 C.F.Gauss: Werke,Band X_1,p.372. Teubner Verlag,Leipzig 1917
2 R.Kuzmin: in Atti del Congresso Internazinale de Matematici(Bologna) vol.6,83-89 (1928)
3 P.Levi: Bull. Soc. Math. France 57,178 -194 (1929)
4 S.Smale: Bull. Amer. Math. Soc. 73,747-817 (1967)
5 D.Ruelle: Amer. J. Math. 98,619-654(1976)
6 R.Bowen,D.Ruelle: Invent. Math. 29,181-202(1975)
7 M.Shub: Amer.J. Math.91,175-199(1969)
8 K.Krzyzewski,W.Szlenk: Studia Math. 33, 83-92(1969)
9 A.Lasota,J.Yorke: Trans. Amer. Math. Soc. 186,481-488(1973)
10 F.Hofbauer,G.Keller: Math.Z. 180,119-140(1982)
11 R.Bowen: Israel J. Math. 28,298-314(1978)
12 D.Ruelle: Bull. Amer. Math. Soc. 5,29-42(1981)

13 R.Williams: Pub. IHES 43,169-203(1974)
14 J.Kaplan,J.Yorke: in "Functional difference equations and approximations of fixed points",Lect. Notes Math. 730,228-237(1979)
15 R.Jensen,C.Oberman: Phys. Rev.Lett. 46,1547-1550(1981)
16 D.Mayer,G.Roepstorff: "Strange attractors and asymptotic measures of discrete time dissipative systems",to appear In J.Stat.Phys. (1983)
17 G.Keller: "Stochastic perturbations of strange attractors" in Proc. of the Sitges Conference on Dynamical Systems (1982)
18 R.Graham: to appear in Phys. Rev. (1983)
19 S. Smale: in "Turbulence Seminar",Lect. Notes Math. 615,48-70(1977)
20 D.van Dantzig: Fund. Math. 15,102-125(1930)
21 G.Keller: C.R.Acad. Sci. Paris 289A, 625-627(1979)
22 D.Ruelle: Commun. Math. Phys. 9, 267-278(1968)
23 D.Ruelle: "Thermodynamic Formalism", Addison-Wesley,Reading,Mass. (1978)
24 D.Mayer: "The Ruelle-Araki Transfer Operator in Classical Statistical Mechanics", Lect. Notes Phys. 123 (1980)
25 A.Renyi: Acta Math. Acad. Sci. Hung. 8,472-493(1957)
26 W.Parry: ibid. 11,401-416(1960)
27 M.Smorodinski: ibid. 24,273-278 (1973)
28 I.Kubo,H.Murato,H.Totoki: Publ. RIMS Kyoto 9,279-303(1974)
29 D.Mayer: Bull. Soc. Math. France 104, 195-203(1976)
30 F.Schweiger: "The Metrical Theory of the Jacobi-Perron Algorithm", Lect. Notes Math. 334 (1973)
31 M.Gordin: Sov. Math. Dokl. 12,331-335(1971)
32 F.Schweiger:in"Ergodic Theory", Lect. Notes Math. 729,199-202(1979)
33 S.Waterman: Z.Wahrsch. verw. Geb. 16,77-103(1970)
J. Math. An. and Appl. 59,288-300(1977)
34 W.Parry,S.Tuncel: "On the classification of Markov chains by finite equivalence", Univ. of Warwick preprint (1981)
35 G.Keller: private communication

Complex Behaviours in Macrosystems Near Polycritical Points

P. Coullet

Equipe de Mechanique Statistique, Université de Nice
F-06100 Nice, France

Abstract

A new concept to analyse complex dynamical systems in the neighbourhood of polycritical points is presented. Two- and three-dimensional singularities are discussed in detail. Particular attention is given to the occurrence of chaotic dynamics.

1. Introduction

Many interesting macrosystems are described by non-linear equations of the form

$$\partial_t U = L_\lambda U + N_\lambda (U) \tag{1}$$

where $\lambda=(\lambda_1, \lambda_2,\dots,\lambda_p)$ represents a set of p parameters, and $U = (U_1,\dots,U_n)$ are functions of time and space. L_λ and N_λ are respectively linear and non-linear operators. We assume that neither operator depends on time or involves ∂_t, but that each generally contains spatial derivatives. When the equations describe the evolution of a system in a "box" Ω, we have to prescribe boundary conditions on $\delta\Omega$. These are written as

$$P_\lambda U + Q_\lambda (U) = 0 \text{ on } \delta\Omega \tag{2}$$

where P_λ is a linear operator and Q_λ is a non-linear one.
We assume that the spectrum of L_λ, with the linear part of (2) is discrete. To simplify the presentation, we shall consider only non-degenerate spectra, though the extension to cases of finite degeneracy is easily made. Because of the form chosen for (1) and (2) U=0 is a steady solution of the problem. Linear stability of this solution is investigated by seeking solutions of the form

$$U = e^{st}\Phi(k,x) \tag{3}$$

for the linear version of (1) and (2). $\Phi(k,x)$ can be written as

$$\Phi(k,x) = \begin{bmatrix} \Phi_1(k)W_1(k,x) \\ \vdots \\ \Phi_n(k)W_n(k,x) \end{bmatrix} \tag{4}$$

where k is a set of parameters characterizing the spatial structure functions $W_i(k,x)$; in the simplest cases, k is a wave vector. For each k, the linear theory reduces to the characteristic value problem

$$L_\lambda \Phi(k,x) = s\,\Phi(k,x) \tag{5}$$

with

$$P_\lambda \Phi(k,x) = 0. \tag{6}$$

The characteristic value equation that results from (5) and (6) is, in the simplest cases, of the form

$$C(s;\lambda) = \sum_k C_k(s;\lambda) \tag{7}$$

where $C_k(s;\lambda)$ is a polynomial of degree n:

$$C_k(s;\lambda) = s^n + a_{n-1}(k;\lambda)\, s^{n-1} + \ldots\, a_o(k;\lambda). \tag{8}$$

For each root of

$$C_k(s;\lambda) = 0 \tag{9}$$

we find the modal amplitudes $\Phi_i(k)$ and the corresponding $\Phi(k,x)$ is a normal mode.

Suppose now that for some value of $\lambda=\lambda^*$ we have l root $s=0$ and m pairs with $\mathrm{Re}\, s=0$ and $\mathrm{Im}\, s \neq 0$ for the characteristic equation

$$C_k(s;\lambda^*) = 0, \tag{10}$$

that is we have $q = l+2m$ roots with Re $s=0$. We describe such roots and the associated modes as critical. We assume that in a neighbourhood Δ of λ^* we have q roots with small $|\mathrm{Re}\, s|$ and that for all the other roots of (10) Re $s<s_o<0$, where s_o is a real constant. The point λ^* in the parameter space is termed a polycritical point. Such points belong to surfaces of codimension l+m called polycritical surfaces. Ordinary bifurcations correspond to codimension one surfaces where either $l=1$, $m=0$ (stationary bifurcation) or $l=0$, $m=1$ (Hopf bifurcation). The problem addressed here is that of extracting the time dependence of the solution of (1) near the polycritical surfaces, that is when one can reduce the problem to that of studying ordinary differential equations. We are first going to show the connection between the polycritical situations and the problem of the singularities of vector fields. Near a polycritical surface of codimension l+m, the dynamic can be reduced to the study of an ordinary differential equation of order 1+2m, the normal form of the perturbation of a singular vector field. A method to perform such a reduction is proposed. The complex behaviours associated with some normal forms are then reviewed. We end this paper showing some application of these ideas to situations of geophysical and astrophysical interest.

2. From the Normal Form of the Perturbations of a Singular Vector Field to the Amplitude Equations Describing Polycritical Phase Transitions in Open Systems

Polycritical situations as defined below are closely related to the singularities of vector fields [1,2]. For sake of simplicity we are going, in the following, to consider differential equations. The extension of the results to partial derivative equations is straightforward. The system of differential equations we are interested in can be written as

$$\partial_t U = LU + N(U,U) \tag{11}$$

where $U=(U_1,..U_n)$, L is an n x n real matrix, N is a n vector whose components are homogeneous real polynomials, at least quadratic in the U_i's. In writing (11) we assumed that U=O was a stationary solution. This assumption is not really restrictive since such a solution can always be chosen to be zero by an appropriate translation of the origin. A difficulty occurs if we consider families of differential equations. In this case a stationary solution can disappear, a possibility not taken into account in (11). In that respect the form chosen for (11) is special. For the sake of simplicity we are still going to consider such restricted systems. Our first task is to construct the normal form of a singular vector field. Let us remark [3] that an invertible, non-linear change of variables

$$U = V+U_2(V,V)+\ldots, \tag{12}$$

where U_2 is quadratic in V, transforms the linear equation

$$\partial_t U = LU \tag{13}$$

into a non-linear one:

$$\partial_t V = LV + (L-LV\cdot \frac{\partial}{\partial V})U_2(V,V)+\ldots \tag{14}$$

where

$$LV\cdot \frac{\partial}{\partial V} = \sum_{i,j} L_{ij}\, V_j \frac{\partial}{\partial V_i}\ . \tag{15}$$

A quadratic perturbation $N_2(V,V)$ of the linear vector field LV lies in a space of dimension $n^2(n+1)/2$. Let α_i be the eigenvalues of L. It is not difficult to show that if $\alpha_i \neq 0$ and $\alpha_k \neq \alpha_i + \alpha_j \quad \forall i,j,k$ the operator

$$L = (L\dot{V} \frac{\partial}{\partial V} - L) \tag{16}$$

is invertible in this space, so that its range has also the dimension $n^2(n+1)/2$ (dimension of the "orbit" of the linear vector field through the non-linear, near identity quadratic change of variables). This is easily checked in the diagonal case $L = \mathrm{diag}(\alpha_1, \alpha_2 \ldots \alpha_n)$ for example, and the result holds as well for non-diagonalizable L. This means in this case that any quadratic perturbation of LV can be removed, since it corresponds to a vector field which is equivalent to LV through the near identity quadratic change of variables. This procedure can now be applied to cubic terms where the non-resonance condition becomes $\alpha_l = \alpha_i + \alpha_j + \alpha_k$, $\forall i,j,k,l$. From the previous discussion two different types of singular situations emerge: the non-hyperbolic cases which occur, when for some i, Re $\alpha_i = 0$ and the hyperbolic resonant cases when for some $i, i_1, \ldots i_p$

$$\alpha_i = \sum_{i=1}^{p} \alpha_{il}, \quad \mathrm{Re}\ \alpha_i \neq 0.$$

Only the non-hyperbolic case will be considered, because of its physical interest. Hyperbolic resonant situations are not interesting in the sense that they describe either stable or unstable equilibrium solutions and that the intuitive feeling is that no topological change is expected in the neighbourhood of such solutions when we cross a resonant condition. This feeling is justified by the theorem of HARTMAN and GROBMAN [4,5] which states that a vector field near a hyperbolic (resonant or not) equilibrium solution is topologically equivalent to its linear part. We first consider the problem of "fully non-hyperbolic" systems.

Let

$$\partial_t U = JU + N_2(U,U) + \ldots \tag{17}$$

and

$$\partial_t V = JV + R_2(V,V) + \ldots \tag{18}$$

be two non-linear perturbations of the same linear singular vector field. J is an n x n matrix with only pure imaginary eigenvalues, written for sake of simplicity in its Jordan form.

Let

$$U = V + U_2(V,V) \ldots \tag{19}$$

be a non-linear near-identity change of variables. At the order k, the condition for the equivalence between the two vector fields through the change of variables defined by (19) is given by

$$(JV\frac{\partial}{\partial V} - J)U_k = I_k - R_k \tag{20}$$

where I_k depends on V through $U_2,\ldots,U_{k-1}$; $R_2,\ldots,R_{k-1}$; $N_2,\ldots N_k$. The basic ides is to try to use the change of variable to simplify as much as possible the expressions of the R_i's. Since

$$L_J = JV\frac{\partial}{\partial V} - J \tag{21}$$

is a non-invertible linear operator, in order to solve (20) we need to impose some solvability conditions. Formally this proceeds as follows. Defining a scalar product for the U_k's, we have to compute L_J^+, the adjoint of L_J in respect to this scalar product and to find its kernel:

$$L_J^+ Z_k = 0 . \tag{22}$$

The solvability condition of (20) is then

$$\langle I_k - R_k, Z_k \rangle = 0 . \tag{23}$$

The prescription chosen to simplify as much as possible the form of (18) is that the R_k not involved in the solvability condition are chosen to be zero. As a trivial example if J was a hyperbolic (non-resonant) matrix, L would be invertible so that no R_k's would be needed to solve (20). In this case they all could be chosen to be zero so that (17) would be equivalent to a linear vector field. Let us give some examples which will be useful in what follows:

ξ singularity $J = (0)$ $\quad \dot{V} = k\,V^2 + \ldots$

ξ^2 singularity $J = \begin{pmatrix} 0 & 1 \\ 0 & 0 \end{pmatrix}$ $\quad \ddot{V} = k_1 V^2 + k_2 V\dot{V} + \ldots$

ξ^3 singularity $J = \begin{pmatrix} 0 & 1 & 0 \\ 0 & 0 & 1 \\ 0 & 0 & 0 \end{pmatrix}$ $\quad \dddot{V} = k_1 V^2 + (k_2\dot{V} + k_3 V)\dot{V} + k_4 V\ddot{V} + \ldots .$

The notation used here to classify the singularities follows at distance Arnold's notation for singularities of matrices [6].
Another example is given by the ω singularity (Hopf bifurcation):

ω singularity $J = \begin{pmatrix} i\omega & 0 \\ 0 & -i\omega \end{pmatrix}$ $\quad \dot{V} = i\omega V + k_1 |V|^2 V$

$\xi\omega$ represents the non-hyperbolic Jordan matrix $J = \begin{pmatrix} 0 & 0 & 0 \\ 0 & i\omega & 0 \\ 0 & 0 & -i\omega \end{pmatrix}$

$\xi^2\xi$ is associated with $J = \left(\begin{array}{cc|c} 0 & 1 & 0 \\ 0 & 0 & 0 \\ \hline 0 & 0 & 0 \end{array}\right)$ and so on.

Let us consider the case of partially non-hyperbolic systems

$$L = \begin{pmatrix} J & O \\ O & H \end{pmatrix} \tag{24}$$

where J and H are respectively an n x n "fully non-hyperbolic" and an m x m "fully hyperbolic" matrix. In this case the operator $L_{J.H}$ is

$$L_{J.H} = JV.\frac{\partial}{\partial V} + HV.\frac{\partial}{\partial V} - L. \tag{25}$$

It is convenient to introduce the following notation

$$\vec{A} = (V_1 \ldots V_n), \quad \vec{B} = (V_{n+1}, \ldots, V_{n+m}). \tag{26}$$

$\mathbb{R}_{n+m}$ decomposed in $\mathbb{R}_n \oplus \mathbb{R}_m$ where $\mathbb{R}_n$ is generated by vectors having only components on the n first components and $\mathbb{R}_m$ is complementary. It is not difficult to show that the space generated by vector constants in $\vec{B}$, that is of the form $W_k(A)$ and having only components on $\mathbb{R}_m$, is invariant under $L_{J.H}$, and that $L_{J.H}$ is invertible on this space. The meaning of this is that the corresponding non-linear terms can be removed from the equation for the B_i's, so that $\vec{B}=0$ is an invariant manifold for the normal form. Since this manifold is locally stable in applications (the spectrum of H has to belong to the left part of the complex plane), we are interested in the study of the dynamics in the manifold (normal form of the so-called central manifold [7]). The reduction of the problem, to the central manifold, can be done by considering only the operator

$$L'_{J.H} = JA\cdot\frac{\partial}{\partial A} - L. \tag{27}$$

We have

$$L_{J.H}U_k = I_k - R'_k \tag{28}$$

where R'_k has only components on $\mathbb{R}_n$. This procedure replaces the change of variables (19) by a reduction change of variables.

$$(U_1, \ldots, U_n, U_{n+1}, \ldots, U_{n+m}) = (A_1, \ldots, A_n, 0 \ldots 0) + U_2(\vec{A}, \vec{A}) + \ldots \tag{29}$$

where U_2 has components on $\mathbb{R}_{n+m}$. Only two slight modifications are necessary to transform this approach into a systematic method [8] to compute the amplitude equations near a polycritical situation as defined in the introduction of this paper.

First the linear change of variable needed to write L in its Jordan normal form can be performed at the same time as the reduction change of variables.

$$U = MA+U_2(\vec{A},\vec{A})+\ldots = \sum_i A_i\Phi_i + U_2(\vec{A},\vec{A}) + \ldots \tag{28a}$$

where $\Phi_i = Me_i \quad e_i^t = (0,0..,1,..0)$ and

$$L\Phi_i = J_{ij}\Phi_j \,.$$

The second modification is the possibility to account for slight perturbations from the singular situation. This is achieved simply by expanding all the quantities in the parameters ε_i's characterizing the deviation from the critical situation. Let us now summarize all that in the following way.

Let

$$\partial_t U = L_\lambda U+N_\lambda(U), \text{ where } \lambda = (\lambda_1,\ldots,\lambda_p) \tag{29a}$$

be a set of differential equations in $\mathbb{R}_{n+m}$.(No major differences occur with partial differential equations if L_λ has a discrete spectrum). Let us assume that for some $\lambda=\lambda^*$ the solution U=0 becomes non-hyperbolic. If the other eigenvalues of L_{λ^*} all have negative real parts, the point λ^* corresponds to a multiple instability (polycritical point in the sense of the introduction). Near such a point the dynamics is dominated by the evolution of the amplitude of critical modes defined by

$$L_{\lambda^*}\Phi_i = J_{ij}\Phi_j \,. \tag{30}$$

A solution of (29) is then sought in the form

$$U=\sum_i A_i\Phi_i + \sum_{i,\{j_k\}} A_i\Phi_i^{(j_1\ldots j_p)}\varepsilon_1^{j_1}\ldots\varepsilon_p^{j_p}+U_2(\vec{A},\vec{A};\varepsilon)+\ldots \tag{31}$$

$$U_k(\vec{A},\ldots,\vec{A};\varepsilon) = \sum_{\{j_k\}} \varepsilon_1^{j_1}\ldots\varepsilon_p^{j_p}\, U_k^{(j_1\ldots j_p)}(\vec{A},\ldots,\vec{A}) \tag{32}$$

together with the amplitude equations

$$\partial_t A = JA+\sum_{\{j_k\}} \varepsilon_1^{j_1}\ldots\varepsilon_p^{j_p}\, J^{(j_1,\ldots,j_p)}A+R_2(\vec{A},\vec{A},\varepsilon)+\ldots \tag{33}$$

$$R_k(\vec{A},\ldots,\vec{A};\varepsilon) = \sum_{\{j_k\}} \varepsilon_1^{j_1}\ldots\varepsilon_p^{j_p}\, R_k^{(j_1,\ldots,j_p)}(\vec{A},\ldots,\vec{A}) \tag{34}$$

where $\varepsilon= \lambda-\lambda^*$.

The equations for the U's and R's are

$$\mathcal{L}\, U_k^{(j_1 \ldots j_p)} = I_k^{(j_1 \ldots j_p)} - R_k \tag{35}$$

for $k \geqslant 2$

where $\mathcal{L} = JA \frac{\partial}{\partial A} - L, \quad L = L_{\lambda *}$.

For k=1 we recover Arnold's theory of perturbations of singular matrices [6, 20]. The method summarized here is reminiscent of the Bogoliubov and Metropolski [9] (B-M) analysis of non-linear oscillations. The reason for this analogy is that the B-M method is by itself a variant of singular perturbation procedures, and that the procedure described here can reproduce all the variants of singular perturbation theories by choosing appropriate scaling relations between the A's and the ε_i's. This procedure works as well for partial differential equations with the condition that the spectrum of J corresponds to an isolated part of the spectrum of $L_{\lambda *}$. Two examples will be explicitly shown in the last section of this paper. The generalization to cases with continua in the spectrum (large systems) is the object of present work.

3. Complex Behaviours Associated with Polycritical Points and the Unfolding of Singular Vector Fields

We are going to consider the unfolding of some singularities, assuming that the normal form near the polycritical situation is not singular by itself. By this, it is meant that none of the coefficients of the lowest order truncation of the normal form are zero. The strategy will be first to discuss some aspects of the qualitative dynamics associated with the lowest order truncation of the amplitude equations, and then to indicate briefly what could be the effects of the higher order terms, not written in their normal form but treated as a perturbation. A typical dynamical behaviour is associated with each singularity. The elementary singularities are the ξ and ω singularity. The quadratic normal form of the ξ singularity (stationary bifurcation [10])

$$\dot{A} = \mu A + k_1 A^2 \tag{36}$$

describes two stationary solutions which collide and exchange their stability at $\mu=0$. This corresponds generally to the appearence of a dissipative structure in an open system like the connective cells in the Rayleigh Benard problem for example. The effects of higher order terms are not significant for $\mu \simeq 0$ if $k_1 \neq 0$. The ω-singularity (Hopf bifurcation [10]), whose lowest order normal form is

$$\dot{A} = (i\omega+\mu)A + k_1 |A|^2 A \tag{37}$$

describes the appearance of a periodic solution. Again the effects of higher order terms do not change the qualitative picture. The linear unfolding of ξ^2 is the matrix [6]

$$\begin{pmatrix} 0, & 1 \\ -\mu, & -\nu \end{pmatrix} \tag{38}$$

whose characteristic polynomial is: $s^2 + \mu s + \nu$.
The corresponding amplitude equation is

$$\ddot{A} = \mu A + \nu\dot{A} + k_1 A^2 + k_2 A\dot{A} \,. \tag{39}$$

Clearly the unfolding of ξ^2 contains ξ and ω. A typical behaviour associated with ξ^2, not present in either ξ or ω, is then the interaction between a periodic solution and a steady solution. This interaction leads to a collision between these two kinds of solution. A closed orbit passing through an unstable stationary solution is called a homoclinic orbit. The unfolding of the ξ^2 singularity is represented in Fig.1. Around such a double point we have very rich scenarios: for example the transition from a homogeneous state to a dissipative structure followed by the birth of an oscillation of this structure whose frequency decreases as we move some control parameter. The

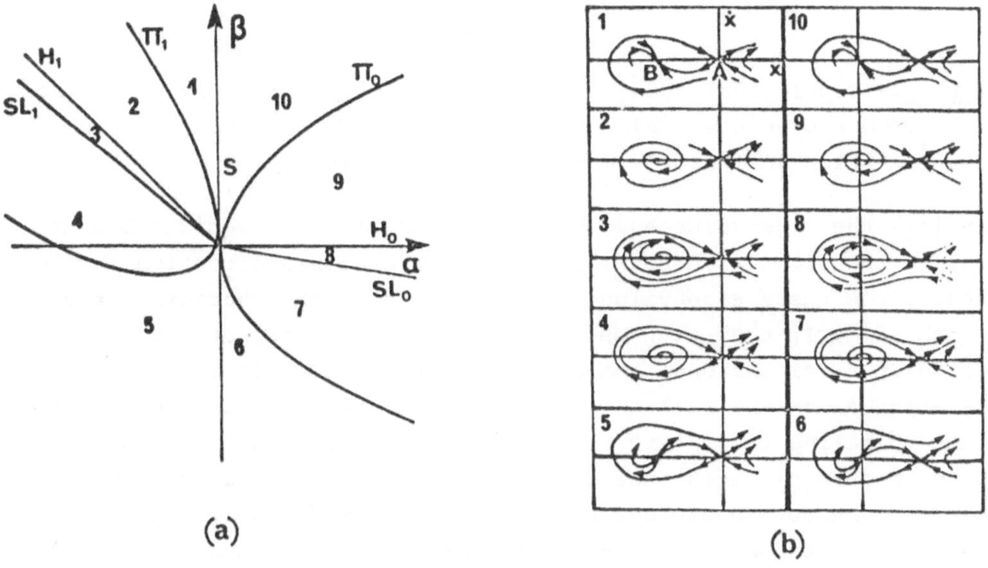

Figure 1. Unfolding of the ξ^2 singularity in the μ-ν plane

striking behaviour near such a double point is the possibility in the oscillatory phase to come back to the unstable initial state through the homoclinic process. When k_1 and k_2 are different from zero, the the higher order terms do not contribute to qualitative changes in the phase portrait of ξ^2. The situation is quite different for the triple instability ξ^3, whose linear unfolding corresponds to the matrix [6]

$$\begin{pmatrix} 0, & 1, & 0 \\ 0, & 0, & 1 \\ -\mu, & -\nu, & -\eta \end{pmatrix} \tag{40}$$

with its associated characteristic equation

$$s^3 + \eta s^2 + \nu s + \mu = 0 . \tag{41}$$

The unfolding of ξ^3 contains ξ^2, so that we can ask the question: what is qualitatively new in ξ^3? A simple answer is the existence of homoclinic orbits of a topological type that ξ^2 cannot have. The reduction from ξ^3 to ξ^2 (which occurs when $|\eta| \gg |\mu|, |\nu|$) allows one to understand in particular the existence of a planar homoclinic orbit as illustrated in Fig.2.

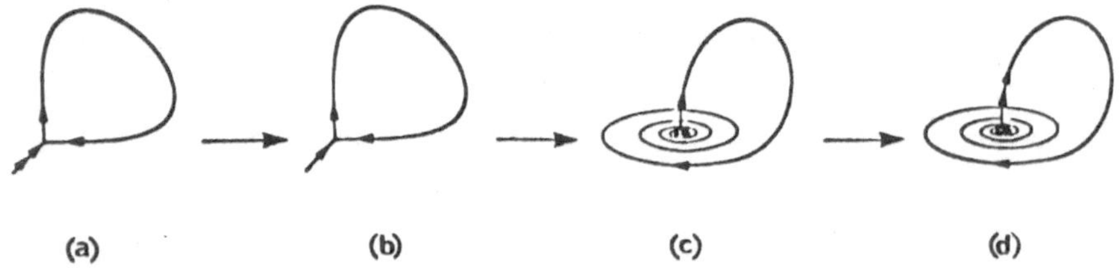

Figure 2. Topological change of the homoclinic orbit when we go from ξ^2 to ξ^3. (a) represents a planar orbit, while (d) corresponds to an orbit for which Shilnikov's conditions are satisfied.

When $|\eta|\downarrow$ the eigenvalue associated with the third direction becomes equal to the eigenvalue corresponding to the stable direction in the plane. Since these eigenvalues are the roots of the characteristic polynomial (41), they then become complex and, we expect the appearance of a fully tridimensional homoclinic orbit as shown in Fig.2. When the dilatation along the real eigendirection is greater than the contraction in the spiraling trajectory toward the origin, a theorem of SHILNIKOV [11,12] proves the existence of infinitely many instable periodic orbits and chaotic behaviour corresponding to infinitely many horseshoes in an appropriate Poincaré map of the third-order amplitude equation [13,14,15] (see Fig.3).

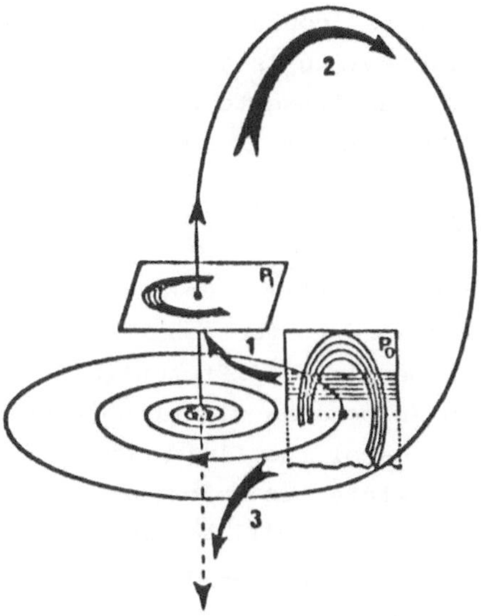

Figure 3. A model Poincaré map for the flow described by the unfolding of ξ^3. It consist of the composition of a linear local motion near the origin (1) and a non-local rigid transport along the homoclinic orbit (2). (3) corresponds to a divergent motion along the lower branch of the unstable manifold of the origin.

These conditions are shown to be numerically satisfied by the homoclinic orbit of the quadratic normal form associated with the ξ^3 singularity when one approaches the triple point $\mu=\nu=\eta=0$ with an appropriate path. The question is now, since ξ^3 contains ξ^2 which is known to display only simple behaviour, what kind of transitions or scenarios to chaos are likely to be observed in the unfolding of ξ^3? Explicit models of situations near a Shilnikov's homoclinic orbit display, as expected, the cascade of period doubling bifurcations as a way to create horseshoes. When the Shilnikov situation is expected, this pathway to chaos has always been observed in numerical simulations. The question of the effects of the higher order terms is very delicate since the unfolding of ξ^3 appears to describe a structurally unstable system. Nevertheless our feeling is that close enough to the triple point, these terms affect only fine details of the dynamics. The chaotic behaviours are preserved since horseshoes are structurally stable objects. The period doubling cascade, as a route to chaos, thanks to its codimensions one character through a renormalization group analysis [16-19], will also be present in the perturbed system. It is clearly not possible to describe with "words" the behaviour of a macrosystem near such a triple point. Roughly speaking a scenario in which the oscillation of the dissipative structure loses its stability and undergoes a sequence of period doubling bifurcations leading to weak turbulence is expected to occur. We have seen some examples of polycritical situations in which we suggested the existence of complex behaviours. We are tempted to conjecture that any kind of complicate non-linear behaviour can be seen near some polycritical point, that is, could appear in the unfolding of some singular vector field. This feeling is justified by some other examples [20]. For example we could envisage to study hyperchaos and hierarchies leading to it [24] with the ξ^n singularity.

4. Some Applications: Multiple Convection

In the following we are going to consider a macrosystem modelling some astrophysical and geophysical situations. The word convection refers mainly to fluid dynamical activity driven primarily by differential buoyancy forces. By multiple convection we mean motions with important buoyancy forces when several physical properties of the fluid may separately influence the stability of a given state. Examples of bi- and triconvection are considered. The system studied will be in every case a horizontal layer of fluid of finite thickness, submitted to a vertical gravitational field. Several other constraints will be successively applied to it: a gradient of temperature, a gradient of salinity, a rotation of the layer. For the sake of simplicity two basic basic assumptions are made. The motion of the fluid is supposed to be bidimensional. The velocity field and the temperature field will depend only on two coordinates: the vertical coordinate and one of the horizontal coordinates. The other assumption is the choice of idealized boundary conditions which generalize those of Rayleigh in his famous study of linear convection. The first example 8 allows us to illustrate the behaviour of a macrosystem near a ξ^2 singularity. Two constraints are then applied to the fluid: a vertical gradient of temperature, and a vertical gradient of salt. The appropriate Boussinesq equations for bidimensional convection, in suitable units, are

$$\begin{aligned}
\partial_t \nabla^2 \psi &= \sigma \nabla^4 \psi - \sigma R \partial_x \Theta + \sigma \tau S \partial_x \Sigma + \frac{\partial(\psi, \nabla^2 \psi)}{\partial(x,z)} \\
\partial_t \Theta &= - \partial_x \psi + \nabla^2 \Theta + \frac{\partial(\psi,\Theta)}{\partial(x,z)} \\
\partial_t \Sigma &= \partial_x \psi + \tau \nabla^2 \Sigma + \frac{\partial(\psi,\Sigma)}{\partial(x,z)} .
\end{aligned} \tag{42}$$

Here Θ and Σ are the deviations of the temperature and salinity from their static values. ψ is the stream function, x and z are the spatial coordinates, ∇^2 is the Laplacian and

$$\frac{\partial(\psi,\Theta)}{\partial(x,z)} = \partial_x \psi \partial_z \Theta - \partial_z \psi \, \partial_x \Theta . \tag{43}$$

The four parameters, R,S, σ and τ that appear in the equations are respectively called the thermal and saline Rayleigh numbers, the Prandtl number, and the Lewis number. We shall consider these equations in the spatial domain

$$0 \le x \le \frac{2\pi}{a} \quad ; \; 0 \le z \le 1 . \tag{44}$$

At the boundaries of this domain we apply the kinematic conditions

$$\psi = 0; \quad \nabla^2\psi = 0 . \tag{45}$$

On the top and bottom plates we impose zero temperature and salinity fluctuations:

$$\Theta = 0, \; \Sigma = 0, \quad z = 0,1 \tag{46}$$

and we allow no fluxes through the side walls

$$\partial_x\Theta=0, \; \partial_x\Sigma=0, \; x=0 , \; \frac{2\pi}{a} . \tag{47}$$

Introducing

$$U = \begin{pmatrix} \psi \\ \Theta \\ \Sigma \end{pmatrix} \qquad E = \begin{pmatrix} 1 \\ 0 \\ 0 \end{pmatrix} \quad \text{and} \tag{48}$$

$$M_\lambda = \begin{pmatrix} \sigma\nabla^4 & -\sigma R\partial_x & \sigma\tau S\partial_x \\ -\partial_x & \nabla^2 & 0 \\ \partial & 0 & \tau\nabla^2 \end{pmatrix}, \; D=\text{diag}(\nabla^2,1,1) \tag{49}$$

the equations become

$$\partial_t DU = M_\lambda U+N(U) \tag{50}$$

where

$$N(U) = (D\partial_z U.E^+)\partial_x U-(D\partial_x.E^+)\partial_z U . \tag{51}$$

We do not indicate any dependence of D and N on λ since such dependence arises only through boundary conditions. The appearance of D on the left side of (51) makes that equation different from (29). The modification required to bring them into agreement is slight. We could write $L_\lambda = D^{-1}M_\lambda$ and $N_\lambda = D^{-1}N$, but we prefer to leave (51) in the form it appears naturally. The double point is defined by

$$R_c = -R_o \frac{\sigma+\tau}{\sigma(\tau-1)}$$

$$S_c = R_o \frac{\tau(1+\sigma)}{\sigma(\tau-1)} \tag{52}$$

where

$$R_o = \frac{(a^2+\pi^2)^3}{a^2} \quad . \tag{53}$$

At this point we have a double zero for the characteristic equation. We can define Φ_1 and Φ_2 (generalized critical modes) by

$$M_{\lambda*}\Phi_1=0 \qquad M_{\lambda*}\Phi_2=D\Phi_1 \tag{54}$$

where λ^* represents any point in the parameter space on the polycritical surface defined by $R=R_c$, $S=S_c$.(Codimension two surface in the 5-dimensional space of the parameters R,S,σ,τ,a). A solution of (50) is then sought of the form

$$U = A_1(t)\Phi_1(x,z)+A_2(t)\Phi_2(x,z)+U_2(A,B)+... \tag{55}$$

together with an equation for the amplitudes A_1, A_2

$$\begin{aligned} \partial_t A_1 &= A_2+F_2(A_1,A_2)+F_3(A_1,A_2)+... \\ \partial_t A_2 &= G_2(A_1,A_2)+G_2(A_1,A_2)+... \end{aligned} \tag{56}$$

where U_k,F_k,G_k are homogeneous polynomials of degree k in the amplitudes.
Inserting (55), (56) into (50) we get U_2. F_2 and G_2 are identically zero because of some intrinsic symmetry of the system. F_3, the coefficients of the terms $A_1A_2^2$ and A_2^3 in G_3, can be chosen equal to zero. The resulting amplitude equation can be written

$$\ddot{A} = k_1A^3+k_2A^2\dot{A} \tag{57}$$

where k_1,k_2 are explicitly known coefficients, when λ is close to λ^*.
The corresponding amplitude equation is then

$$\ddot{A} = \nu\dot{A}+\mu A+k_1A^3+k_2A^2\dot{A} \tag{58}$$

where μ and ν are small parameters computed from the linearized theory. (58) differs from (39) by the presence of cubic non-linear terms instead of quadratic ones. The consequence of the dynamical behaviour represented by (39) is the existence of symmetric homoclinic and heteroclinic solutions as illustrated in Fig.4.

The second example [22,23] allows to illustrate the behaviour near a ξ^3 singularity. We consider the same system as before, but we apply to it a vectorial rotation. The equations become

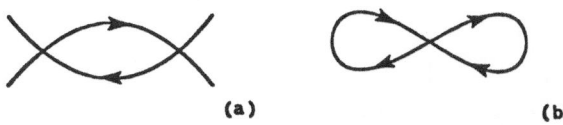

Figure 4. (a) Heteroclinic orbits present in the phase portrait of (58) when $k_1 > 0$; (b) homoclinic orbits displayed when $k_1 < 0$.

$$\partial_t \nabla^2 \psi = \sigma \nabla^4 \psi - \sigma R \partial_x \Theta + \sigma \tau S \partial_x \Sigma + \sigma^2 T \partial_z \gamma + \frac{\partial(\psi, \nabla^2 \psi)}{\partial(x,z)}$$

$$\partial_t \Theta = \nabla^2 \Theta - \partial_x \psi + \frac{\partial(\psi,\Theta)}{\partial(x,z)} \tag{59a}$$

$$\partial_t \Sigma = \tau \nabla^2 \Sigma + \partial_x \psi + \frac{\partial(\psi,\Sigma)}{\partial(x,z)}$$

$$\partial_t \gamma = \sigma \nabla^2 \gamma - \partial_z \psi + \frac{\partial(\psi,\gamma)}{\partial(x,z)} \tag{59b}$$

where γ represents the y velocity , T is the Taylor number. The additional boundary conditions are

$$\partial_z \gamma = 0 \quad \text{for } z=0 \text{ and } 1$$

$$\gamma = 0 \quad \text{for } x=0 \text{ and } \frac{\pi}{a} . \tag{60}$$

(59) can be written in the form

$$\partial_t DU = M_\lambda U + N(U) . \tag{61}$$

The tricritical surface is defined by

$$R = R_c = R_o \frac{\tau+2\sigma}{\sigma(1-\sigma)(1-\tau)}$$

$$S = S_c = -R_o \tau^3 \frac{(1+2\sigma)}{\sigma\tau(\tau-\sigma)(1-\tau)} \tag{62}$$

$$T = T_c = R_o \frac{a^2\sigma(\tau-1)(1+\sigma+\tau)}{\pi^2(\tau-\sigma)(1-\sigma)(1-\tau)}$$

for $\lambda=\lambda^*$ we can define the generalized null vectors

$$M_{\lambda^*} \Phi_1 = 0$$

$$M_{\lambda^*} \Phi_2 = D \Phi_1 \tag{63}$$

$$M_{\lambda^*} \Phi_3 = D \Phi_2 .$$

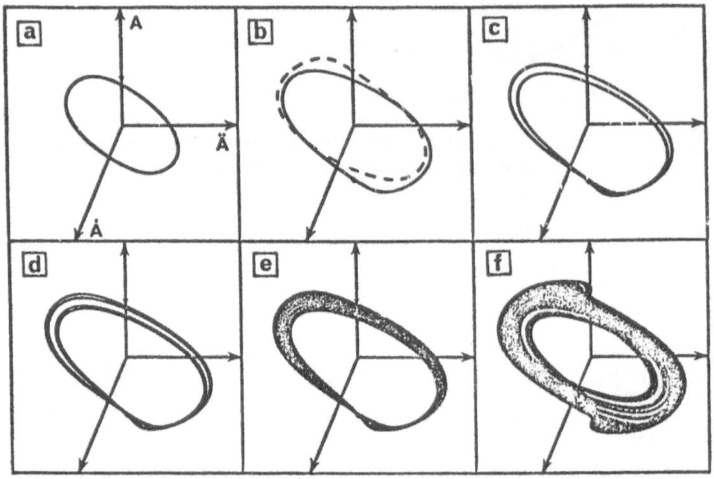

Figure 5. A scenario by which the dynamics of (65) becomes chaotic when $k_1 > 0$. One observes a symmetry breaking in (b), then a cascade of period doubling bifurcations leading to chaos (e). (f) shows a symmetric chaos solution. One goes from 5a to 5f by increasing $-\mu$, keeping fixed ν, η.

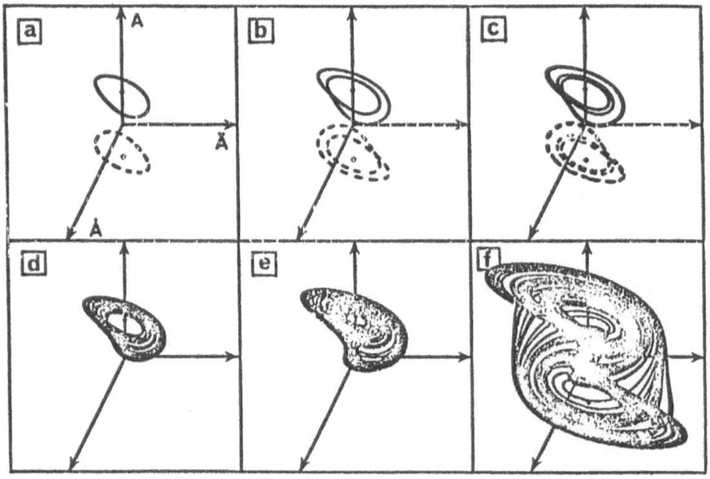

Figure 6. Scenario when $k_1 < 0$. Here the symmetry is broken by the initial stationary bifurcation. A cascade of period doubling is then observed when $-\mu$ increases. (f) illustrates a symmetric chaotic solution.

On the tricritical surface a solution of (61) can be sought of the form

$$U = A_1(t)\Phi_1(x,z) + A_2(t)\Phi_2(x,z) + A_3(t)\Phi_3(x,z) + U_2(A_1,A_2,A_3) + \ldots \quad (64)$$

The corresponding equation for the amplitude, when λ is close to λ^*, is given by

$$\dddot{A} = \eta\ddot{A}+\nu\dot{A}+\mu A+k_1A^3+k_2A^2\dot{A}+k_3A^2\ddot{A}+k_4A\dot{A}^2+k_5\dot{A}^3+k_6A\ddot{A}^2 \qquad (65)$$

where μ,ν,η and the k_i's are known functions. The equation (65) exhibits as expected chaotic behaviour describing weak turbulence for the geophysical system. The nature of the chaos can be understood as for the quadratic case by the presence of some homoclinic and heteroclinic orbits [23].

References:

1 F.Takens: Pub.IHES 43, 47 (1973)
2 V.I.Arnold: Russ.Math.Survey 27, 54 (1972)
3 H.Poincaré: Thèse 1879, Oeuvre 1 (Gautier Villars Paris 1928) 69
4 P.Hartman: Ordinary Differential Equations (Wiley, N.Y. 1964)
5 D.Grobman: Dokl.Akad.Nauk.USSR 128,880 (1965)
6 V.I.Arnold: Russ.Math,Survey 26, 29 (1971)
7 A.Kelley: J.Diff.Eqns. 3, 546 (1967)
8 P.H.Coullet, E.A.Spiegel: Amplitude Equations for Systems with Competing Instability (S.I.A.M. to be published)
9 N.N.Bogoliubov, Y.A.Mitropolsky: Asymptotic Methods in the Theory of Nonlinear Oscillations (Gordon and Breach, N.Y. 1961)
10 G.Iooss, D.D.Joseph: Elementary Stability and Bifurcation Theory (Springer Verlag, N.Y. 1980)
11 L.P.Shilnikov: Soviet Math. 6, 163 (1965)
12 L.P.Shilnikov, Math.USSR Sbornik 10, 91 (1970)
13 A.Arneodo, P.H.Coullet, C.Tresser: Comm.Math.Phys. 79, 573 (1981)
14 A.Arneodo, P.H.Coullet, C.Tresser: J.Stat.Phys. 27, 171 (1982)
15 A.Arneodo, P.H.Coullet, C.Tresser: Asymptotic Chaos (Physica D) to be published
16 M.J.Feigenbaum: J.Stat.Phys. 19, 25 (1978)
17 M.J.Feigenbaum: J.Stat.Phys. 21, 669 (1979)
18 P.H.Coullet, C.Tresser: J.Phys.Paris C5, 5 (1978)
19 C.Tresser, P.H.Coullet: C.R.A.S.Paris 287, 577 (1978)
20 P.H.Coullet: Chaotic Behaviours in the Unfolding of Singular Vector Fields, to appear in the Proceedings of the "Workshop on Common Trends in Particle and Condensed Matter Physics" at Les Houches (1983)
21 Rayleigh, Lord: Phil.Mag.(6) 32, 529 (1916)
22 A.Arneodo, P.H.Coullet, E.A.Spiegel: Phys.Lett. 92A, 369 (1982)
23 A.Arneodo, P.H.Coullet, E.A.Spiegel: The Dynamics of Triple Convection (Preprint 1983)
24 O.E.Roessler, this volume

Part III

Chaotic Dynamics-Real Systems and Experimental Verification

Chaos in Classical Mechanics: The Double Pendulum

P.H. Richter and H.-J. Scholz
Fachbereich Physik der Universität Bremen
D-2800 Bremen 33, Fed. Rep. of Germany

The dynamics of a double pendulum is presented in terms of Poincaré sections. It is shown that the simple classical textbook example displays all the complexity of non-integrable Hamiltonian systems.

1. Introduction

The following note describes an experiment. As far as we know the experiment has never been performed but that is not important: nobody questions the validity of Newtonian mechanics for systems with velocities small compared to c, and actions much larger than ħ. In fact, it may not even be wise to do the actual experiment in order to understand the double pendulum: there is always noise and dissipation present which both interfere strongly with the dynamics of the ideal system; if friction is compensated for by driving then an extra dimension is introduced which further complicates the analysis. And finally, given that numerical computation is by now much faster than classical experimentation—why should anybody put an effort in the real thing?

So let us try to solve the equations of motion for the ideal double pendulum (Fig.1). The Lagrangian can be looked up in any textbook of classical mechanics [1].

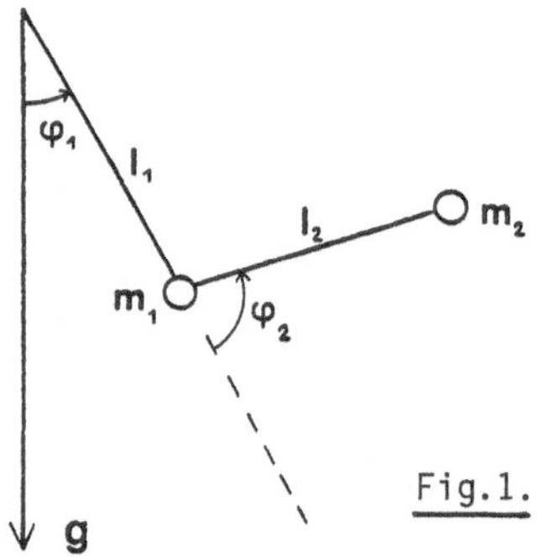

Fig.1. The double pendulum

Those books, however, tend to stop short of the interesting part of the analysis. They seem to tacitly assume that deriving the equations of motion is the hard part. Nowhere except in *Arnold*'s book [2] have we found a hint to the bewildering complexity that this simple system exhibits. The aim of this presentation is to demonstrate some of it. There is a growing number of examples in the literature that show complex dynamical behaviour in Hamiltonian systems: they are from celestial mechanics [3,4], from accelerator physics [5], plasma physics [6], or solid state physics [7-10]. We have found it remarkable that the seemingly simplest textbook systems, such as the double pendulum, give rise to just as complex a picture.

2. The Physical Background

The equations of motion are derived in the standard way from the Lagrangian $L = T - V$ where kinetic and potential energy are given as follows

$$T/m_1\ell_1^2 = \frac{1}{2}(1+\mu)\dot{\varphi}_1^2 + \mu\ell\cos\varphi_2\dot{\varphi}_1(\dot{\varphi}_1+\dot{\varphi}_2) + \frac{1}{2}\mu\ell^2(\dot{\varphi}_1+\dot{\varphi}_2)^2 \tag{1}$$

$$V/m_1g\ell_1 = (1+\mu)(1-\cos\varphi_1) + \mu\ell(1-\cos(\varphi_1+\varphi_2)) \quad . \tag{2}$$

Here we have introduced the two parameters

$$\mu = m_2/m_1 \quad , \quad \ell = \ell_2/\ell_1 \quad . \tag{3}$$

From (1) we obtain the angular momenta $L_i = \partial T/\partial\dot{\varphi}_i$

$$L_1/m_1\ell_1^2 = (1+\mu+2\mu\ell\cos\varphi_2+\mu\ell^2)\dot{\varphi}_1 + \mu\ell(\ell+\cos\varphi_2)\dot{\varphi}_2 \tag{4}$$

$$L_2/m_1\ell_1^2 = \mu\ell(\ell+\cos\varphi_2)\dot{\varphi}_1 + \mu\ell^2\dot{\varphi}_2 \quad . \tag{5}$$

It is easy to verify that L_1 is the total angular momentum of the system whereas L_2 is formally conjugate to φ_2 but without an obvious interpretation.

Using the fact that the total energy $E = T + V$ is a conserved quantity we introduce dimensionless variables by scaling

$$t\sqrt{E/m_1\ell_1^2} \rightarrow t \quad ; \quad L_i/\sqrt{Em_1\ell_1^2} \rightarrow \lambda_i \quad . \tag{6}$$

This leaves us with the Hamiltonian

$$h \equiv \frac{T+V}{E} = \frac{1}{1+\mu\sin^2\varphi_2}\left\{\frac{1}{2}\lambda_1^2 - \frac{\ell+\cos\varphi_2}{\ell}\lambda_1\lambda_2 + \frac{1+\mu+2\mu\ell\cos\varphi_2+\mu\ell^2}{2\mu\ell^2}\lambda_2^2\right\}$$

$$+ \gamma\{(1+\mu)(1-\cos\varphi_1) + \mu\ell(1-\cos(\varphi_1+\varphi_2))\} \quad . \tag{7}$$

The parameter γ measures the strength of gravity relative to the total energy

$$\gamma = m_1g\ell_1/E \quad . \tag{8}$$

The Hamiltonian (7) shows that the physics of the double pendulum depends on the three parameters μ, 1, and γ. The equations of motion have the canonical form

$$\dot{\varphi}_i = \frac{\partial h}{\partial \lambda_i} \quad , \quad \dot{\lambda}_i = - \frac{\partial h}{\partial \varphi_i} \quad . \tag{9}$$

Because of energy conservation they describe a motion on the three dimensional energy surface $h = 1$ in four-dimensional phase space $(\varphi_1, \lambda_1, \varphi_2, \lambda_2)$. The energy surface has two obvious symmetries: reflection invariance in space $(\varphi_{1,2} \to -\varphi_{1,2})$ and time $(\lambda_{1,2} \to -\lambda_{1,2})$. In case of vanishing gravity $\gamma = 0$ (or very high energy E), there is also invariance with respect to rotation, $\varphi_1 \to \varphi_1 + \alpha$. Then angular momentum λ_1 is a second conserved quantity besides energy, and the system is integrable. When $\gamma \to \infty$ the system is again integrable because the Hamiltonian becomes quadratic in its variables (see next section). But except for those limiting cases, total energy conservation is the only physical restriction to the motion.

To get a feeling for the potential consider Fig.2 where the equipotential lines are shown for three typical cases. As can be seen from (7) the potential assumes its minimum value (zero) for $\varphi_1 = \varphi_2 = 0$. The backfolded situation $\varphi_2 = \pi$ corresponds to a saddle point of the potential if $\varphi_1 = 0$ $(V/\gamma E = 2\mu\ell)$ or $\varphi_1 = \pi (V/\gamma E = 2(1+\mu))$. The two saddles become degenerate when the center of gravity for $\varphi_2 = \pi$ coincides with the suspension point, $\ell = (1+\mu)/\mu$, Fig.2b. When the second pendulum is shorter or longer the potential looks as in Fig.2a,c respectively. The motion is restricted to angles $|\varphi_{1,2}| < \pi$ when the potential energy exceeds E at the lower saddle point, i.e. for $2\gamma \cdot \min(\mu\ell, 1+\mu) > 1$. Both angles are unrestricted when the potential top is below E, i.e. for $2\gamma(1+\mu+\mu\ell) < 1$.

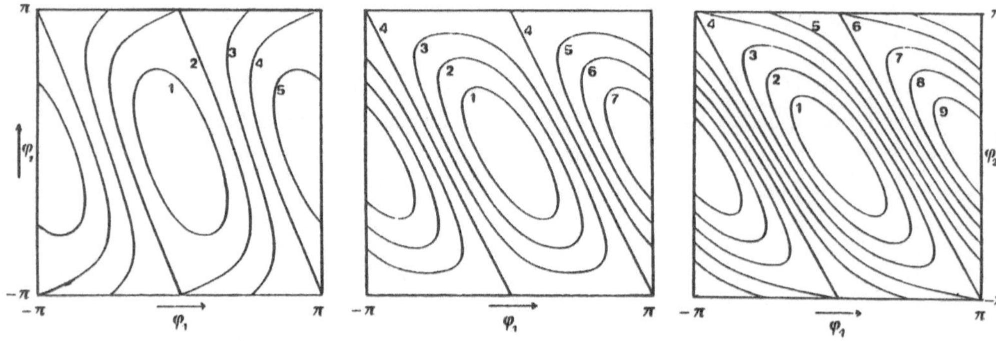

Fig.2a-c. Equipotential lines $V/\gamma E = 1,2,3,\ldots$ (as indicated) for equal masses $\mu = 1$ and length ratios $\ell = 1$ (*left*), $\ell = 2$ (*center*), and $\ell = 3$ (*right*)

3. The Integrable Limiting Cases

We mentioned already that the cases $\gamma = 0$ and $\gamma \to \infty$ are integrable: there is a second conserved quantity besides the total energy. The motion is thus confined to a two-dimensional submanifold of the energy surface which turns out to be topologically equivalent to a torus.

a) $2\gamma \gg \max(1/\mu\ell,\ 1/(1+\mu))$. This is the case of strong gravity or small energy E. The motion is confined to small values of φ_1, φ_2 so that the Hamiltonian (7) may be approximated by

$$h = \frac{1}{2}\lambda_1^2 - \left(1 + \frac{1}{\ell}\right)\lambda_1\lambda_2 + \frac{1 + \mu(1+\ell)^2}{2\mu\ell^2}\lambda_2^2$$

$$+ \frac{1}{2}\gamma\left((1 + \mu\ (1 + \ell))\varphi_1^2 + 2\mu\ell\varphi_1\varphi_2 + \mu\ell\varphi_2^2\right) \quad . \tag{10}$$

As shown in [1] this quadratic form can be diagonalized. The two independent harmonic oscillators have frequencies

$$\omega_{1,2}^2 = \frac{\gamma}{2\ell}\left\{(1+\mu)(1+\ell) \pm \sqrt{(1+\mu)^2(1+\ell)^2 - 4\ell(1+\mu)}\right\} \quad . \tag{11}$$

Each of these oscillations can be described in terms of action-angle variables (I_i, Θ_i), and their connection to the old variables (λ_i,φ_i) be given explicitely. Suffice it to mention that the Hamiltonian reads

$$h = I_1\omega_1 + I_2\omega_2 \tag{12}$$

from which it follows that $I_i = \text{const.}$ and $\Theta_i = \omega_i t$. The toroidal nature of the motion is thereby obvious: it is the direct product of two circular motions. The frequency ratio ω_1/ω_2 or "winding number" tells us how many turns along the circle the 2-motion completes while the 1-motion goes once around. For $\mu = \ell = 1$ we find $\omega_1/\omega_2 = \sqrt{2}+1$ (= [2,2,2,...] in terms of continued fractions). In general we infer from $\omega_i = \partial h/\partial I_i$ that

$$W \equiv \frac{\omega_1}{\omega_2} = -\left.\frac{\partial I_2}{\partial I_1}\right|_h \quad . \tag{13}$$

For the Hamiltonian (12) this ratio is independent of I_1 throughout the range $0 \leq I_1 \leq 1/\omega_1$; it only depends on the parameters μ and ℓ. The lines of constant winding numbers W in the ℓ,μ-plane are given by

$$\mu = \frac{\ell}{(1+\ell)^2}\,\frac{(1+W^2)^2}{W^2} - 1 \quad . \tag{14}$$

In terms of the old variables φ_i, λ_i the motion is a superposition of the two eigenmodes. For irrational winding numbers this means it is quasiperiodic and the φ_i versus λ_i phase portraits are filled densely if the complete motion is recorded. We thus follow Poincaré's prescription and consider φ_1,λ_1 only after full periods of the φ_2-motion ($\varphi_2 = 0$, $\dot{\varphi}_2 > 0$), and similarly for the φ_2,λ_2 values. The phase portraits so generated can be viewed as deformed sections across the invariant tori mentioned above: The successors to an initial point must all lie on the curve corresponding to the invariant torus. That curve will eventually be filled in case of irrational ω_1/ω_2, whereas for rational winding numbers a finite number of points on it will be generated periodically. The last picture in the sequence of Fig.8 is an example of this kind of representation (to the extent that $2\gamma = 4$ is large compared to 1).

b) $\gamma = 0$. In the extreme case of vanishing gravity (which may be realized by putting the double pendulum flat on a table), the energy surface no longer depends on φ_1, and so the total angular momentum λ_1 is conserved during the motion. Again this implies the existence of invariant tori, but this time the picture is more involved because the φ_2-motion can be of two types: it is libration (oscillation, $|\varphi_2| < \pi$) in case λ_1 is near its maximum possible value, and rotation ($\dot{\varphi}_2$ not changing sign) when λ_1 is small which means the angular momentum of the φ_2-motion compensates that of the φ_1-rotation. To verify these statements consider (4) through (7) from which we find

$$\lambda_2 = \frac{\mu\ell(\ell + \cos\varphi_2)\lambda_1 \pm \mu\ell^2(1 + \mu\sin^2\varphi_2)\dot{\varphi}_2}{1 + \mu + 2\mu\ell\cos\varphi_2 + \mu\ell^2} \tag{15}$$

$$\dot{\varphi}_2 = \pm\sqrt{\frac{2(1 + \mu + 2\mu\ell\cos\varphi_2 + \mu\ell^2) - \lambda_1^2}{\mu\ell^2(1 + \mu\sin^2\varphi_2)}} \quad . \tag{16}$$

The maximum possible value of $|\lambda_1|$ is seen to be

$$\lambda_1^{max} = \sqrt{2(1 + \mu(1 + \ell)^2)} \quad , \tag{17}$$

and occurs when the double pendulum is stretched out ($\varphi_2 = 0$). From (16) we see that $\dot{\varphi}_2$ cannot be zero if $|\lambda_1| < \lambda_1^{sep}$ where

$$\lambda_1^{sep} = \sqrt{2(1 + \mu(1 - \ell)^2)} \quad . \tag{18}$$

We conclude that the φ_2-motion is rotation for $|\lambda_1| < \lambda_1^{sep}$, and libration for $|\lambda_1|$ between λ_1^{sep} and λ_1^{max}.

A way to represent this situation is in terms of action-angle variables. Because of angular momentum conservation, λ_1 is itself the action variable I_1 (although φ_1 is not the corresponding angle Θ_1!). The second action I_2 is therefore obtained by integrating (15) over a period of the motion

$$I_2 = \frac{1}{2\pi}\oint \lambda_2 d\varphi_2 \quad .$$

Figure 3 shows a few representative plots of I_1 versus I_2, for the energy surfaces $h = 1$. The separatrices $I_1 = \pm\lambda_1^{sep}$ show up as discontinuities of the action I_2. Taking the derivatives according to (13) we see that the winding numbers W depend strongly on I_1. Figure 4 displays them for the case $\mu = \ell = 1$. It can be shown that W diverges logarithmically at the separatrix. This is reminiscent of the critical

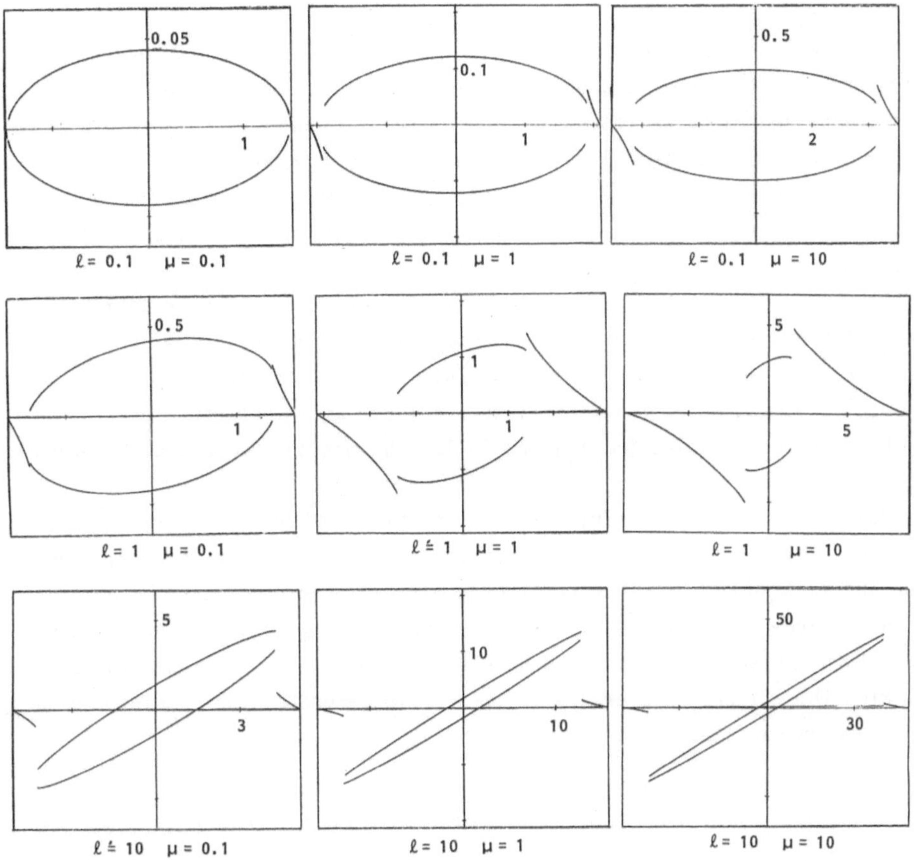

Fig.3. Energy surfaces in action variable representation. The plots show $I_1 = \lambda_1$ versus I_2 at constant energy $h = 1$, for nine different sets of parameters μ, ℓ, as indicated. At large values of I_1, $|I_1| < \lambda_1^{sep}$, the φ_2-motion is libration whereas for $-\lambda_1^{sep} < I_1 < \lambda_1^{sep}$ the angle φ_2 rotates in either of the two possible directions. At fixed μ, the ratio $\lambda_1^{sep}/\lambda_1^{max}$ tends towards 1 for $\ell \to 0$ and $\ell \to \infty$; it is constant at lines $\mu = 1/(\ell/\mu_1 - (\ell - 1)^2)$, $0 < \mu_1 < \infty$

slowing down at phase transitions, and indeed, a symmetry is being broken at the transition from libration to rotation, namely that of time reversal invariance.

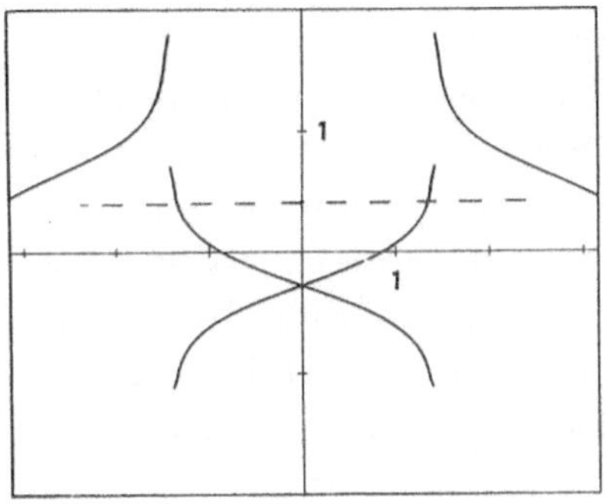

Fig.4. Winding number $W = \omega_1/\omega_2$ as a function of $\lambda_1 = I_1$; $\mu = \ell = 1$. The dashed line shows the winding ratio $\sqrt{2} - 1$ which characterizes the motion at $\gamma \to \infty$, independent of initial conditions

4. Energy Surfaces

The problem of understanding the double pendulum for values of γ between zero and infinity might be described as follows: find the transition between the pictures of Fig.3 ($\gamma = 0$) and the straight lines $\omega_2 I_2 = 1 - \omega_1 I_1$ ($\gamma \to \infty$, see (12)). The catch is that orbits are not necessarily confined to tori any more: when a torus breaks up the action variable is no longer defined. We shall see in the next section that the connection between the two integrable limits is far from being smooth. Rather we find that upon increasing γ, more and more invariant tori disappear until no one is left; the new invariant tori and corresponding action variables gradually emerge from chaos as γ is further increased. This implies that action-angle variables are not very suitable for intermediate γ-values. We therefore return to the original phase space variables (φ_i, λ_i) and follow the motion by means of Poincaré sections.

Before doing so it may be helpful for the intuition to have a picture of the energy surface. This is easily obtained when $\gamma = 0$ since then there is no φ_1-dependence, and any section $\varphi_1 = \text{const.}$ gives the same two-dimensional surface in λ_1, φ_2, λ_2-space. Figure 5 illustrates this surface for $\mu = \ell = 1$. The lines are levels of constant λ_1 which should be imagined as a third coordinate pointing towards the reader. Taking this figure once around the φ_1-circle generates the complete three-dimensional energy surface.

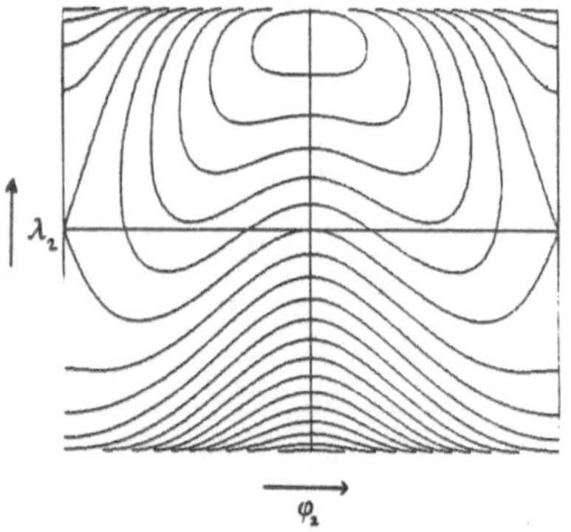

Fig.5. Energy surface for $\mu = \ell = 1$, $\gamma = 0$. The abscissa is φ_2, the ordinate λ_2. Level lines $\lambda_1 = \text{const.}$ are shown

When $\gamma \neq 0$ the energy also depends on φ_1. The sections $\varphi_1 = \text{const.}$ are then all different. Figure 6 shows them at eight angles $\varphi_1 = n\pi/4$, again using λ_1 as a level coordinate. For $\gamma = 0.1$ (Fig.6a) the energy surface still extends to all φ_1-values whereas for $\gamma = 0.4$ (Fig.6b) it does not go much beyond $|\varphi|_1 = \pi/2$. In contrast to the case $\gamma = 0$, the level lines $\lambda_1 = \text{const.}$ no longer represent invariant tori; however, for small γ it will be interesting to observe how the dynamics deviates from angular momentum conservation, and how with growing γ it eventually ignores the level lines completely. An alternative view on the same two energy surfaces as in Fig.6 is presented in Fig.7 where the level lines $\lambda_2 = \text{const.}$ are given in the φ_1, λ_1-plane, for different sections $\varphi_2 = n\pi/4$.

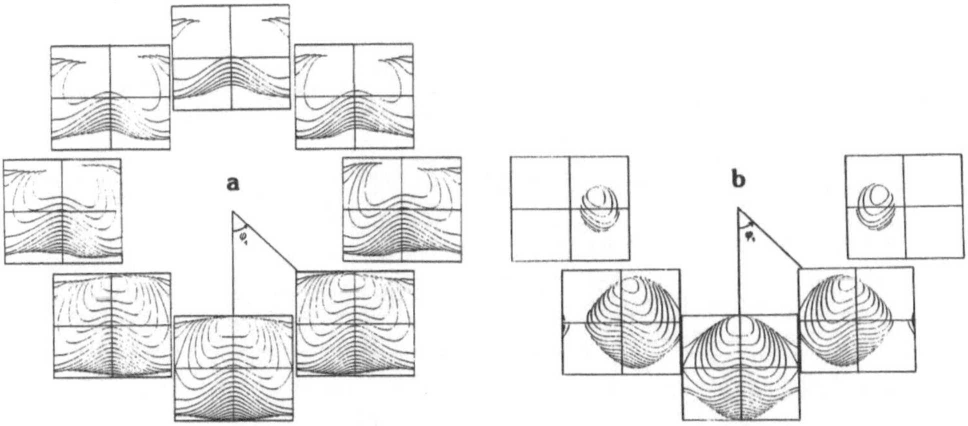

Fig.6. (a): Energy surface for $\mu = \ell = 1$, $\gamma = 0.1$. The φ_2 versus λ_2-phase portraits are shown for eight angles $\varphi_1 = 0, \pm\pi/4, \pm\pi/2, \ldots$. The lines show levels of constant λ_1. (b) Energy surface as in (a), except for $\gamma = 0.4$

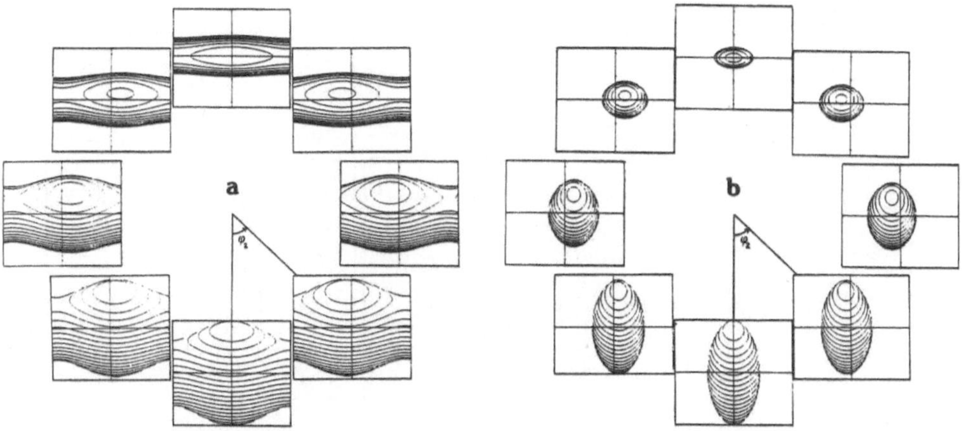

Fig.7. (a): The same energy surface as in Fig.6a, from a different perspective: φ_1 versus λ_1-phase portraits at angles $\varphi_2 = 0, \pm\pi/4, \ldots$. The level lines are taken at constant λ_2. (b) The same energy surface as in (b), from the perspective of (a)

5. Chaos and Order

We are now ready to take a glance at the complexity of the double pendulum. For the special case $\mu = \ell = 1$ we consider two series of Poincaré sections that show how the dynamics changes as γ is increased. Figure 8 collects 12 φ_1 versus λ_1-phase portraits ($\varphi_2 = 0$, $\dot{\varphi}_2 > 0$). First in the series is the integrable case $\gamma = 0$; all orbits are confined to straight lines ($\lambda_1 = \text{const.}$). 600 image points have been calculated for each of 16 initial conditions. Most of these orbits are quasi-periodic (or of very high period), except for one orbit of period 1 in the upper part

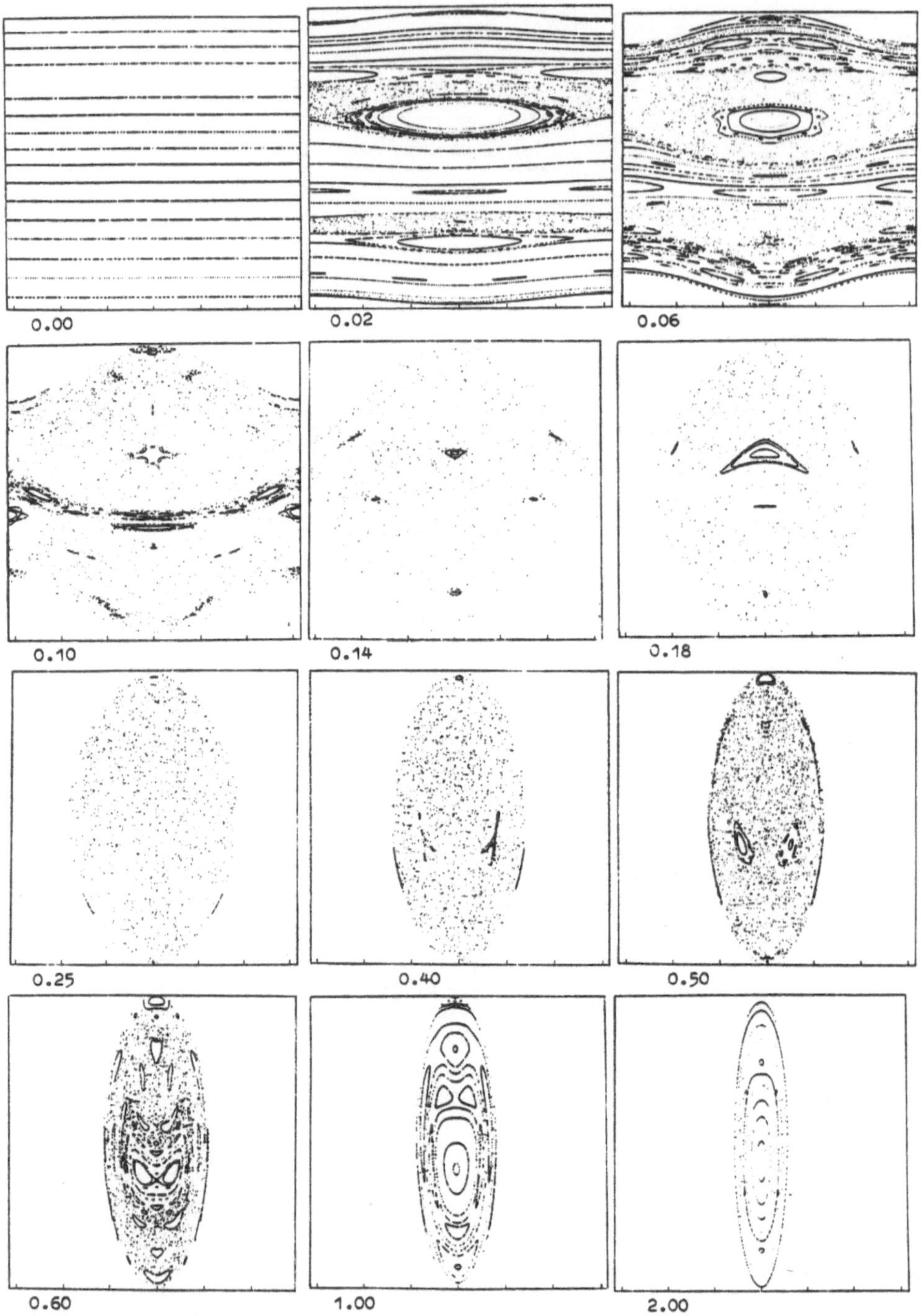

Fig.8. Poincaré sections φ_1 versus λ_1 at $\varphi_2 = 0$, $\dot{\varphi}_2 > 0$. $\mu = \ell = 1$, γ as indicated

of the figure. The next picture shows the case $\gamma = 0.02$. We observe a number of typical features of this kind of Poincaré maps. (i) As the nonintegrable perturbation is fairly small a lot of invariant tori continue to exist, albeit slightly deformed with respect to the case $\gamma = 0$. From the Kolmogoroff-Arnold-Moser (KAM) theorem [6] we know that these tori must have "very irrational" winding numbers. (ii) The tori with rational winding numbers for $\gamma = 0$ have broken up into "chains of islands" along which there is a succession of elliptic and hyperbolic periodic points. This behaviour reflects the Poincaré-Birkhoff theorem [6]. (iii) Some hyperbolic points have become centers of rather broad "chaos bands" where the images of an initial point do not lie on a curve any more but fill a whole area densely. This is an obvious manifestation of non-integrability; the dynamics has become ergodic on part of the energy surface. (iv) The homogeneity of points within the chaos bands reflects that the Poincaré maps are area preserving. For a proof of this statement we refer to the book by *Siegel* and *Moser* [11].

The next pictures ($\gamma = 0.06$, 0.1, 0.14) describe how more and more invariant tori are eaten up by the expanding chaos bands. The last KAM torus that is about to disappear at $\gamma = 0.1$ has winding number $W = 0.3820 = [0,2,1,1,1...]$. It is no accident that this is just the golden mean ratio: no continued fraction expansion converges slower than this succession of 1's which characterizes the golden mean as "the most irrational" number. It is in line with the KAM theorem that the corresponding torus should survive longest again nonintegrable perturbations, and experiments by *Greene* [12] as well as theoretical work by *Escande* [13] and *MacKay* [14] have confirmed this expectation. The figure for $\gamma = 0.1$ also shows that the golden mean torus is approached from alternating sides by periodic orbits of periods 2,3,5,8,13,... - one recognizes the well known Fibonacci series. Their winding numbers are rational approximations to the limiting value of 0.3820: 1/2, 1/3, 2/5, 3/8, 5/13, ...

For $\gamma = 0.18$, 0.25, 0.40, 0.50 the dynamics is largely dominated by chaos, i.e., except for a few small elliptic island the energy surface is densely covered by chaotic trajectories. When γ is further increased both angles φ_1, φ_2 are confined to values less than π, and more and more regularity emerges as the Hamiltonian becomes effectively quadratic in its variables.

To complement this series of views on the dynamics along the energy surface, Fig.9 presents some corresponding Poincaré sections at $\varphi_1 = 0$, $\dot{\varphi}_1 > 0$ (φ_2 versus λ_2-phase portraits). At small values of γ the lines of constant λ_1 are still meaningful for the dynamics, as comparison with Figs.5 and 6a reveals. The strongest violation of angular momentum conservation occurs near the separatrix between libration and rotation. At larger values of $\gamma (\gamma \gtrsim 0.1)$ the motion ignores the λ_1-levels. In agreement with Fig.8 the dynamics is overwhelmingly chaotic for $0.1 \lesssim \gamma \lesssim 0.5$, and gains regularity at further increased γ.

Summing up, it may be said that the integrable limits $\gamma = 0$ and $\gamma \to \infty$ are so different in character that the transition between these two kinds of order has to first completely destroy one of them before the other can emerge. We have not at-

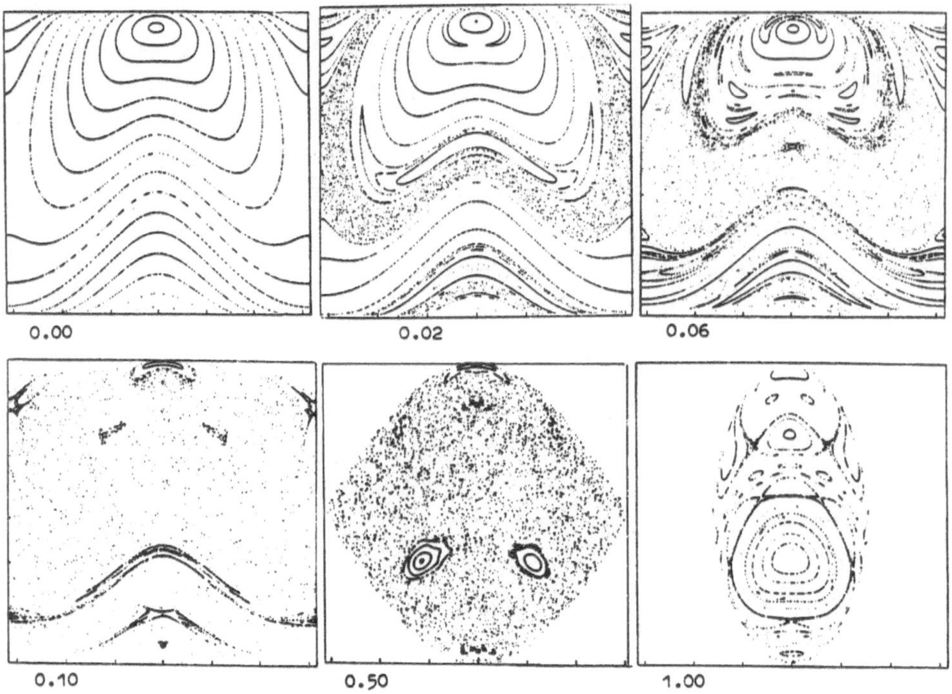

Fig.9. Poincaré sections φ_2 versus λ_2 at $\varphi_1 = 0$, $\dot{\varphi}_1 > 0$. $\mu = \ell = 1$, γ as indicated

tempted here to give an anlytic treatment of this very complex dynamics. An analysis on the basis of the theory developed in [13] and [15] can, however, be performed and will be presented elsewhere: on the basis of a renormalization group approach the width of the chaotic bands will be calculated as well as the critical -values where KAM tori break up. The survey on the double pendulum that we gave in this little note was only meant to illustrate the beautiful complexity of one of the simplest physical systems.

References

1. L.D. Landau, E.M. Lifshitz: *Course of Theoretical Physics*, Vol.1, Mechanics, 3rd ed. (Pergamon, Oxford 1980)
2. V.I. Arnold: *Mathematical Methods of Classical Mechanics* (Springer, Berlin, Heidelberg, New York 1978)
3. J. Moser: *Stable and Random Motions in Dynamical Systems* (Princeton University Press 1973)
4. M.V. Berry: Regular and Irregular Motion. In: *Topics in Nonlinear Dynamics*. Amer. Inst. Phys. Conf. Proc. **46**, S. Jorna (ed.) 1978
5. R.H.G. Helleman: One Mechanism for the Onset of Large-Scale Chaos in Conservative and Dissipative Systems. In: *Long-Time Prediction in Dynamics*, ed. by W. Horton et al. (Wiley, New York 1982)
6. A.J. Lichtenberg, M.A. Lieberman: Regular and Stochastic Motion (Springer, Berlin, Heidelberg, New York 1983)
7. S. Aubry: The New Concept of Transitions by Breaking of Analyticity in a Crystallographic Model. In: *Solutions and Condensed Matter*, ed. by A.R. Bishop and T. Schneider (Springer, Berlin, Heidelberg, New York 1979)

8. P. Bak: Commensurate Phases, Incommensurate Phases, and the Devil's Staircase, Rep. Prog. Phys. **45**, 587-629 (1982)
9. H. Büttner, H. Bilz: Nonlinear Excitations in Ferroelectrics. In: *Recent Developments in Condensed Matter Physics*, Vol.1, J.T. Devreese (ed.) (Plenum, New York 1981)
10. M.E. Fisher, W. Selke: Low Temperature Analysis of the Axial Next-Nearest Neighbour Ising Model Near its Multiphase Point. Phil. Trans. R. Soc. **302**, 1-44 (1981)
11. C.L. Siegel, J.K. Moser: *Lectures on Celestial Mechanics* (Springer, Berlin, Heidelberg, New York 1971)
12. J.M. Greene: A Method for Determining a Stochastic Transition. J. Math. Phys. **20**, 1183-1201 (1979)
13. D.F. Escande: Large-Scale Stochasticity in Hamiltonian Systems. Physica Scripta T2/1, 126-141 (1982)
14. R.S. MacKay: Renormalization in Area Preserving Maps. Ph.D. Thesis, Princeton 1982
15. M.S. Mohamed-Benkadda: Calcul de Seuil de Stochasticité et Universalité pour des Systèmes Hamiltoniens: Applications à la Physique des Plasmas. Thèse, Orsay 1983

Chaos in Continuous Stirred Chemical Reactors

J.L. Hudson and J.C. Mankin

Department of Chemical Engineering, Thornton Hall, University of Virginia
Charlottesville, VA 22901, USA

O.E. Rössler

Institute for Physical and Theoretical Chemistry, University of Tübingen
D-7400 Tübingen, Fed. Rep. of Germany

Oscillations and chaos in open, well-mixed chemical reactors are discussed. We first review experimental results obtained with the Belousov-Zhabotinsky reaction in an isothermal reactor, and describe the series of oscillation types which have been obtained with variation of a single parameter, the reactor residence time. We then describe simulated behavior of two reactors, coupled by means of heat exchange, in which a first order, exothermal reaction is taking place.

1. Introduction

Experimental evidence of chaotic behavior in chemical systems is relatively new. SCHMITZ, GRAZIANI, and HUDSON [1] published data obtained with the Belousov-Zhabotinsky reaction which showed evidence of chaotic states. OLSON and DEGN [2] presented results on chaos in a biochemical system, the horseradish peroxidase reaction. These studies were guided by the work of RÖSSLER [3] who had shown that simple sets of three ordinary differential equations could produce chaos, thus indicating that chaos was most likely to be found in laboratory chemical reactors.

Experimental research on the Belousov-Zhabotinsky reaction continued. HUDSON et al. [4] showed that an entire sequence of states, some periodic and some chaotic, could be obtained with variation of a single parameter, the flow rate or residence time. ROUX et al. [5] and TURNER et al. [6] have since shown similar behavior.

The investigation of chaotic behavior in chemical reactors is now quite active. For example, SCHMITZ et al. [7] have obtained chaotic behavior in a system of three ordinary differential equations governing a surface reaction. KAHLERT et al. [8] and JORGENSON and ARIS [9] have shown that chaos is possible with two consecutive non-isothermal reactions in a CSTR. Even higher forms of chaos (more than one positive Lyapunov characteristic exponent) is likely to be found in chemical systems (RÖSSLER and HUDSON [10]). There have also been several studies in heterogeneous systems governed by partial differential equations; these problems will not be discussed in this paper.

We start by reviewing experimental evidence of chaos in the Belousov-Zhabotinsky reaction. We will mention a few models which have been developed to describe the oscillations seen in the laboratory, but will not discuss them in detail since Professor Noyes discusses the relationship between models and experiment elsewhere in this volume. Rather, we would like to introduce another simple model which is capable of producing chaotic behavior, viz., that of a single exothermic reaction taking place in two stirred tank reactors which are coupled by heat transfer. This model is of interest since it shows that complicated behavior is possible even with the simplest chemical reaction, if transport (diffusion or transfer) occurs; furthermore, this system of four ordinary differential equations shows behavior which is possibly more complicated than that obtained in the three variable systems mentioned above.

A more detailed discussion of recent work on chaos in stirred chemical systems (governed by ordinary differential equations) can be found in HUDSON and RÖSSLER [11].

2. Experiments with the Belousov-Zhabotinsky Reaction

The experiments were carried out in an isothermal continuous stirred tank reactor of volume 25.4 ml. The reactants are fed by means of peristaltic pumps and most recently by means of precise constant volume pumps. Data are taken with a platinum wire electrode and a bromide ion electrode which are connected to a digital computer.

Data have been obtained as a function of flow rate, temperature, and feed concentration. We limit our discussion here to results obtained at a single temperature (25°C) and mixed feed concentrations (malonic acid, sodium bromate, sulfuric acid, and cerous ion) of 0.3, 0.14, 0.2, and 0.001 M, respectively.

We discuss here the results obtained with variation of one parameter, the flow rate to the reactor. Further information on the apparatus and observations can be found in [4,12,13].

With increase in flow rate to the reactor from 2.9 ml/min to 5.5 ml/min the following sequence of oscillations has been obtained:

$\Pi(1)$, $\Pi(1,2)$, $\Pi(2)$, $\chi(2,3)$, $\Pi(2,3)$, $\chi(2,3)$, $\Pi(3)$, $\chi(3,4)$, $\Pi(4)$,
$\chi(4,5)$, $\Pi(5)$, $\Pi(m)$, $\Pi'(1)$, s.s.

$\Pi(n)$ denotes a periodic oscillation having n peaks, $\Pi(n, n+1)$ denotes a periodic oscillation which alternates between n and n+1 peaks, and $\chi(n, n+1)$ denotes a chaotic behavior in which there is irregular switching between n and n+1 peaks. $\Pi(n)$ denotes a large-amplitude relaxation oscillation, whereas $\Pi'(1)$ is a small amplitude harmonic oscillation; s.s. denotes steady state. $\Pi(m)$ denotes a region of oscillations having a large number of peaks, the exact number which is difficult to obtain exactly. The sequence shown is not complete, but is presented in order to demonstrate the richness of the dynamic behavior. For example, periodic behavior of types $\Pi(1,3)$ and $\Pi(1,4)$ as well as quasiperiodic beating have been observed. In addition the periodic types $\Pi(3,4)$ and $\Pi(4,5)$ probably could be found but have not been looked for in detail. Other types of oscillations are certainly possible.

A portion of the series of oscillations is shown in Fig. 1. All recordings shown were taken with the bromide electrode. The single peak relaxation oscillations obtained at residence times (τ) in the neighborhood of eight minutes are not shown. The oscillations in Fig. 1a are alternating single and double peaked (τ=6.67 min) and those in Fig. 1b (τ=6.26 min) are double peaked. Both oscillations are periodic and stable. At τ=5.89 minutes (Fig. 1c) chaotic behavior is observed. This behavior is reproducible and continues until the external conditions are changed. Note that this chaos is primarily an (irregular) mixture of two and three peaks. Two other regions of chaotic behavior were found (Figs. 1e and 1g).

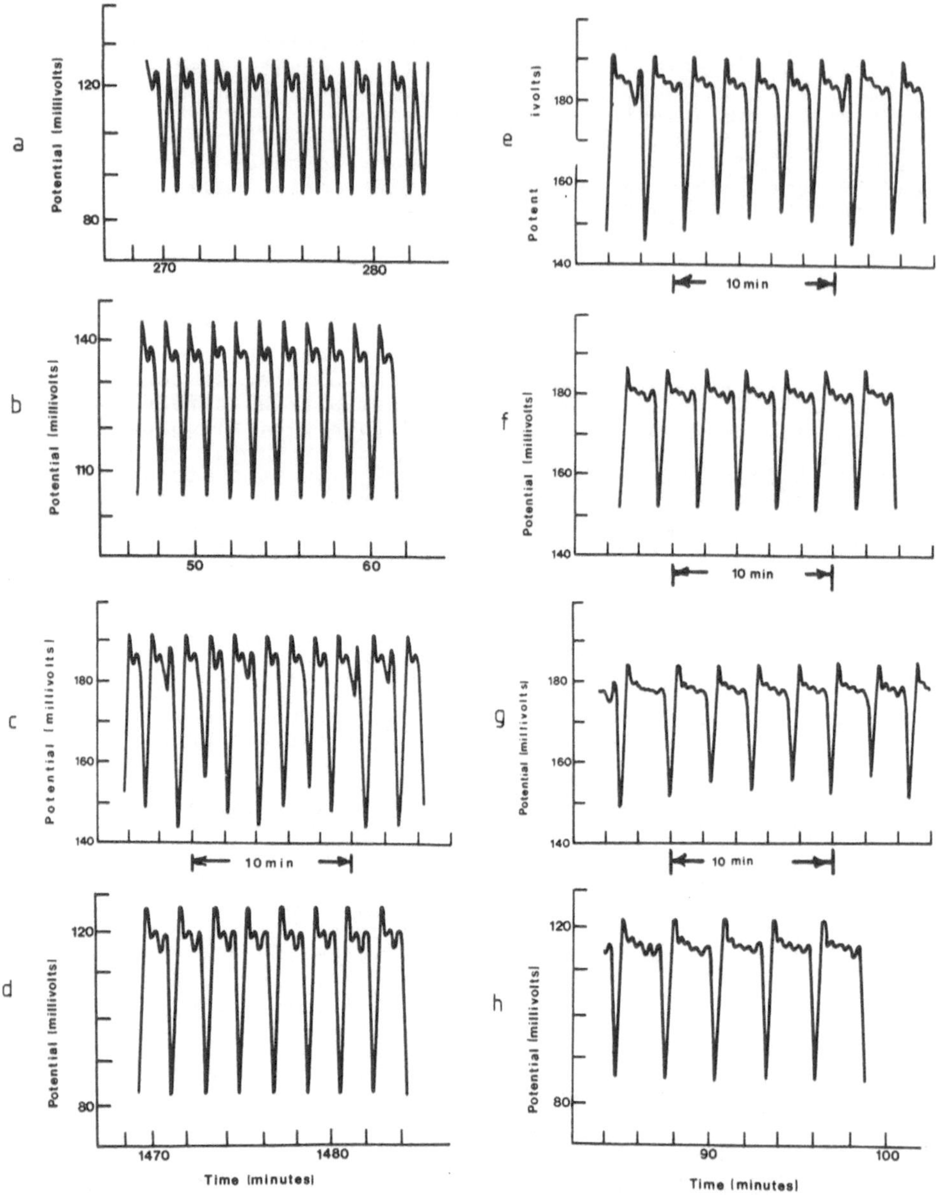

Fig. 1 Summary of the behavior of the Belousov-Zhabotinsky-Reaction with variation of a single parameter; T=25°C; (a) τ=6.76 min; (b) τ= 6.26 min; (c) τ=5.89 min; (d) τ=5.85 min; (e) τ=5.63 min; (f) τ=5.50 min; (g) τ=5.34 min; (h) τ=5.28 min [4]

A next amplitude map constructed from the data of Fig. 1c is shown in Fig. 2 (HUDSON and MANKIN [12]). Such a map was first obtained by TOMITA and TSUDA [14] using the data of HUDSON et al. [4]. Similar maps have been obtained with the Belousov-Zhabotinskii reaction by VIDAL [15] and ROUX et al. [5].

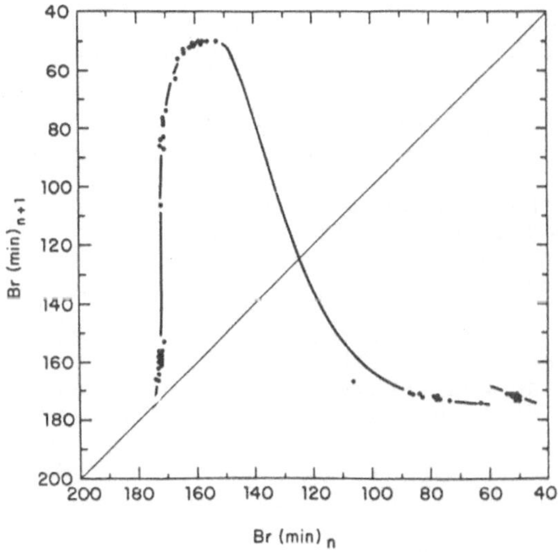

Fig. 2 Return map from experimental data. Minimum in bromide ion electrode potential vs. the value at the previous minimum [12]

A Lyapunov characteristic exponent for the results of Fig. 1c has been calculated to be 0.62 per iterate, which is equivalent to 1.0 min^{-1}.

Several models have been developed to describe the experimentally observed behavior. Many of these are based on variations of the original FIELD, KÖRÖS, and NOYES [16] reaction scheme. Since Professor Noyes discusses models in a contribution to this meeting, we make only a few brief comments. The ability of realistic Belousov-Zhabotinsky reaction models to generate chaos is not yet completely clear. NOYES and his coworkers (e.g., SHOWALTER et al., [17]) have looked carefully at such models, and have seen only periodic solutions. However, TOMITA and TSUDA [18] and TURNER et al. [6] have obtained chaos in modifications of these equations. PISMEN [19] considered flow on a folded slow manifold and showed that such a manifold can produce an infinite number of periodic and chaotic regimes interspersed in parameter space. TOMITA and TSUDA [14] and TSUDA [20] have shown that the entire sequence of experimental findings can be generated by varying the form of a map (Fig. 2) obtained from a single experiment. PIKOVSKY [21] showed that a folded manifold can be used to formulate a set of three ordinary differential equations which model the experimental findings. Finally, it should be pointed out that sets of differential equations derived from mass action kinetics do exist that produce chaotic behavior (WILLAMOWSKI and RÖSSLER [22]).

3. Chaos in Coupled Exothermic Reactions

We now present simulated results of chaos in a realistic chemical system, viz., a single, irreversible, exothermic reaction. Such a reaction has been shown to produce sustained oscillations in a non-adiabatic continuous stirred reactor by means of both experiments and simulation (e.g., HAFKE and GILLES [23], CHANG and SCHMITZ [24], WIRGES [25]). However, since such a system is governed by two ordinary differential equations, an energy balance and a material balance, any resulting oscillations must be periodic.

Consider now two such reactors coupled through the transport of heat. We now have a system of four equations, capable of more complicated behavior. The coupling of two limit cycles to produce more complicated behavior has been considered by several authors. For example, FUJISAKA and YAMADA [26] have shown chaos in coupled Belousov-Zhabotinsky reactions. RÖSSLER [27] has shown chaos through linear coupling of Turing oscillators.

The behavior of the coupled system depends, of course, on the parameters chosen. For example, when one reactor is much larger than the other, the system reduces to a two-variable system (the smaller reactor) with forcing, which is mathematically equivalent to a three-variable system. In this case, chaos is obtained via a series of period doubling bifurcations with variation of a parameter, the heat transfer coefficient between the two reactors.

We consider here a situation in which the two reactors are almost identical and in which only one parameter, the heat transfer coefficient governing heat flow between the two reactors (β_m), is varied. For the case chosen, the governing differential equations are:

$$\begin{aligned}
\dot{C}_1 &= -C_1 + (1-C_1)0.085\exp(T_1) \\
\dot{T}_1 &= -(3.45+\beta_m)T_1 + (1-C_1)(22)(0.085)\exp(T_1) \\
&\quad +\beta_m T_2 - 2.805 \\
\dot{C}_2 &= -C_2 + (1-C_2)0.087\exp(T_2) \\
\dot{T}_2 &= -(3.45+\beta_m)T_2 + (1-C_2)(22)(0.087)\exp(T_2) \\
&\quad +\beta_m T_1 - 2.805
\end{aligned} \tag{1}$$

In these equations C and T are the dimensionless concentration and temperature, respectively, the subscripts 1 and 2 refer to each of the two reactors, each reactor has a (dimensionless) cooling temperature of -1.1449 and heat transfer coefficient of 2.45, and the Damköhler numbers in the two reactors are 0.085 and 0.087 respectively. The parameters are chosen such that without coupling each reactor oscillates independently.

We consider now the behavior as the parameter β_m is varied. For $\beta_m=0.57$ the system is periodic; transient trajectories spiral around the final limit cycle, and approach it very slowly. The parameter β_m is now lowered systematically. For $\beta_m=0.55$ the three-dimensional projections of the four-dimensional trajectories appear to be quite complicated. Theyare, however, quasiperiodic and confined to the surface of a torus; a cross section of the flow is a closed circle. (Such cross sections are obtainable in several ways. For example, T_1, T_2, C_1, is plotted at the minima of T_2.

For $\beta_m=0.525$ a stereoplot is shown (Fig. 3) with the variables T_1, T_2, and C_1. Although the exact nature of this flow is not yet known, it is certainly chaotic with a positive Lyapunov characteristic exponent. (The remaining three have not as yet been calculated.)

A next amplitude map is given in Fig. 4, where successive values of the temperature in reactor two at the minima in this temperature are shown. The cross section of this flow is no longer a closed circle but rather a more complicated pattern. The minima fall mostly on the curve in the lower left corner of the figure, with occasional jumps

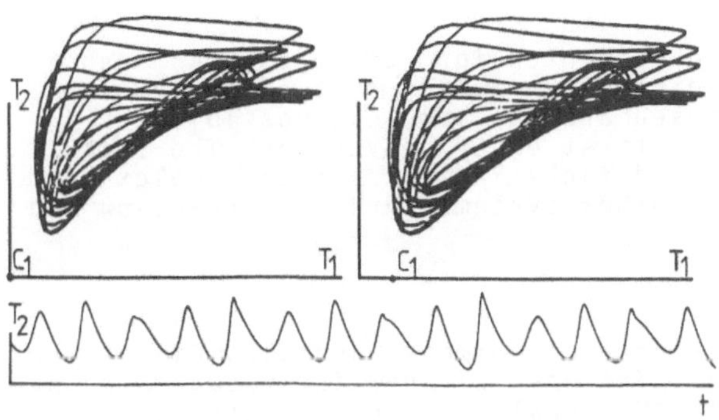

Fig. 3 Trajectories calculated with (1) with β_m=0.525. The axes are: C_1 : 0.8 to 1.0; C_2 : 0.88 to 0.94; T_1 : 3.5 to 8; T_2 : 4.0 to 5.5; time 0 to 20. Corresponding I.C.: 0.934, 0.9325, 4.94, 4.98

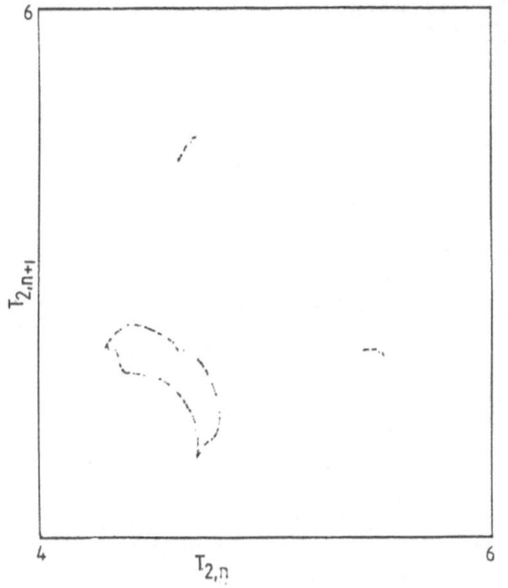

Fig. 4 Next amplitude map obtained from the results of Fig. 3 at the minima of T_2

to the two small islands. The flow appears to have the topology of a folded torus such as that found with the driven van der Pol oscillators (e.g., SHAW [28]).

4. Discussion

We have presented results demonstrating chaotic behavior in two systems.

Experimental evidence for chaotic behavior of the Belousov-Zhabotinsky reaction is now plentiful. The chaos occurs in a reproducible region of parameter space, and is surrounded by periodic behavior. An (almost) one-dimensional next amplitude map can be constructed from the data and from it a positive Lyapunov characteristic exponent calculated. Several models are available which reproduce the series of periodic and chaotic oscillations observed in the experiments. Improvements in these models will perhaps next come from an even closer correspondence to a detailed reaction mechanism.

Chaos, perhaps even a more complicated form of chaos, was seen in numerical simulations of an exothermic reaction in two coupled reactors. The reaction kinetics chosen are the simplest possible for a non-isothermal reaction, viz., a first order conversion. The fact that chaos is found in two coupled tanks indicates that complex behavior may be prevalent in many other systems involving reaction and transport.

Acknowledgements

We thank Norman Packard for his helpful comments and E.D. Gilles for his hospitality and stimulation. The work was partially supported by the Fulbright Commission and the National Science Foundation, grant CPE 80.21950.

1 R.A. Schmitz, K.R. Graziani and J.L. Hudson: J. Chem. Phys. 67, 3040-3044 (1977)
2 L.F. Olsen and H. Degn: Nature 271, 177-178 (1977)
3 O.E. Rössler: Z. Naturforsch. 31a, 259-264 (1976a)
4 J.L. Hudson, M. Hart and D. Marinko: J. Chem. Phys. 71, 1601-1606 (1979)
5 J.C. Roux, J.S. Turner, W.D. Mc Cormick and H.L. Swinney: "Experimental Observations of Complex Dynamics in a Chemical Reactor", in Nonlinear Problems, Presence and Future (North-Holland, Amsterdam 1981)
6 J.S. Turner, J.C. Roux, W.D. Mc Cormick and H.L. Swinney: Phys. Letters 85A, 9 (1981)
7 R.A. Schmitz, G.T. Renola and A.P. Ziondas: "Strange Oscillations in Chemical Reactions - Observations and Models", in Dynamics and Modelling of Reactive Systems (Academic Press, New York 1980)
8 C. Kahlert, O.E. Rössler and A. Varma: "Chaos in a Continuous Stirred Tank Reactor with Two Consecutive Reactions, One Exo-, One Endothermic", in Modelling of Chemical Reaction Systems (Springer Verlag, New York-Berlin 1981)
9 D.V. Jorgensen and R. Aris: "On the Dynamics of a Stirred Tank with Consecutive Reactions", Preprint (1982)
10 O.E. Rössler and J.L. Hudson: "Higher Chaos in Simple Reaction Systems", in Symposium on Chemical Applications of Topology and Graph Theory (Elsevier, Amsterdam 1983; to appear)
11 J.L. Hudson and O.E. Rössler: "Chaos and Complex Oscillations in Stirred Chemical Reactors", in Dynamics of Nonlinear Systems (Gordon & Breach, to appear 1983)
12 J.L. Hudson and J.C. Mankin: J. Chem. Phys. 74, 6171-6177 (1981)
13 J. Mankin, P. Lamba and J.L. Hudson: "Transitions Between Periodic and Chaotic States in a Continuous Stirred Reactor", in American Chemical Society 196 (1982)
14 K. Tomita and I. Tsuda: Prog. Theor. Phys. 64, 1138 (1980)
15 C. Vidal: "Chemical Kinetics as an Experimental Field for Studying the Onset of Turbulence", in Nonlinear Phenomena in Chemical Dynamics (Springer-Verlag, N.Y., Berlin 1981)
16 R.J. Field, E. Körös and R.M. Noyes: J. Amer. Chem. Soc. 94, 8649-8664 (1972)
17 K. Showalter, R.M. Noyes and K. Bar-Eli: J. Chem. Phys. 69, 2514 (1978)
18 K. Tomita and I. Tsuda: Physics Letters 71a, 489 (1979)
19 L.M. Pismen: Phys. Lett. 89 A, 59 (1982)
20 I. Tsuda: Phys. Lett. 85 A, 4 (1981)
21 A.S. Pikovsky: Phys. Lett. 85 A, 13-16 (1981)
22 K.D. Willamowski and O.E. Rössler: Z. Naturforsch. 35a, 317-318 (1980)

23 C. Hafke and E.D. Gilles: Messen-Steuern-Regeln 11, 204-208 (1968)
24 M. Chang and R.A. Schmitz: Chem. Eng. Sci. 30, 21 (1975)
25 H.-P. Wirges: Chem. Eng. Sci. 35, 2141-2146 (1980)
26 H. Fujisaka and T. Yamada: Z. Physik B 37, 265-275 (1980)
27 O.E. Rössler: Z. Naturforsch. 31a, 1168-1172 (1976c)
28 R. Shaw: Z. Naturforsch. 36a, 80 (1981)

The Interface Between Mathematical Chaos and Experimental Chemistry

R.M. Noyes

Max-Planck-Institut für Biophysikalische Chemie, Am Fassberg
D-3400 Göttingen-Nikolausberg, Fed. Rep. of Germany and

Department of Chemistry, University of Oregon, Eugene, OR 97403, USA

If the mechanism of a chemical reaction is assumed, presently available computer technology is adequate to model the dynamic behavior of a real system by means of deterministic equations. Our own computations have not yet discovered significant ranges of parameter space within which such behavior is chaotic. If experimental measurements exhibit chaotic responses under carefully controlled conditions, those responses need not reflect intrinsic chaos in the chemical mechanism unless sensitivity to uncontrolled fluctuations can be ruled out. The measuring apparatus itself is usually sensitive enough that it need not be a cause of concern, and the quantum-mechanical uncertainty principle is not important in macroscopic systems. However, fluctuations in experimentally controlled parameters and in statistical-mechanical populations of species may both be amplified many orders of magnitude if they occur when the system is making a transition between manifolds. It appears very difficult to rule out the possibility that such fluctuations are responsible for any specific example of experimentally observed chaotic behavior.

1. Introduction

Strange attractors and chaotic behavior are often studied as problems of pure mathematics and of theory of equations independent of any applicability to real systems. A considerable body of literature has also developed describing experimental studies of chemical systems whose behaviors are so aperiodic as to satisfy criteria defining "chaos". The transition between mathematical theory and experimental measurement is not always easy to accomplish, and sequence of procedures depends upon the direction in which that transition occurs.

In order to pass from theory to experiment, it is appropriate to propose a mechanism with intrinsic capacity for chaotic behavior, to model that system by computations with selected parameter values, and to make measurements to test the predictions of the computations.

In order to pass from experiment to theory, it is appropriate to observe chaotic behavior in a laboratory system, to eliminate possible causes other than those intrinsic to the mechanism, and to develop an explanation consistent with the observations.

Any effort to make such transitions will be complicated by the fact that observations in any laboratory system and also efforts to model such a system by numerical computations are impacted by an irreducible level of random noise. Such noise may be associated with the measurement and control of physical quantities, may be due to fluctuations inherent in the behavior of a finite number of independent particles, and may be due to round-off error in numerical computations. Each of these sources of noise must be analyzed and its potential contributions must be assigned with confidence before any residual random effects can be ascribed legitimately to the kind of behavior of interest to this conference. The object of the present paper is to examine the probable contributions to chaotic behavior due to each of these possible sources of noise. We shall then discuss the confidence with which we can ascribe experimental aperiodicities to chaotic behavior inherent in the deterministic equations defining the evolution of the system.

2. Fluctuations Due to Measurement and Control of Experimental Quantities

2.1. Measurement of Response

The experimental record of the behavior of the system will often, but not necessarily, be in the form of a recorder trace of voltage against time. That record will have a certain irreducible level of fluctuation inherent in the measuring instrument. However, a good experimenter will understand the sensitivity of that instrument and will not draw definite conclusions from experiments unless the signal-to-noise ratio is sufficiently large. Instrumental sensitivity is not a problem for many of the systems which have been studied.

2.2. Control of Constraints

Description of any chemical system requires specification of a number of quantities such as temperature, pressure, concentration of reagent, residence time in reactor, etc. The values of these constraints must be controlled either to remain constant or to vary in known ways. The precision of that control is often known with the same sort of confidence that can be assigned to the measured response discussed in Section 2.1.

However, we are often less confident about the true sensitivity of the system to small fluctuations in the controlled quantities. An illustration of the potential for difficulty is provided by Fig. 1 taken from a previous publication [1]. The plot shows concentration of species Y for an Oregonator [2] model having dynamic and stoichiometric parameters chosen so as to generate a locally stable but excitable stationary state.

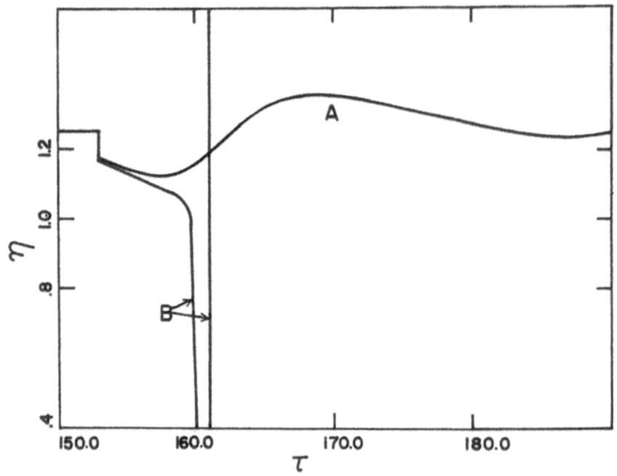

Figure 1. Plot of behavior of η (dimensionless variable representing Y or Br⁻) for an Oregonator model of the Belousov-Zhabotinsky system with stoichiometric factor f = 1.5. The stationary state is locally stable but excitable. At time 153 units, the system was perturbed discontinuously by reducing η, and the trajectory for return to the stationary state was computed. Curve A describes the effect of a perturbation of 6.0%. Curve B (which ranged approximately from 0.01 to 1000) describes the effect of a perturbation of 6.5%. Reprinted with permission from Reference 1.

Curve A illustrates the behavior if the concentration of species Y (normalized as η) is suddenly reduced by 6.0% and the system is then allowed to evolve according to the model. It decays to the original stationary state with damped oscillations having a maximum amplitude of about 10% corresponding to about 0.04 units in log η.

Curve B illustrates the behavior if the concentration is reduced by 6.5%. The concentration then drops to about 0.01 of the stationary-state value, rises to about 1000 times that value, and eventually decays to the original stationary state. The amplitude of this excursion is about 5 units in log η.

Figure 1 illustrates the effect of a single instantaneous perturbation and shows that a variation of less than a percent in a quantity being perturbed can make a difference of five orders of magnitude in the range of the resulting excursion! The parameters were by no means selected to illustrate a maximum sensitivity to perturbation, and the effects could be truly "chaotic" if the system were subjected to repeated perturbations of random magnitude and temporal separation.

Figure 1 shows that it is comparatively simple to devise systems exhibiting extremely nonlinear responses to small perturbations. If an experimental system has the misfortune to exhibit such hypersensitivity to perturbation, there does not seem to any way to insure that observed "chaotic" behavior is ultimately due to factors inherent in the mechanism rather than to effects of unavoidable fluctuations in experimental conditions. The only exception to this pessimistic conclusion would arise if the chemical mechanism were so well understood that it could be asserted with confidence that the sensitivity was not of such a pathological type.

3. Fluctuations Due to Molecular Effects

3.1. Quantum-Mechanical Effects

One of the most important developments of twentieth-century science has been the recognition that quantities like position, momentum, energy, and time can not be measured simultaneously with indefinitely high precision. The uncertainty principle imposes an ultimate limit on the accuracy with which the behavior of any real system can be described by deterministic equations.

Fortunately, the magnitude of Planck's constant is so small that it is doubtful the observable behavior of any real chemical system (except perhaps at temperatures near 0 K) will ever be impacted by this type of quantum effects. The possibility is mentioned only to make the analysis of this paper complete.

3.2. Statistical-Thermodynamic Effects

The detailed state of any chemical system should presumably differ little from the most probable state in an ensemble of systems identical in all observable properties. However, fluctuations from the most probable state will occur even if the system is at equilibrium. If that most probable state contains N_i molecules of a particular energy and chemical configuration, the real system may have deviations of the order of $(N_i)^{1/2}$ from that value, and the fractional variation in the concentration of that species will be of the order of $(N_i)^{-1/2}$.

The magnitude of Avogadro's number is so large that these variations are of negligible significance for the major components of any macroscopic chemical system. However, some minor chemical species may be present at concentrations as low as 10^6 molecule cm^{-3} or even less, and a biological cell may have only a few molecules of a particular enzyme. Concentrations in such systems may fluctuate by significant fractional amounts.

These fluctuations are probably never of importance for any study of the properties of an equilibrium system during a significant period of time. However, the situation can become more complicated if the system is far removed from equilibrium. If a molecule of a particular species is destined to react further in a time short compared to that for evolution of the system as a whole, the concentration of that transient intermediate will attain a steady state in which the number of molecules can be expected to fluctuate by a fraction at least as large as the $(N_i)^{-1/2}$ predicted for a system at equilibrium.

Fortunately, the steady states in most chemical systems are stable to perturbation, and fluctuations decay rapidly with little influence on observable properties. However, even locally stable stationary states may be so sensitive that fluctuations are amplified far beyond the magnitudes predicted by equilibrium statistical mechanics.

An example of such a situation is provided by the mechanism of equations 1 to 4.

$$A \rightarrow X \tag{1}$$

$$B + X \rightarrow P + Q \tag{2}$$

$$C + X \rightarrow 2X \tag{3}$$

$$2X \rightarrow R + S \tag{4}$$

The temporal dependence of the concentration of X is given by equation 5.

$$d[X]/dt = k_1[A] - k_2[B][X] + k_3[C][X] - 2k_4[X]^2 \tag{5}$$

If any X formed in steps 1 and 3 is destined to react rapidly by step 2 or 4, we may use a steady-state approximation that d[X]/dt is nearly zero during any time scale of experimental interest. To the validity of that approximation, we can write equation 6.

$$[X] = \frac{k_3[C] - k_2[B] + \sqrt{\{k_3[C] - k_2[B]\}^2 + 8k_1k_4[A]}}{4k_4} \tag{6}$$

As long as $\{k_3[C] - k_2[B]\}^2 >> 8k_1k_4[A]$, the concentration of X can be approximated very well by either equation 7 or equation 8, whichever is appropriate.

$$[X]_{small} = \frac{k_1[A]}{k_2[B] - k_3[C]} \qquad \text{if } k_2[B] > k_3[C] \tag{7}$$

$$[X]_{large} = \frac{k_3[C] - k_2[B]}{2k_4} \qquad \text{if } k_3[C] > k_2[B] \tag{8}$$

Because species B and C are both major components of the reacting system and therefore subject to truly infinitesimal fluctuations, the concentration of X calculated by these deterministic equations will not fluctuate by much more than would be anticipated for an equilibrium system having the same net concentration of this species.

However, if $k_3[C] \cong k_2[B]$, neither equation 7 nor 8 will be a valid approximation, and the concentration of X may become *extremely* sensitive to very small fluctuations in the concentrations of species B and C. If the chemical mechanism is such that the quantity $k_3[C] - k_2[B]$ repeatedly changes sign, such changes may be accompanied by significantly different trajectories in the concentration of X and in net reaction during the change. These differences could be manifested by "chaotic" components in the dynamic behavior.

The above discussion used "concentrations" of chemical species and thereby tacitly assumed the composition of the macroscopic system was uniform even when it was undergoing statistical-mechanical fluctuations. The real situation is much more complicated, and *local* fluctuations in a sensitive system may influence surrounding regions. Drs. ROUX, DE KEPPER, and BOISSONADE [3] at the University of Bordeaux have recently obtained evidence which suggests that such local fluctuations may exert observable effects on transitions in a system exhibiting bistability.

It should be emphasized that these statistical-mechanical effects are separate from questions as to how well various experimental parameters can be controlled and measured. Although the theory of fluctuations is an important area of current research, most of the examination of very sensitive systems seems to have been concerned with whether phase transitions do or do not occur rather than with the detailed dynamic trajectories in the fluctuating system. It is these detailed trajectories which are important to the question of whether statistical-mechanical effects are contributing significantly to observed "chaotic" behavior.

4. Fluctuations Due to Computational Effects

One of the significant tests of understanding of a scientific phenomenon is the ability to model it by computations based on current theories. If "chaotic" behavior is observed in the laboratory, and if the chemical mechanism is understood in detail, then computations based on the assumed dynamic behavior should reproduce the observations.

Figures 2 to 4 illustrate such a test based on strictly deterministic equations. Figure 2 is the experimental trace of electrode potential for a Belousov-Zhabotinsky system studied in a flow reactor by HUDSON and MANKIN [4]. The system exhibits oscillations of both large and small amplitude. Between each maximum and subsequent minimum in a large-amplitude oscillation, there are either one or two

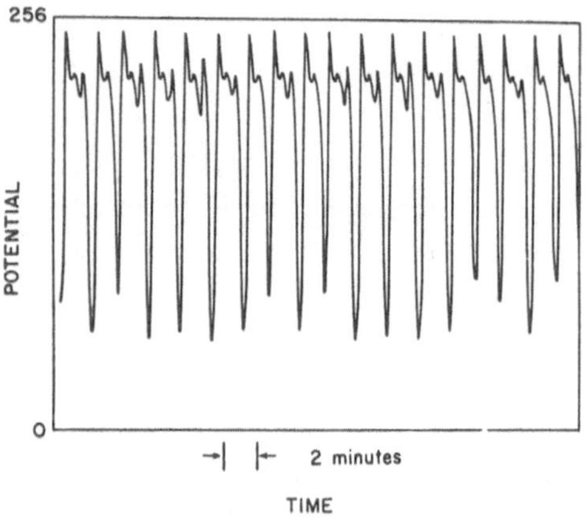

Figure 2. Trace of evolution of potential of an electrode specific to bromide ion measured experimentally in a stirred flow reactor with residence time 6.01 min. Reprinted with permission from Reference 4.

oscillations of small amplitude. However, there is no pattern to the occurrence of single and double small-amplitude oscillations. The trace in Fig. 2 can fairly be called an experimental example of "chaotic" behavior.

When the residence time in the flow reactor was varied from the conditions in Fig. 2, the system could be made to exhibit repeated single or repeated double small-amplitude oscillations. The region of chaotic behavior corresponded to a few percent of the residence time, and it could be entered or left from either direction without evidence of hysteresis. The residence time could be controlled considerably more accurately than the range of the region of chaotic behavior.

The chemical mechanism is thought to be known, and Figures 3 and 4 illustrate an effort by GANAPATHISUBRAMANIAN and NOYES [5] to model the experimental system of Fig. 2. Differences in the detailed form of the traces indicate either that the chemical mechanism is not completely understood or else that the associated rate constants are not known exactly, but each pair of extrema is separated by either one or two small-amplitude oscillations just as in the experimental system. However, Figures 3 and 4 are both regular. In Fig. 3, each example of a single small-amplitude oscillation is followed by two double oscillations. In Fig. 4, each is followed by three double oscillations. Yet these two cases of regular behavior differ by only about one part per million in residence time!

Of course the bifurcation in Figures 3 and 4 can not be infinitely sharp for any computations carrying a finite number of significant figures. SHOWALTER, NOYES, and BAR-ELI [6] showed for similar computations that regular behavior became "chaotic" if the error parameters were relaxed for the matching of derivatives in successive increments of time. Figures 3 and 4 each show sixteen small-amplitude os-

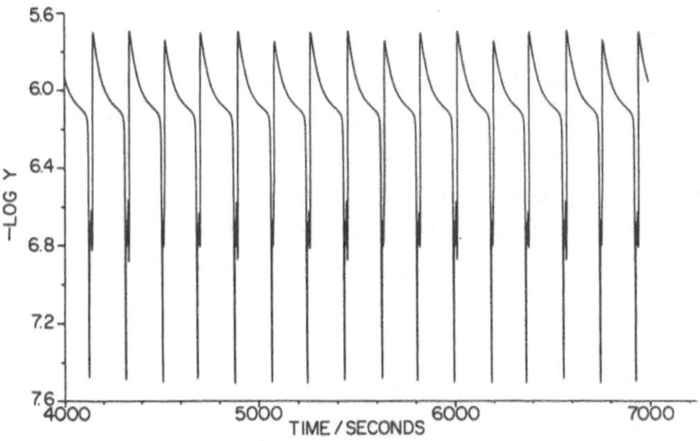

Figure 3. Computed trace of -log Y (supposedly proportional to the potential of an electrode specific to bromide ion) for a model of the reaction of Fig. 2. The calculations used a reciprocal residence time of 7.80884 x 10^{-3} sec^{-1} corresponding to a residence time of 2.134333 min. Reprinted with permission from Reference 5.

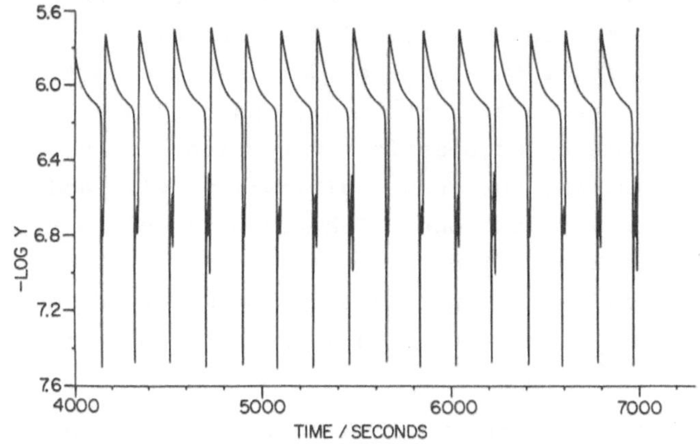

Figure 4. Computed trace like that in Figure 3 except that reciprocal residence time of 7.80885 x 10^{-3} sec^{-1} corresponded to a residence time of 2.134330 min. Reprinted with permission from Reference 5.

cillations. If the computations were continued for several hundred such oscillations, occasional variations in the regular pattern might appear. Similarly, an effort to identify the transition still more accurately must ultimately show a range of residence times within which the behavior is chaotic. All we can say is that such a range is less than 10^{-6} residence time and many orders of magnitude narrower than the range observed experimentally.

The computations in Figures 3 and 4 were done in double precision carrying 16 significant figures. Dr. John RINZEL [7] at the National Institutes of Health reproduced them for the same parameters except that the computations carried 20 significant figures. He also observed sharp transitions between the behaviors of Figures 3 and 4, but they came at slightly different residence times. The difference suggests that the computational results can be observably impacted if the precision of individual computations is changed by factors approaching the ratio of unity to Avogadro's number! It would be unrealistic to expect regular behavior in experimental systems exhibiting such a high level of sensitivity.

5. Conclusions

As was pointed out in the Introduction, the interface between mathematical theory and experimental measurement can, in principle, be crossed in either direction.

The transition from theory to experiment is more likely to be feasible. The plots in Figures 3 and 4 show that if a mechanism is no more complicated than that presently believed for the Belousov-Zhabotinsky reaction, then the computer technology and software exist to model that mechanism numerically at the level of deterministic differential equations. It may also be possible to modify such computations to incorporate the effects of stochastic fluctuations at the levels anticipated in real systems, but such modifications will add considerable complexity to the computations. Chaotic behavior computed with deterministic equations will necessarily be confined to a narrower range of parameter values than will chaotic behavior in the equivalent experimental system which is inevitably subject to fluctuations.

The transition from experiment to theory is likely to be more difficult to accomplish. There certainly are experimental systems which exhibit "chaotic" behavior in spite of heroic measures to control all recognized parameters. Such behavior does not prove that the chaos is inherent in the mechanism itself rather than due to unavoidable stochastic fluctuations. Before such a proof were possible, the chemical mechanism would have to be known with certainty, which the philosophy of science tells us is impossible. Even if a mechanism is assumed, it is necessary to develop objective procedures to evaluate the sensitivity to be anticipated for the observed trajectory because of the inevitable fluctuations in control parameters and in statistical-mechanical populations of transient intermediates. Such sensitivity may not merely be averaged over a period but must be integrated along the trajectory of evolution of the system in phase space. Any system capable of exhibiting intrinsic chaos will undergo repeated transitions between manifolds. The discussion in Section 3.2 indicates that during such transitions the system will be particularly sensitive to fluctuations. Potential sensitivities are so great that it appears impossible at present to demonstrate that any experimentally observed "chaos" could not possibly have been caused by stochastic fluctuations.

Acknowledgment

This manuscript was prepared at the Max-Planck-Institut für Biophysikalische Chemie while the author was on leave from the University of Oregon. Much of the work described here was supported in part by grants from the National Science Foundation and by a Senior American Scientist Award from the Alexander von Humboldt Stiftung.

References

1. R. J. Field, R. M. Noyes: Faraday Symp. Chem. Soc. 9, 21 (1974)

2. R. J. Field, R. M. Noyes: J. Chem. Phys. 60, 1877 (1974)

3. J. C. Roux, P. De Kepper, J. Boissonade: (private communication)

4. J. L. Hudson, J. C. Mankin: J. Chem. Phys. 74, 6171 (1981)

5. N. Ganapathisubramanian, R. M. Noyes: J. Chem. Phys. 76, 1770 (1982)

6. K. Showalter, R. M. Noyes, K. Bar-Eli: J. Chem. Phys. 69, 2514 (1978)

7. J. Rinzel: (private communication)

The Enzyme and the Strange Attractor – Comparisons of Experimental and Numerical Data for an Enzyme Reaction with Chaotic Motion

L.F. Olsen

Institute of Biochemistry, Odense University, Campusvej 55
DK-5230 Odense M, Denmark

Experimental results showing periodic and apparently aperiodic (chaotic) motions in the oscillating peroxidase-oxidase reaction are presented. These results are compared with computer simulations of a model of the reaction. On the basis of these comparisons it is argued that the system displays chaotic motion of a higher order than previously observed in chemical systems.

1. Introduction

In 1976 RÖSSLER proposed that oscillating chemical reactions in the homogenous phase and with external supplies of the reactants may under some conditions exhibit chaotic motion [1]. Since then several attempts have been made to provide experimental support for this idea [2-6]. Experimental verification, however, has proven extremely difficult, mainly due to the inevitable existence of small fluctuations in the inlet and mixing systems [7,8]. It is known that small periodic perturbations may turn an otherwise periodically oscillating system into a chaotic system [6]. Nevertheless, evidence for chaotic behaviour in real chemical systems is accumulating, chiefly because of the formulation of realistic models capable of reproducing the various types of behaviours found in the experimental systems [9-12].

The present work is about the oscillating peroxidase-oxidase reaction in an open system. In this reaction reduced nicotinamide adenine dinucleotide (NADH) is oxidised with molecular oxygen as electron acceptor:

$$2H^+ + 2NADH + O_2 \longrightarrow 2H_2O + 2NAD^+.$$

The reaction is catalysed by peroxidase (EC 1.11.1.7). Previous work by YAMAZAKI *et al.* [13-16], DEGN [17,18] and others [19-22] has demonstrated the existence of bistability, damped oscillations and sustained oscillations when one or both substrates are continuously supplied to the reaction mixture. The oscillating peroxidase-oxidase reaction is also the system for which experimental evidence for chaotic behaviour was first presented [2]. As for the Belousov-Zhabotinskii reaction [23], the oscillating peroxidase-oxidase reaction can be described by a simple model capable of reproducing the various waveforms observed in the experimental system [19]. A recent modification of this model has been shown to contain a strange attractor [12].

Here I aim at a semiquantitative comparison of experimental results with numerical results in order to provide further evidence for chaos in this system.

2. The experimental system

Experiments were performed in a hexagonal glass cuvette, fitted with a stirrer for efficient mixing. The stirring shaft was provided with a cone in order to stabilise the surface of the liquid (4.5 ml). The reaction mixture was in contact with a N_2/O_2 gas phase, the composition of which could be regulated by the help of a digital gas mixer [24]. The rate of diffusion, V_d, of oxygen into the liquid is then determined by the equation [25]:

$$V_d = K([O_2]_{eq} - [O_2])$$

where $[O_2]$ is the oxygen concentration in the liquid and $[O_2]_{eq}$ is the concentration at equilibrium between the gas and the liquid. K is a constant at a fixed temperature and a fixed surface area of the liquid (constant stirring rate). The cuvette is inserted into a brass block adapted for circulation of water from a thermobath. NADH was supplied by infusion of a concentrated (0.2 to 0.25 M) aqueous solution into the reaction mixture through a capillary whose tip was below the surface of the liquid. A Harvard Apparatus Co. model 971 high precision infusion pump was used for the supply of NADH. The whole setup was inserted into a Hitachi-Perkin Elmer 356 spectrophotometer. O_2 was measured with a Clark electrode (Radiometer, Copenhagen). NADH was measured spectrophotometrically at 380 nm.

3. Experimental results

A typical example of periodic motion in the peroxidase-oxidase reaction is illustrated in Fig. 1. The oscillations are of the relaxation type with critical limits of the reactants at which the reaction is turned on and turned off [19-21]. Figure 2 shows the effect of varying the enzyme concentration on the oscillating mode. For clarity only the NADH traces are shown. The critical limit of NADH at which the reaction is turned on increases with decreasing concentrations of peroxidase. It is also to be noted that whereas the oscillations in Fig. 2a and c

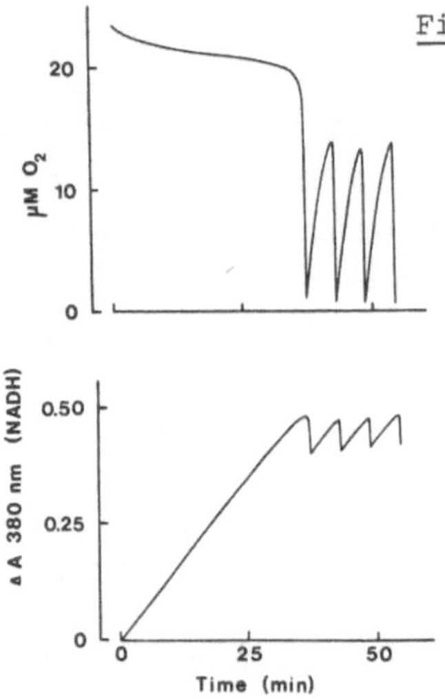

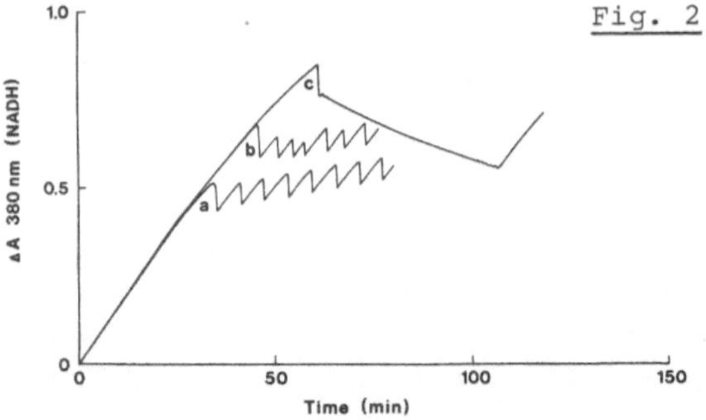

Fig. 1 Periodic oscillations of O_2 and NADH. 1.32 μM horseradish peroxidase, 0.2 μM methylene blue and 10 μM 2,4-dichlorophenol in 0.1 M Na-acetate buffer, pH 5.1. O_2 content in the gas phase was 1.9% (v/v). 0.2 M NADH was pumped into the liquid at a rate of 10 μl per hour. Temperature 30°C

Fig. 2 Dependence of oscillatory waveform of NADH on the concentration of peroxidase; (a) 0.9 μM, (b) 0.54 μM, (c) 0.42 μM. Other conditions as in Fig. 1 except for the NADH flow rate (11 μl per hour) and the temperature (28°C)

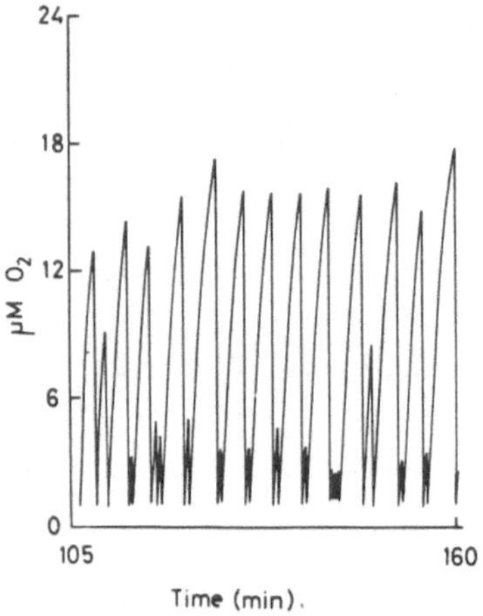

Fig. 3 Aperiodic oscillations of O_2 • 0.5 μM horseradish peroxidase, 0.2 μM methylene blue and 20 μM 2,4-dichlorophenol in 0.1 M Na-acetate buffer, pH 5.1. O_2 content in gas phase 2.15% (v/v). 0.25 M NADH was pumped into the liquid at a rate of 15 μl per hour. Temperature 28°C

show regular periodic motion, the oscillations in Fig. 2b seem to be 'wobbly'. The behaviour in Fig. 2b illustrates the onset of aperiodic motion in the reaction. Figure 3 shows the time course of oxygen during the aperiodic motion. Analyses of traces similar to that shown in Fig. 3 have been used to provide evidence for chaotic behaviour by construction of a next-amplitude map [2] and using LI and YORKE's theorem [26].

However, as will be discussed below the behaviour found here may be more complex than previously expected.

4. Reaction schemes and a minimal model

The peroxidase-oxidase reaction is overall an autocatalytic reaction [18,19,27] involving 10-20 individual reaction steps [19]. These reaction steps may be combined in several ways to form various overall autocatalytic schemes. Two examples are shown in Fig. 4. Ideally each of these schemes should have a branching factor of 3 but the side reaction [28]:

$$2H^+ + 2O_2^{-\cdot} \longrightarrow O_2 + H_2O_2$$

results in a reduction of this factor in scheme 4b. It has previously been argued that a combination of two mutually coupled autocatalytic schemes is necessary to account for the experimentally observed oscillatory behaviours [19,20]. This is achieved with the following minimal model, which is an extension of a model proposed earlier [29] to explain the oscillations in the Bray reaction:

$$B + X \xrightarrow{k_1} 2X \qquad (1)$$

$$2X \xrightarrow{k_2} 2Y \qquad (2)$$

$$A + B + Y \xrightarrow{k_3} 3X \qquad (3)$$

$$X \xrightarrow{k_4} P \qquad (4)$$

$$Y \xrightarrow{k_5} Q \qquad (5)$$

$$X_o \xrightarrow{k_6} X \qquad (6)$$

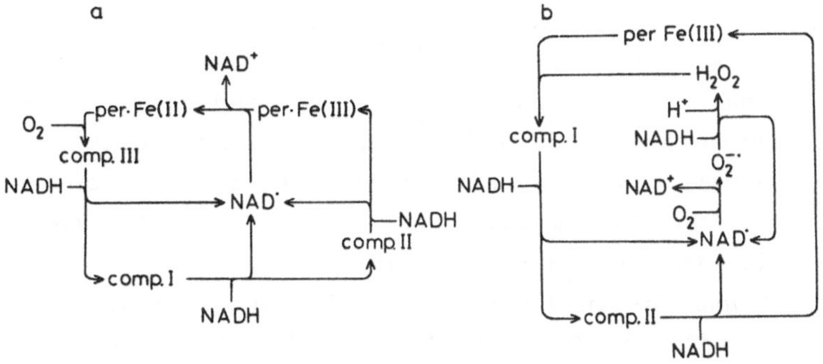

Fig. 4 Examples of autocatalytic schemes derived from known or proposed reaction steps [19]. per·Fe(III), ferriperoxidase; per·Fe(II), ferroperoxidase; comp., compound. Compounds I, II and III are enzyme intermediates

$$A_o \underset{k_7}{\overset{k_7}{\rightleftharpoons}} A \tag{7}$$

$$B_o \xrightarrow{k_8} B. \tag{8}$$

Here A denotes O_2, B denotes NADH and X and Y represent intermediate free radical species equivalent to those shown in Fig. 4. Reactions (7) and (8) simulate the inputs of O_2 and NADH from their respective sources. It is to be noted that reaction (1) and reaction (2) plus reaction (3) constitute two autocatalytic reactions and the quadratic branching step in (2) is the coupling of these. The enzyme itself has been omitted from the minimal scheme. However, there is strong evidence that reaction (1) involves the enzyme [20] in such a way that we can assume k_1 to be linearly dependent on the enzyme concentration.

The model was simulated on a computer by numerical integration of the differential equations:

$$\dot{A} = k_7\ (A_o - A) - k_3\ ABY \tag{9}$$

$$\dot{B} = k_8\ B_o - k_1\ BX - k_3\ ABY \tag{10}$$

$$\dot{X} = k_1 BX - 2k_2\ X^2 + 3k_3\ ABY - k_4\ X + k_6\ X_o \tag{11}$$

$$\dot{Y} = 2k_2\ X^2 - k_3\ ABY - k_5\ Y. \tag{12}$$

The oscillatory waveformes of A and B obtained with this model were essentially identical to those of O_2 and NADH in the experimental system. Furthermore the changes in behaviour resulting from changes in k_1 followed those obtained by changing the enzyme concentration in the experimental system [12]. A critical point in the test of the model is to obtain chaotic motion of species A and B. An example of such behaviour of A is shown in Fig. 5. The pattern of oscillations of A was followed for several tens of thousands of excursions finding no repetition. Furthermore, the solutions of differential equations with parameters as in Fig. 5 were highly sensitive to initial conditions. Figure 6 shows a bifurcation diagram obtained by varying k_1 over the interval $0.3 \leq k_1 \leq 0.41$. The shaded region indicates the interval where chaotic solutions are obtained - chaotic solutions in this context are defined as aperiodic solutions that are highly unstable to small changes in initial conditions [30-32]. It should be noted that the transitions to chaotic motion are different when the chaotic regime is approached either from the left or from the right. From the

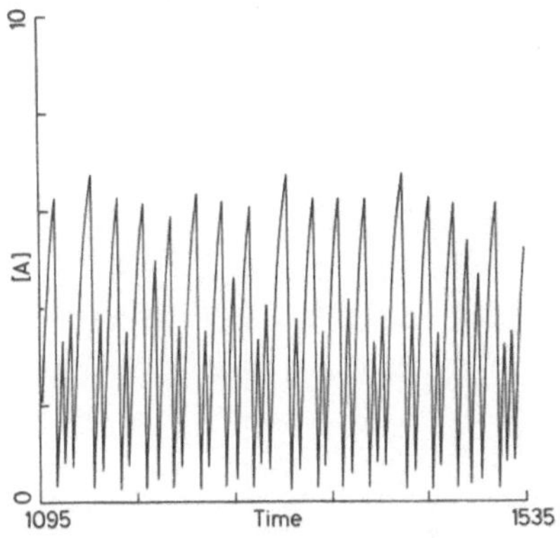

Fig. 5 Aperiodic oscillations of A obtained by numerical integration of eqs. (9) - (12). Parameters: $k_1 = 0.35$, $k_2 = 2.5 \times 10^2$, $k_3 = 3.5 \times 10^{-2}$, $k_4 = 20$, $k_5 = 5.35$, $k_6 X_0 = 10^{-5}$, $k_7 = 0.1$, $A_0 = 8$, $k_8 B_0 = 0.825$. Units of time and concentration are dimensionless

right the transition occurs via the well-known route of period doubling bifurcations [33,34] whereas from the left the order of transitions, starting at $k_1=0.3$, is $P_4 \rightarrow P_7 \rightarrow P_3$, where the index indicates the number of excursions per period, P. At $k_1 > 0.338$ period multiplication and chaotic motion begin.

5. Comparisons of experimental and numerical data - next-amplitude plots

Despite the close resemblance of the experimental results with the corresponding numerical data there are differences - quantitative as well as qualitative - between the model and the experimental system. The two most important are listed below:

1) The k_1 values in the model giving chaotic solutions are relatively closer to the k_1 values producing simple relaxation oscillations (P_1) compared to the corresponding experimentally observed enzyme concentrations.
2) The k_1 interval in the model giving chaotic solutions appears to be narrower compared to the experimentally observed enzyme concentration interval.

However, these discrepancies can probably be overcome by a choice of different sets of rate constants in the model. It was found that changing k_2 resulted in a shift in the position relative to P_1 and a change in the width of the chaotic regime in the k_1 bifurcation diagram (Fig. 6). A k_2 versus k_1 phase diagram has yet to be made.

Another way to compare numerical data with experimental data is to construct a one-dimensional (next-amplitude) map for both. Originally the map obtained from experimental data was believed to define a single-valued curve [2] for which evidence for chaotic behaviour could be obtained by using LI and YORKE's theorem [26]. However recent numerical

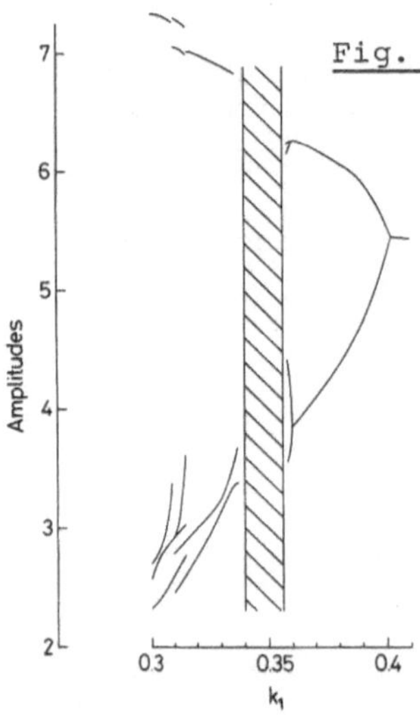

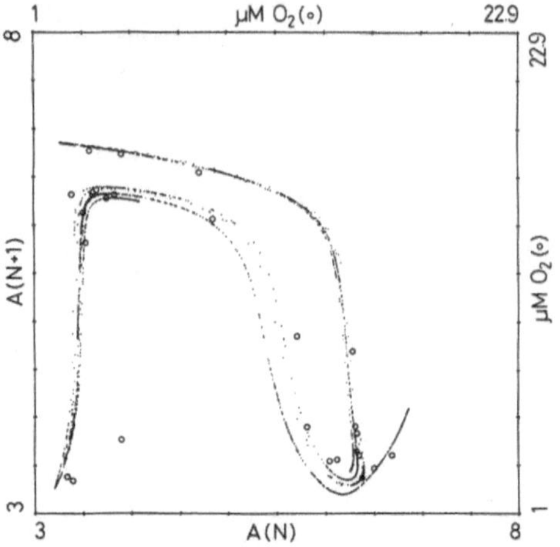

Fig. 7

Fig. 6 Bifurcation diagram obtained by varying k_1 and keeping all other parameters constant (values as in Fig. 5). The ordinate indicates the amplitudes of excursions of A. Only periods with less than 8 excursions are indicated. The shaded regime corresponds to oscillations with no apparent period

Fig. 7 Next-amplitude plots constructed from experimental (o) and numerical (·) data (Figs. 3 and 5)

results have suggested that the map is far from being single valued [12]. Figure 7 shows a comparison of the numerical and experimental next-amplitude plots constructed from the data shown in Figs. 3 and 5. The map obtained from numerical data was shown to have a 'Cantor set like' or 'fractal' transversal structure [12] reminiscent of that found in HENON's map [35]. The good agreement between numerical and experimental data suggests that the experimental system has a similar strange attractor with a fractal transversal structure. If so then the peroxidase-oxidase reaction reveals chaotic motion of a higher order than that reported recently for the Belousov-Zhabotinskii reaction [9,10].

This research was supported by the Danish Natural Science Research Council.

1. O.E. Rössler: Z. Naturforsch. 31a, 259 (1976)
2. L.F. Olsen and H. Degn: Nature 267, 177 (1977)
3. R.A. Schmitz, K.R. Graziani and J.L. Hudson: J. Chem. Phys. 67, 3040-3044 (1977)
4. O.E. Rössler and K. Wegmann: Nature 271, 89 (1978)
5. V. Hlavacek and P. Van Rompay: Chem. Eng. Sci. 36, 1587 (1981)
6. K. Tomita: Phys. Rep. 86, 113 (1982)

7. K. Showalter, R.M. Noyes and K. Bar-Eli: J. Chem. Phys. 69, 2514 (1978)
8. N. Ganapathisubramanian and R.M. Noyes: J. Chem. Phys. 76, 1770 (1982)
9. J.L. Hudson and J.C. Mankin: J. Chem. Phys. 74, 6171 (1981)
10. R.H. Simoyi, A. Wolf and H.L. Swinney: Phys. Rev. Lett. 49, 245 (1982)
11. J.S. Turner, J.-C. Roux, W.D. McCormick and H.L. Swinney: Phys. Lett. 85A, 9 (1981)
12. L.F. Olsen: Phys. Lett. 94A, 454 (1983)
13. I. Yamazaki, K. Yokota and R. Nakajima: Biochem. Biophys. Res. Commun. 21, 582 (1965)
14. I. Yamazaki and K. Yokota: Biochim. Biophys. Acta 132, 310 (1967)
15. S. Nakamura, K. Yokota and I. Yamazaki: Nature 222, 794 (1969)
16. I. Yamazaki and K. Yokota: Mol. Cell. Biochem. 2, 39 (1973)
17. H. Degn: Nature 217, 1047 (1968)
18. H. Degn: Biochim. Biophys. Acta 180, 271 (1969)
19. L.F. Olsen and H. Degn: Biochim. Biophys. Acta 523, 321 (1978)
20. L.F. Olsen: Biochim. Biophys. Acta 527, 212 (1978)
21. H. Degn, L.F. Olsen and J.W. Perram: Ann. N.Y. Acad. Sci. 316, 623 (1979)
22. V.R. Fed'kina, T.V. Bronnikova and F.I. Ataullakhanov: Studia Biophysica 82, 159 (1981)
23. R.J. Field and R.M. Noyes: J. Chem. Phys. 60, 1877 (1974)
24. J. Lundsgaard and H. Degn: IEEE Trans. Biomed. Eng. BME 20, 384 (1974)
25. H. Degn, J. Lundsgaard, L.C. Petersen and A. Ormicki: in Methods of Biochemical Analysis (D. Glick, ed.) Vol. 26 (John Wiley & Sons, New York 1979) pp 47-77
26. T.-Y. Li and J.A. Yorke: Am. Math. Month. 82, 985 (1975)
27. H. Degn and D. Mayer: Biochim. Biophys. Acta 180, 291 (1969)
28. I. Fridovich: in Free Radicals in Biology (W.A. Pryor, ed.) Vol.1 (Academic Press, London 1976) pp. 239-277
29. P. Lindblad and H. Degn: Acta Chem. Scand. 21, 791 (1967)
30. R. Shaw: Z. Naturforsch. 36a, 80 (1981)
31. H. Degn: Phys. Rev. A26, 711 (1982)
32. J. Ford: Phys. Today April 1983, 40 (1983)
33. R.M. May: Nature 261, 459 (1976)
34. M.J. Feigenbaum: J. Stat. Phys. 19, 25 (1978)
35. M. Hénon: Commun. Math. Phys. 50, 69 (1976)

Nonuniform Information Processing by Strange Attractors of Chaotic Maps and Flows

J.S. Nicolis[(1)], G. Mayer-Kress, and G. Haubs

Institut für Theoretische Physik, Universität Stuttgart, Pfaffenwaldring 57
D-7000 Stuttgart 80, Fed. Rep. of Germany

Summary

We introduce a new parameter -the " nonuniformity factor" (NUF)- by way of estimating and comparing the deviation from average behavior (expressed by such factors as the Lyapunov characteristic exponent(s) and the information dimension) in various strange attractors (discrete and chaotic flows). Our results show for certain values of the control parameters the inadequacy of the above averaging properties in representing what is actually going on - especially when the strange attractors are employed as dynamical models for information processing and pattern recognition. In such applications (like for example visual pattern perception or communication via a burst-error channel) the high degree of adherence of the processor to a rather small subset of crucial features of the pattern under investigation or the flow, has been documented experimentally: Hence the weakness of concepts such as the entropy in giving in such cases a quantitative measure of the information transaction between the pattern and the processor. We finally investigate the influence of external noise in modifying the NUF.

1. Introduction

Taking averages in physical sciences in general and in communication theory in particular results always in some selective loss of detail. If it happens that a few details account practically for the whole pattern then the averaging process simply "washes out" all the essential information. In statistical mechanics for one the pursuit of evolution of the microscopic probability density function and its moments through the formalism of the Master equation and Fokker-Planck-equation in systems far from equilibrium and near bifurcation points manifest the "break down" of the law of large numbers; this has been amply demonstrated in recent years [1,2] - together with the ensuing invalidation of the "mean field regime". The entropy for example is just the mean value of the distribution $-\ln p(x)$ where $p(x)$ stands for the a priori probability density distribution of a (finite or infinite) set of elements constituting a certain pattern. Some of the elements of the set may be extremely improbable vis-a-vis a certain observer or prone to deliver upon reception a disproportionally large amount of information. Of particular interest is the case where the median value of $p(x)$ is the least probable. In such cases the usual expression(s) for the entropy:

$$S = - \sum_i p_i(x) \log_2 p_i(x) \qquad \text{or} \tag{1}$$

$$S = -\int p(x) \log_2 p(x)dx \qquad \text{(in bits)} \tag{2}$$

is perhaps inadequate in characterizing quantitatively the information transaction.
In this paper we intend to treat dynamical systems where the variety production or information dissipation are given by the dynamical analogs of the entropy and are couched in terms of the Lyapunov-exponents of the flow or discrete map concerned.

---

(1) On leave of absence form the Dpt. of Electrical Engineering, University of Patras, Greece

In the following we do three things. First we briefly review some experimental evidence about the dynamics of visual pattern perception and recognition (what are the "crucial features" of the pattern in such a case and how is the processor dealing with them?) as well as the irrelevance of the "law of averages" in certain "coin tossing" and communication problems. Second we discuss the possible use of strange attractors as dynamical models in information processing. Thirdly we calculate how the NUF fares in different attractors as the control parameters change. We provide expressions which under specific circumstances should compliment the Lyapunov exponent(s) and the information dimension of the attractors involved.

2. The Break Down of the "Law of Large Numbers" in Pattern Recognition and Random Walks

I. Review of Some Experimental Results in Visual Pattern Perception [3,4,5]

What are the "crucial features" of a visual pattern and how is the optical (human) cortex (in cooperation with the sensor (fovea) [(2)] and the optical muscular apparatus) dealing with them? In a remarkable series of experiments conducted in the early seventies, Noton and his associates performed a number of investigations which are discussed below. To begin with just look at fig. 1. The left part represents the famous bust of Queen Nefertiti. At right is displayed the trajectory of the eye movement of a human observer (as recorded by A.L. Yarbus of the Institute for problems of Information Transmission in Moscow, USSR) - as it was scanning Nefertiti´s head and neck in the process of perception. The experiments were performed in an attempt to settle the controversy of Gestalt (parallel, global) versus sequential (-step by step - or iterative) pattern recognition. (An essential prerequisite for the meaningfullness of the attempted comparison is of course, that the pattern must be extended enough in space in order to allow for scanning by the sensor.) The result of the experiments heavily support the hypothesis of serial, or piecemeal perception and recognition. Specifically two questions are answered, a) what are the features of the pattern that the optical cortex selects as the key items for identifying the object? and b) how are such features integrated and related to one another to form the complete internal representation of the object?

First of all we have good evidence that the optical system is an hierarchical one [4] and between the successive hierarchical levels mappings take place giving rise to feedforward - feedback loops. The higher levels (cortex) receive information from the lower levels and respond by sending commands to the sensor´s muscular apparatus either to move in order to"phaselock" with the feature under investigation or to move away and assume the next algorithmic step of scanning the pattern. The experiments demonstrate that the above scanning algorithm is far from smooth and "homogeneous". There are parts on the pattern that hold for the processor the most information about the subject. The fixation or the "holding time" of the receptor tend to cluster around the parts characterized by sharp curves (small radius of curvature), angles, or in general around areas where the curve is "nondifferentiable". It appears that the angles are the principal features the brain uses to store and recognize the pattern. (There is also neurophysiological justification for such a preference revealed through the painstaking experiments of D.H. Hubel and T.N. Wiesel; it appears that there are angle-detecting neurons in the frog´s retina and angle-detecting neurons in the visual cortex of cats and monkeys. Recordings obtained from the human visual cortex by E. MARG [3] of the University of California at Berkely give indications, that this result can be extended to the human visual cortex as well.) The scanning of fig. 1 bears witness to the heavy preference of the processor to areas of the face characterized by sharp curves. Such features are complex (they need "many bits" for their most laconic description which would lead to the constructions of a replica through a finite state machine [13]) and are therefore endowed with high information content. The next question concerns how these features are integrated by the brain into a coded internal representation.

(2) The fovea constitutes a small central area of the retina and is characterized by the highest concentration of photoreceptors.

Once again evidence comes from the recording of the eye movements. Dynamical analysis of the sequence of fixation of the receptor at the different "states"-features of the (somewhat coarse-grained) pattern suggest a format for the interconnection of these "states" into the overall internal representation. The result of this experiment reveals that the sensor directed by the brain is essentially involved with two types of "scanning pathways": Regular and irregular. Specifically it appears as NOTON and STARK [3] put it that "(the) eyes usually scanned it (the pattern) following - intermittently but repeatedly - a fixed path which we have termed his "scan path" .." The occurrences of the scan path were separated by periods in which the fixations were ordered in a less regular manner... Scan paths usually occupied from 25 to 35 percent of the subject's viewing time, the rest being devoted to less regular eye movement." We may conclude then following Noton and Stark that the internal representation (or the mapping) of a pattern in the memory system is an assemblage of features (states) mediated essentially by a feedback loop: A sequence of senso-motor traces recording, abstracting and subsequently reporting to the brain a "Markov chain" as it were, of states-features - whereupon the brain directs the next move of the peripheral activity. The time intervals during which the system holds a given state has obviously to do with the excitation of the directing neuronal tissue - which in turn depends on the degree of curvature of the feature-state involved. This Markov chain, this algorithm, is subsequently stored in the brain isomorphically as, say, a pattern of circulating electrical activity. When the observer is subsequently encountering the same pattern again he recognizes it by matching it with the "feature-ring" or the Markov chain which constitutes the internal representation in his memory, state-by-state. Matching then or recognizing consists in calculating the "distance" or the crosscorrelation between two Markov chains: The memorised one and the one which runs as the observer re-examines the pattern. Obviously the first chain directs the steps of the second through the brain-senso-motor loop and so learning takes place: Beyond a few reruns the object becomes "familiar" as the crosscorrelation above tends to one. It would be then, under such circumstances, utterly improper to try to gather the information convoyed from observing the Nefertiti's bust by calculating the "entropy" of this pattern. By the way, this could be done as follows: You partition the two-dimensional pattern in fig. 1 in $n(\varepsilon)$ squares of size ε from each of which the trace of the scanning path passes at least once. Let be a number proportional to the relative frequency with which the scanner visits the specific square i. Then the "entropy" of the pattern could be calculated as

$$-\sum_{i=1}^{n(\varepsilon)} p_i(\varepsilon)\, \log_2 p_i(\varepsilon) \qquad \text{bits} \tag{3}$$

and as the resolution ε becomes finer and finer one may be interested in calculating how the distribution $p_i(\varepsilon)$ scales with resolution. In that case one could obtain the "information dimension", or the bits required for the determination of a point of the scanning curve,

$$D_I = \lim_{\varepsilon \to 0} \left\{ \frac{-\sum_{i=1}^{n(\varepsilon)} p_i(\varepsilon)\, \log_2 p_i(\varepsilon)}{\left|\log_2 \frac{1}{\varepsilon}\right|} \right\} \tag{4}$$

as the asymptotic value of the slope of entropy versus resolution ($\|\log \varepsilon\|$). It is implicitly assumed of course that the Nefertiti's bust constitutes indeed a "strangely attracting object" or, that the scanning path remains bounded in space and continuously iterating, on the basis of a dynamical control algorithm dictated by the brain centers - which centers "fire" more or less frequently depending of the curvature of the encountered last detail (state). In view of the shape of scanning trace on the right side of fig. 1 instead on relying on equ. (4) one could obtain a more meaningful collective property or quantitative measure of the information gathered by investigating the variance of D_I. However to make calculations feasible we should not start from formula (4) but rather from a relationship between the information dimensionality and the spectrum of Lyapunov-exponents of the flow or the discrete map, and calculate the variance of D_I, ΔD_I, from the NUFs of the corresponding Lyapunov-exponents. The NUF of D_I would indicate how much fuzziness enters in the degree of compressibility $(N - D_I)/N\%$ of information realized by the attractor attractor in N-dimensional space.[(3)]

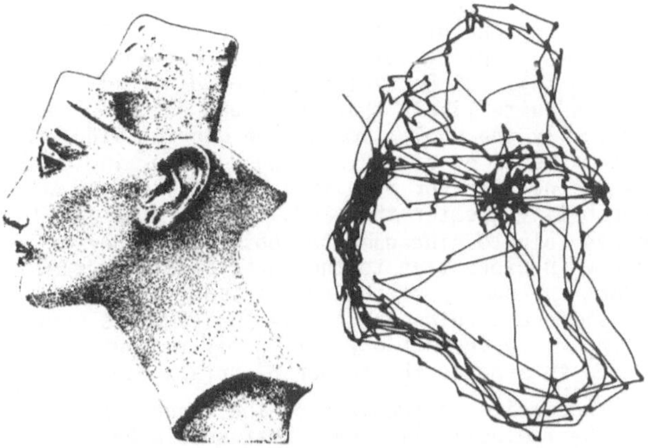

Fig.1: Scanning Nefertiti´s bust (After Noton and Stark, 1971)

II. Information Transfer through Intermittent Channels or "Error Clustering" Media

Let us now change gears and refer to another seemingly unrelated example concerning again the break down of the "law of averages" in Markov chain sequences. The example has essentially to do with coin-tossing games or random walks and has been elevated into the prominence of a new scientific paradigm by Feller [6] and subsequently by Berger and Mandelbrot [7]. According to "widespread beliefs" (the characterization is Feller´s) a so-called "law of averages" should ensure that in a long coin-tossing game each player will be in the winning side for about half the time and that "switching" will take place frequently from one player to the other. Fig. 2, taken from Feller represents the result of a computer experiment simulating 10.000 tosses of a fair coin. The top line contains the graph of the first 550 trials; the

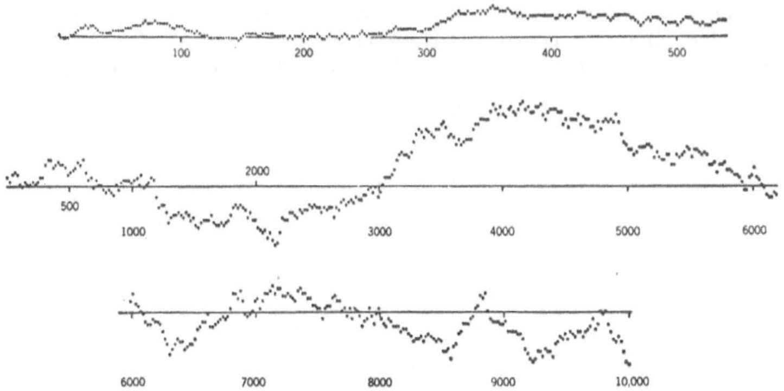

Fig.2: Fluctuations of gains in the game of fair coin tosses (after Feller 1970)

---

(3) In the present work we calculate the NUF for the dominant λ of maps and flows. The variance of D_I will be given in a forthcoming paper.

next two lines represent the entire record. The surprising thing has to do of course with the length of intervals between sucessive crossing of the "zero"-axis. On a isomorphic basis you can imagine these graphs as standing for a one-dimensional random walk without barriers and ask how often the walker is likely to change sides. Because of the walk´s symmetry one expects that in a long walk the man should spend about half of his time on each side of the starting spot. Exactly the opposite is true. Regardless of how long he walks the most probable number of changes from one side to the other is 0, the next probable 1 followed by 2, 3 and so on. If a man walks one step every second say for a year, in one out 20 repetitions of this experiment the walker would expect to go along one direction for more than 364 days and 10 hours. (In such simulations one is not using coins of course but decides on the basis of the upcoming digits of to say, 100 decimals: An even digit is "0" and an odd is "1".) In conclusion then it is quite likely that in a long coin-tossing game one of the players remains practically the whole time on the winning side, the other on the loosing side. (In the experiments of fig. 2 mentioned above, in 10.000 tosses of a perfect coin the lead is at one side for more than 9930 trials and at the other for fewer than 70 with probability greater than 10%.) On the other hand one "should expect" that in a prolonged coin-tossing game the observed number of changes of lead should increase roughly in proportion to the duration of the game. This again is false. Feller proves that the number of changes of lead in n trials increases only as $\sqrt{n}$; so in 100n trials one should expect only 10n times as many changes of lead as in n trials. Putting it in another way, if N_n is the number of changes in lead $N_n/n = \sqrt{n}/n = 1/\sqrt{n}$ goes to zero as the number of trials goes to infinity. So the waiting times between successive changeovers of the winner-looser roles between the contestants are likely to be fantastically long. What should be the pertinent probability density function for such a process? Imagine a huge sample of records of ideal coin tossing games each consisting of 2n tosses. We pick one at random and observe the number of the last trial at which the accumulated numbers of heads and tails were equal - that is the last changeover. This number is even and we denote it by 2K; so that $0 \leq K \leq n$. Frequent changeovers (that is in the lead of one player) would imply that K is likely to be relatively close to n - but this is not so. Interestingly enough, Feller´ s calculation give for the p.d.f. of the variable $x = K/n$ the expression:

$$P(x) \cong \frac{1}{n\pi\sqrt{x(1 - x)}} \tag{5}$$

The symmetry of the distribution implies that the inequalities $K \geq n/2$ and $K \leq n/2$ are equally likely. We see that the probabilities near the end points are greatest; the most probable values of K are then the extremes 0 and n, while the median value $K = n/2$ is least likely to occur. The above probability density function is identical (save the proportionality factor 1/n) to the invariant measure of the logistic map $f(x) = 4rx(1 - x)$ for $r = 1$, i.e. for the maximum value of the control parameter

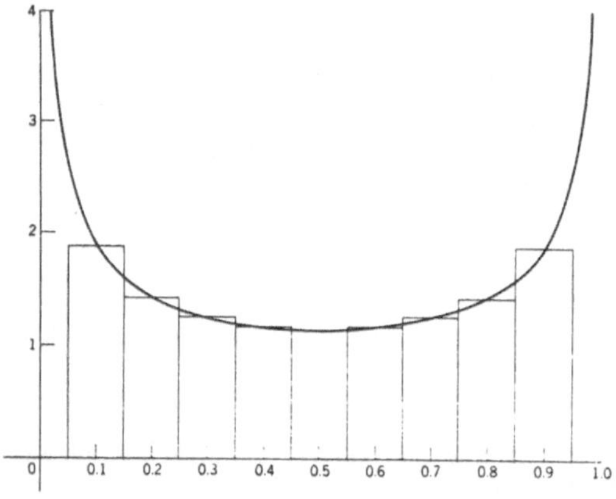

Fig.3: The probability-density-function of x=k/n (see text, after Feller, 1970)

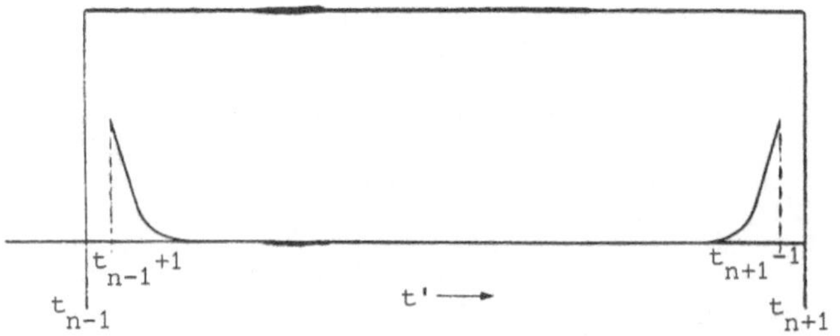

Fig.4: The p.d.f. of the "intermediate error" in a burst noise channel (after Berger and Mandelbrot, 1963)

where the chaotic regime occupies the whole invariant interval and the map becomes exactly two-to-one. Let us now turn to the distribution of the occurence of errors in data transmission on real communication channels, the simplest being the binary symmetric channel (Fig. 4), (Gilbert 1960, Berger and Mandelbrot 1963). The channel is modelled as a Markov chain with two states G (Good = errorless) and B (Bad). The probability of error depends upon the state. State G is associated with zero probability of an error. In state B a (loaded) coin is tossed to decide whether an error will occur or not. The (biased) coin tossing feature is included because actual bursts or clusters do contain "good" digits interspersed with the errors. To simulate burst errors the states G and B must persist; hence the transition probabilities P_{GB} and P_{BG} should be small and the probabilites P_{GG} and P_{BB} should accordingly be large. Berger and Mandelbrot started by considering groups of three successive errors taking the time position t_{n-1} and t_{n+1} fixed and allowing t_n to hover in between; what is the probability density function of the inter-error intervals and the probability density of t_n? When the inter-error intervals are geometrically distributed (like in some cases assumed by the memoryless binary symmetric channel) the distribution of t_n is uniformly distributed between t_{n-1} and t_{n+1}. Berger and Mandelbrot, in order to bring the model as close as possible to the collected data, assumed for the inter-error intervals a Pareto distribution. In that case, for the probability density of t_n the theoretical type of curve best fitting the experimental results is illustrated in fig. 5. It says essentially that the probability of the occurrence of an error increases as we approach t_{n-1} or t_{n+1} symmetrically and becomes greatest in the close neighborhood of the occurrence of another error (the previous or the next). The probability of having an error in between is practically zero. Similarly in the case of a large number of errors a sizeable portion of the total sample length will be found in a few of the longest error intervals, say L of them, thus creating a pattern in which errors are mostly grouped in L clusters. Again the median value of the distribution is the most improbable one while the greatest probability refers to light error clustering in time. The shape of the probability density function is again of the "hyperbolic" type we have already encountered twice above. The channel has his "Good" days and his "Bad" days - occuring in persistent clusters. This persistence and the scarcity of a possible changeover reminds one of the random walk business or the fair tossing games again. In fact one might conjecture that the time axis within the transmission interval displays a fractal property that is the clustering is self-similar down to small scales. The channel spends most likely in the errorless regime, either too long, or too short a time. However the relative number of errors or the average number of digits in errors should be expected to tend to zero as the length of the message n increases to infinity. This conjecture comes for the same reasonsing mentioned pre-

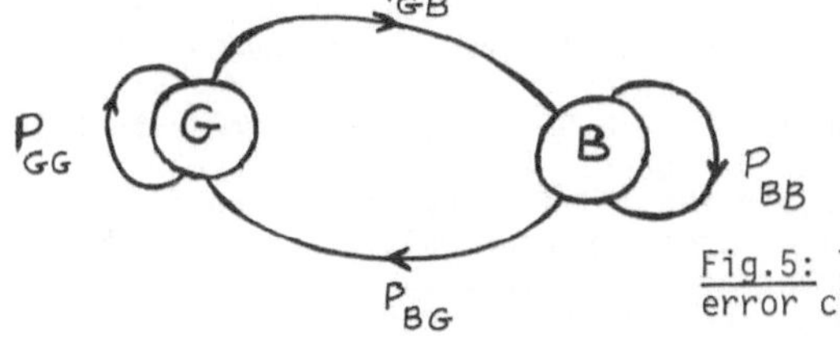

Fig.5: The Markov-chain model for the burst error channel

viously with reference to the average number of changes in the lead of one regime as time outstretches to infinity. It turns out then that as the length of the transmitted message goes to infinity despite the presence of an unbounded number of errors the channel capacity of the burst-error binary symmetric channel tends to one that is the figure one usually gets for the noiseless memoryless channel. Finally in their study of intermittent behavior near a tangent bifurcation of the logistic map, Hirsch, Huberman and Scalapino (1982) derive a similar probability density function for the path lengths that the orbit spends in the laminar (pseudo-limit cycle) regime or the chaotic regime - where it wanders aperiodically before getting a chance of being reinjected in the "channel" formed between the map and the bisector line - that is in the limit cycle regime. In other words in the intermittent regime the orbit spends, say within the limit cycle regime, either too little or too much time. Again the average value of the (symmetrical) distribution function is the least likely to occur.

3. Possible Role of Chaos in Information Processing

It appears therefore that in a broad domain of seemingly unconnected physical phenomena a chaotic dynamics or macroscopic "turbulent noise" prevails - in the absence of microscopic multidimentional noise - and in fact it is responsible for the break down of the "law of large numbers". (No wonder; chaos is manifested even with 3 macrovariables, coupled via quite deterministic non-linear ordinary differential equations.) Is there any compelling reason to believe that in biological organisms in general and man in particular chaos may play any role in information processing? Reliable information processing rests upon the existences of a "good" code (or map) or language: namely a set of recursive rules which generate variety at a given hierarchical level and subsequently compress it thereby revealing information at a higher level. To accomplish this a language - like good music - should strike at every moment an optimum ratio of variety (stochasticity) versus the ability to detect and correct errors (memory). Is there any dynamics available which might emulate this dual objective in state space? The answer is: In principle yes (J.S. Nicolis, 1982, 1983). We remind ourselves that there are two available dynamical ways of producing information: Either by cascading bifurcations giving rise to broken symmetry or via cascading iterations increasing resolution. The last way is simpler. A three-dimensional strange attractor for example creates variety along the direction of his positive Lyapunov exponent λ_+ and constrains variety (thereby revealing information) along the direction of his negative Lyapunov exponent λ_-, ($\|\lambda_-\| > \lambda_+$). The issue then of relevance of strange attractors in information processing can first be debated on parsimonial grounds namely what possible evolutionary advantages may the chaotic mode bestow on an organism. Neurophysiological evidence must of course follow. Discussions on the above two topics have been couched by one of us (J.S. Nicolis) in a recent series of publications, where for example the thalamo-cortical pacemaker of the (human) brain has been tentatively identified as a chaotic processor; we are not going to repeat them here. Enough of emphasising that a chaotic processor even as an artifact ensures a very rich behavioral repertoire (software) with very simple hardware. What we intend to do in the remaining of this paper is to provide a general discussion on chaotic dynamics and report on a number of calculations on strange attractors (both chaotic maps and flows) which calculations justify, we think, mistrust in the intuitive reliance on the "median" values in characterizing the dynamics of these attractors; we thereby propose some more representative parameters. Let us remind ourselves that basic parameters of a strange attractor are a) his Lyapunov exponents determining the average amount of information produced ($\lambda > 0$) or dissipated ($\lambda < 0$) per iteration and b) the information dimension of the attractor D_I which esentially determines the average value of the degree of compressibility ensured by the processor namely the average number of bits one needs to determine any point on the attractor after transients subside. For the three-dimensional Lorenz attractor for example, $D_I = 2.06$ (bits) which means that if it is used as an information processing unit the Lorenz attractor can "save" $3 - 2.06 = 0.94$ bits for any point on its basin that is for any initial condition. Since compressibility of a time series is the necessary prerequisite for subsequent simulation of the dynamical phenomena conveyed by the series, the importance of strange attractors as information processors cannot be overlooked.

4. One-dimensional maps and Markov chains

In the preceding sections we have been referring to a category of dynamical processes - all of them characterized by a probability density function of the "hyperbolic" type; in such a case the median value $\int p(x)x\,dx$ is very small and the variance

$$\sigma = \sqrt{p(x)x^2\,dx - \left[\int p(x)x\,dx\right]^2} \qquad (6)$$

is large.
We now want to test the conjecture that perhaps the median value of the information distribution (that is the entropy) is, under specific circumstances (for example in the vicinity of intermittency) also inadequate in describing the information processing going on by the evolution of a strange attractor. Let us start then with examining the correspondance between one-dimensional maps and Markov chains. Let us consider the asymmetric two-to-one piece-wise linear map of fig. 6.

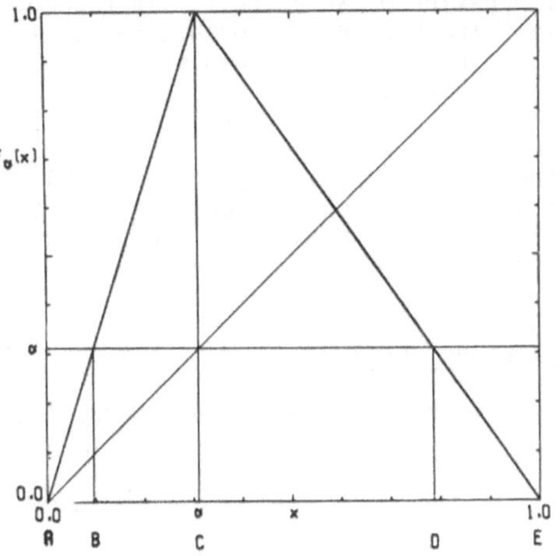

Fig.6: Graph of $f_\alpha(x)$ for $\alpha = 0.31$ (see eq. (21))

It is instrumental to start in this simple case by calculating the elements of the Markov chain which represents the iterative process of the map as an information source. We devide the unit interval AE in two parts of length α and $1 - \alpha$ respectively and every time the trace of the iterating trajectory falls on AC we get the symbol 1 while every time the trace falls on CE we get the symbol 0. So we obtain (for every value of the control parameter) strings of 0010110110110111101... - which do not constitute however just fair coin tosses but rather one-sided Bernouilli shifts they are constrained by the (non-linear) shape of the map f(x). What are the transitional probability elements p_{ij} of the two-state Markov chain and what are the probabilities $u_1 = p(0)$ and $u_2 = p(1)$? From fig. 6 it is clear that $l(AB) = P_{11}$ since all points on the interval within AB which belongs to α are projected by the map on a portion equal to α. From the geometry we see that $P_{11} = \alpha^2$. Likewise we observe that $l(CD) = P_{22}$ since all points within this subinterval which belong to $1-\alpha$ are projected by the map on the portion $1-\alpha$. $P_{12} = l(BC)$ since all the points within it, belonging to α are projected to $1-\alpha$ and $P_{21} = l(DE)$ since all points within it belonging to $1-\alpha$ are projected to α. From the geometry we get: $P_{12}\ \alpha - \alpha^2 = \alpha(1-\alpha)$, $P_{21} = \alpha(1-\alpha)$ and $P_{22} = (1-\alpha) - \alpha(1-\alpha) = (1-\alpha)^2$. (The values above are not normalized; one should have $P_{11} + P_{12} = 1$, $P_{21} + P_{22} = 1$.)

The probabilities u_1, u_2 are calculated from the relations $u_1 = u_1P_{11} + u_2P_{21}$ and $u_1 + u_2 = 1$. We get: $u_1 = \alpha$, $u_2 = 1-\alpha$.

We now intend to forward some general discussion on the nature of Lyapunov exponents in multidimensional flows and maps.

5. The Nonuniformity Factor as a Further Characterization of Dynamical Systems

I. The Concept of the Local Divergence Rate

The spectrum of the Lyapunov-exponents constitutes a way to classify dynamical systems in general as well as particular solutions of a dynamical system. It is nowadays widely used in the literature [14,15,18].

In this section we want to interprete the maximal Lyapunov-exponent as the time average of a statistical variable we call the local divergence rate $Y(\vec{x}(t))$. Talking about time averages it is quite natural to consider higher moments of the statistical variable like the variance. The variance of the local divergence rate ΔY is a measure for the nonuniformity of the dynamical behavior of an attractor in respect to the separation of nearby trajectories. Besides the Lyapunov-exponents the variance ΔY will give additional information about a dynamical system and may turn out to be helpful for further classification of chaotic systems.

In the second part of this section we derive the expression for the statistical variable, whose first moment will be the Lyapunov-exponent. We will discuss the meaning of the corresponding variance and define the Non-Uniformity Factor (NUF) ΔY in the context of dynamics. In a third part we will first consider analytically and numerically the case of different 1-dimensional maps. Then in the last subsections we will give some numerical results for a 2-dimensional map (Hénon-map) and a 3-dimensional flow (Rössler-system).

II. Average Value and Variance of the Local Divergence Rate

We start from a continuous dynamical system which is described by a set of coupled ordinary nonlinear differential equations

$$\dot{\vec{x}} = \vec{F}(\vec{x}) \tag{7}$$

The maximal Lyapunov exponent is defined as [14,15,16,17]:

$$\lambda = \lim_{t\to\infty} \frac{1}{t} \ln \|\vec{u}(t)\|^{4)} \qquad \text{where } \vec{u}(t) \text{ is a solution of:} \tag{8}$$

$$\dot{\vec{u}} = \frac{\delta\vec{F}}{\delta\vec{x}}(\vec{x}(t))\,\vec{u} \tag{9}$$

which is obtained by linearizing eq.(7) along a trajectory $\vec{x}(t)$. Note that eq.(9) is a linear system of differential equations with time dependent coefficients that describes the behavior of a perturbation of the trajectory $\vec{x}(t)$ of system (7). The solution of eq.(9) to a given initial condition $\vec{e}(t=0)$ at time t can be written as:

$$\vec{u}(t) = U_0^t(\vec{x}(0))\,\vec{e}(0) \tag{10}$$

$U_0^t(\vec{x}(0))$ is a time dependent matrix and is sometimes called fundamental solution matrix. The argument $\vec{x}(0)$ shall indicate that $U_0^t(\vec{x}(0))$ depends on the trajectory $\vec{x}(t)$ with initial condition $\vec{x}(0)$. We will now write eq.`s (10) and (8) in a form that is very helpful for the intuitive understanding of the Lyapunov-exponent and more suitable for its numerical computation for flows as well as for discrete maps [17]. To this end we formally discretize time into intervals of length τ, such that $t=n\tau$. A solution (10) to eq.(9) can then be written as:

4) rigorously, lim should be replaced by limsup, see [26]

$$\vec{u}(t) = U^{n\tau}_{(n-1)\tau}\, U^{(n-1)\tau}_{(n-2)\tau} \ldots U^{2\tau}_{\tau}\, U^{\tau}_{0}\, \vec{e}(0)$$

We take τ as the time unit and, changing the notation slightly, we obtain:

$$\vec{u}(n) = U^{n}_{n-1}\, U^{n-1}_{n-2} \ldots U^{2}_{1}\, U^{1}_{0}\, \vec{e}(0) \tag{11}$$

Let us read eq.(11) from right to left. $\vec{u}(1) = U^1_0\, \vec{e}(0)$ denotes the solution of eq.(9) to the initial condition $\vec{e}(0)$ after time τ. It is again a vector with length d_1 and direction $\vec{e}(1)$: $U^1_0\, \vec{e}(0) = d_1\, \vec{e}(1)$, $\|\vec{e}(1)\| = 1$.

Now we apply U^2_1 to $\vec{e}(1)$ and obtain another vector of length d_2 and direction $\vec{e}(2)$. Finally we obtain:

$$\vec{u}(n) = d_n\, d_{n-1} \ldots d_2\, d_1\, \vec{e}(n)\ , \qquad \|\vec{e}(n)\| = 1\ . \tag{12}$$

We have changed a product of operators into a product of real numbers. The Lyapunov-exponent (eq.(8)) now reads:

$$\lambda = \lim_{n\to\infty} \frac{1}{n\tau} \ln \| d_n \ldots d_1\, \vec{e}(n) \| \quad \text{or}$$

$$\lambda = \lim_{n\to\infty} \frac{1}{n} \sum_{k=1}^{n} \frac{1}{\tau} \ln d_k \tag{13}$$

From the definition of $d_k = U^k_{k-1}\ \vec{e}(k-1)$ we see that $\ln d_k$ is the exponential change of the length of $\vec{e}(0)$ during the time interval τ when the system (7) moves along the trajectory between $\vec{x}((k-1)\tau)$ and $\vec{x}(k\tau)$. Normalizing this quantity to one time-unit we define the <u>local divergence rate</u> for discrete or discretized systems:

$$Y_k = \frac{1}{\tau} \ln d_k \tag{14}$$

Equation (13) now suggests to interprete Y_k as a statistical variable. The Lyapunov-exponent λ is the time average (in time-units of τ) of the local divergence rate Y_k along the trajectory $\vec{x}(t)$:

$$\lambda = \langle Y_k \rangle\ . \tag{15}$$

Note that although eq.(13) describes a discrete sampling of Y_k along a continuous flow it is still exact. $Y_k = \frac{1}{\tau}\ln d_k$ is obtained by solving the continuous eq.`s (7) and (9). On the other hand eq.(13) can also be used for disrcete maps. Y_k is then obtained by iterating the map and its linearization. With the interpretation of λ as the time average of the local divergence rate we define the nonuniformity factor (NUF) to a solution $\vec{x}(t)$ as the coresponding variance:

$$\Delta Y = \langle\langle (Y_k - \langle Y_k \rangle)^2 \rangle\rangle^{1/2} = (\langle Y_k{}^2 \rangle - \langle Y_k \rangle^2)^{1/2} \qquad \text{or} \tag{16}$$

$$(\Delta Y)^2 = \lim_{n\to\infty} \left(\frac{1}{n} \sum_{k=1}^{n} \left(\frac{1}{\tau}\ln d_k\right)^2 \right) - \lambda^2 \tag{17}$$

The NUF is a measure for the deviation of the local divergence rate from its mean value, the Lyapunov-exponent. It characterizes the nonuniformity of a solution of a dynamical system in respect to the sensitivity of initial conditions, or in other words, how much the local divergence rate changes along the flow. Chaotic flow, e.g., with uniform turbulent behavior all over the attractor, yields a positive Lyapunov-exponent and a zero variance. Intermittent chaos, however, where regular laminar and chaotic phases alternate irregularly with each other, also give a positive Lyapunov-exponent (possibly the same) but the variance (NUF) should be large, since the local divergence rate varies strongly. So the NUF ΔY can be a very helpful number besides the Lyapunov-exponents to characterize the behavior of dynamical systems, especially chaotic ones. Its computation does not take more effort than that of the Lyapunov-exponents. In the following section we consider some chaotic dynamical systems that are well known in the literature [15].

III) Analytical and numerical results from specific models

a) Maps on the interval

For a discrete-time dynamical system, which is generated bya map f on the one-dimensional interval I, the Lyapunov-ex-ponent λ as defined in eq. (13) is given by:

$$\lambda = \lim_{n\to\infty} \frac{1}{n} \sum_{k=0}^{n-1} \ln\|f'(x_k)\| \tag{18}$$

where $x_k = f^k(x_0)$ and x_0 is some initial value in the basin of the attractor under consideration. In the case of one-dimensional maps we can apply the ergodic hypothesis and replace the time average (18) by the ensemble average of the local divergence rate Y(x) with respect to the probability-density-function p(x) of f [18] . In our case the divergence-rate Y(x) as defined in eq. (14) takes the simple form:

$$Y(x) = \ln\|f'(x)\| \tag{19}$$

The Lyapunov-exponent λ therefore can be expressed as:

$$\lambda = \langle Y(x)\rangle = \int_I p(x)\, Y(x)\, dx \tag{20}$$

Accordingly, we obtain the NUF as the standard-deviation ΔY of the statistical variable Y(x) (see eq (16)). The interpretation of the NUF is especially simple in this case: it just describes the variation of the modulus of the slope of the function f along the attractor. Thus, for the "uniform" maps $\tilde{f}(x) = 2x \bmod 1$ and the "tent"-map:

$$f_{0.5}(x) = \begin{cases} 2x & x\varepsilon\ [0,0.5] \\ 2(1-x) & x\varepsilon\ [0.5,1] \end{cases}$$

we have Y(x) = ln 2 and p(x) = 1. This is why the NUF vanishes, i.e. the separation of trajectories happens uniformly in the whole interval.

We now wish to disturb this uniformity in two different ways by considering:

(i) piece-wise linear but asymmetric maps

(ii) symmetric but nonlinear maps

For perturbations of the type (i) we define the family f_α by:

$$f_\alpha(x) = \begin{cases} \dfrac{x}{\alpha} & x\ \varepsilon\ [0,\alpha] \\ \dfrac{1-x}{1-\alpha} & x\ \varepsilon\ [\alpha,1] \end{cases} \tag{21}$$

The graph of $f_{.32}$ is shown in fig.6. For each α we have constant density p(x) = 1 , and therefore we can calculate both the Lyapunov-exponent λ_α and the NUF ΔY_α analytically. The Lyapunov-exponent is given by:

$$\lambda_\alpha = (\alpha-1)\ \ln(1-\alpha) - \alpha\ \ln\alpha \tag{22}$$

From

$$Y_\alpha^2(x) = \alpha(\ln\alpha)^2 + (1-\alpha)\ (\ln(1-\alpha)^2 \quad \text{and} \tag{23}$$

$$\lambda^2 = \alpha^2(\ln\alpha)^2+(1-\alpha)^2(\ln(1-\alpha\ )^2)+2\alpha(1-\alpha)\ln\alpha\ln(1-\alpha) \tag{24}$$

we deduce the NUF:

$$\Delta Y_\alpha = (\alpha(1-\alpha))^{1/2}\ \ln(\frac{1-\alpha}{\alpha}) \tag{25}$$

In fig.7 we have plotted the resulting Lyapunov-exponent λ_α (solid line) and the NUF ΔY_α (broken line) as a function of the asymmetry parameter α. Note that the NUF vanishes in the uniform cases, for which $\alpha \varepsilon \{0,\frac{1}{2},1\}$. It has two maxima which are given by the transcendental equation:

$$\ln(\frac{1-\alpha}{\alpha}) = \frac{2}{1-2\alpha} \tag{26}$$

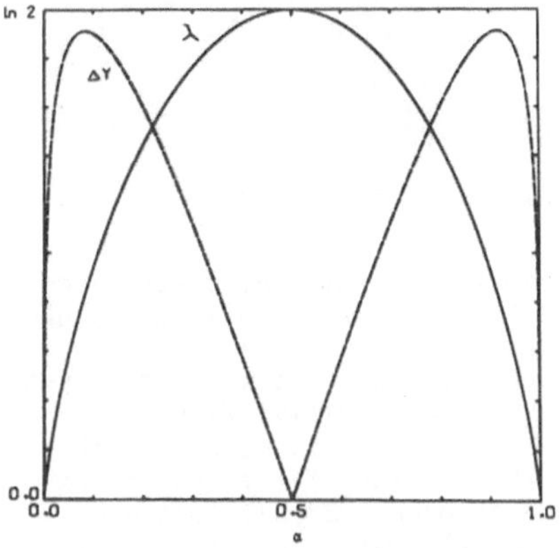

Fig.7: Lyapunov exponents (λ) and NUF (ΔY) versus the asymmetry parameter α for $f_\alpha(x)$ of eq.(21)

We also see from fig.7 that beyond some threshold values of the asymmetry-parameter α, the NUF exceeds the Lyapunov-exponent.
From what we have mentioned above, it is clear that all the results also hold for the family of discontinuous functions $\tilde{f}_\alpha$ defined by:

$$\tilde{f}_\alpha = \frac{x}{\alpha} \bmod 1\ . \tag{27}$$

In order to study non-uniform systems of type (ii), we consider the family of non-linear symmetric functions $g_{\gamma,r}(x)$, defined by [19]:

$$g_{\gamma,r}(x) = r(1 - \|1 - 2x\|^\gamma) \tag{28}$$

In fig.8 the graph of the "tent"-map $g_{1,1}(x) = f_{0.5}(x)$ (solid line), of the "Lorenz"-map $g_{0.5,1}(x)$ (broken line), and of the "logistic" map $g_{2,1}(x)$ (broken dotted line) are presented. The Lyapunov-eponent λ_γ and the NUF have been numerically determined for various values of the exponent γ and fixed parameter $r = 0.99999$. The results are shown in fig.9, where λ_γ (solid line) and ΔY_γ (broken line) are plotted versus the exponent γ. Note that λ_γ is constant except in a neighborhood of $\gamma = 0.5$, where intermittent chaos is observed, which gives rise to the vanishing Lyapunov-exponent [9,21,22]. Again we find $\Delta Y_\gamma = 0$ for the uniform case $\gamma = 1$ and two crossings of the λ_γ- and ΔY_γ-curves close to the "Lorenz"- and the "logistic" functions. An interesting feature in the intermittency case γ - 0.5 is the singularity of the relative standard-deviation $\Delta Y_\gamma/\lambda_\gamma$, which indicates the strong non-uniformity, stemming from the alternation of long laminar phases and chaotic bursts. This observation is confirmed in fig.10, where we have plotted λ (solid line) and ΔY (broken line) as a function of the height-parameter r in the "logistic" system

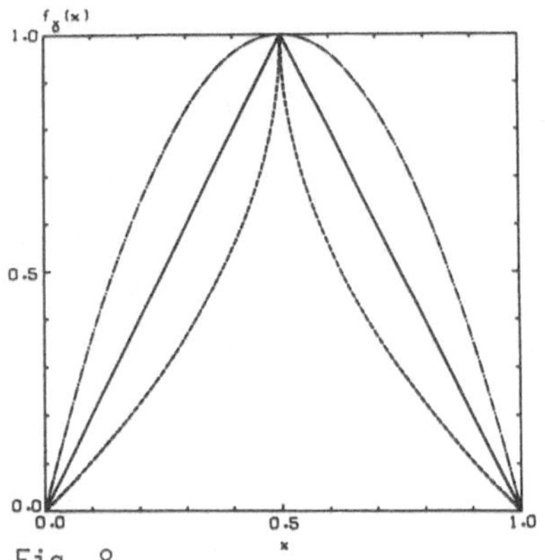

Fig. 8

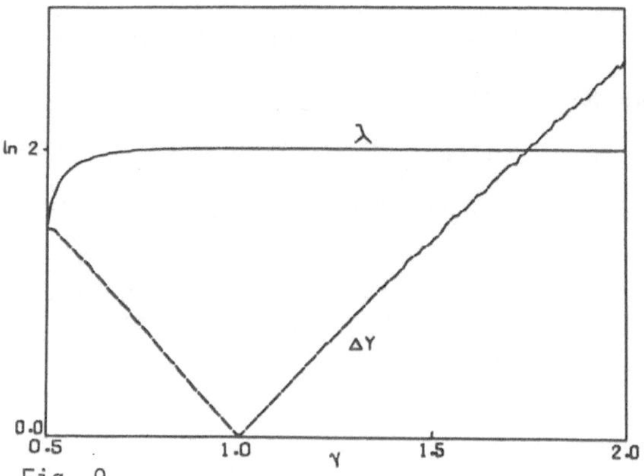

Fig. 9

Fig.8: Graph of $g_{\gamma,r}(x)$ for r=1. and $\gamma=\frac{1}{2}$ (---), γ=1 (——), γ=2 (-••-)

Fig.9: Lyapunov exponent (λ) and NUF (ΔY) of $g_{\gamma,r}(x)$ vs. γ for r=1

$g_{2,r}(x)$. At each value of r at which the system exhibits intermittent chaos, i.e. where λ_r drops from positive to negative values [18,20,9,22], we find large values of the NUF, just like in the case of the "Lorenz"-map.

We have also performed numerical simulations at parameters r, at which the system is fully chaotic, and where we have perturbed the system by additive fluctuations. Our results indicate that the NUF does not change very much (a few percent), when the noise-level is increased to its maximal value. This insensitivity against external noise has also been observed for the Lyapunov-exponent [22,23], although the probability-density-function becomes more and more flat with increasing noise-level.

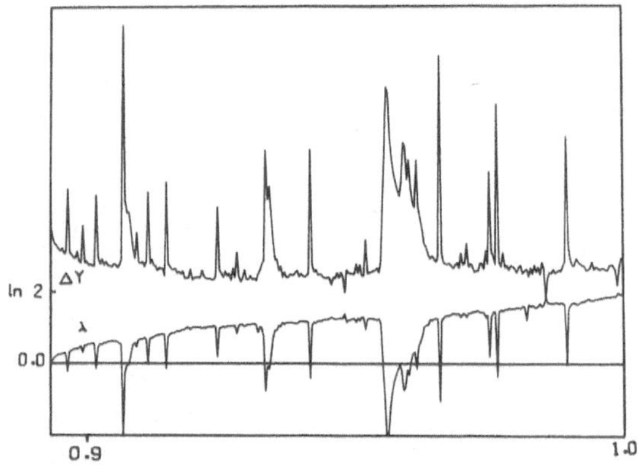

Fig.10: Lyapunov exponent (λ) and NUF (ΔY) of $g_{\gamma,r}(x)$ vs. r for γ=2

b) The Hénon-Map

We used eq.`s(13) and (17) to compute the maximal Lyapunov-exponent and the NUF for the Hénon-map [24,15]:

$$x_{n+1} = y_n + 1 - ax_n^2 \qquad y_{n+1} = bx_n$$

We picked the usual parameter choice for b = 0.3 and varied the parameter a from 0.0 to 1.42 . Figure 11 shows the largest Lyapunov-exponent (lower curve) and the corresponding NUF (upper curve). Figure 12 shows a magnification of the chaotic regime of fig. 11.

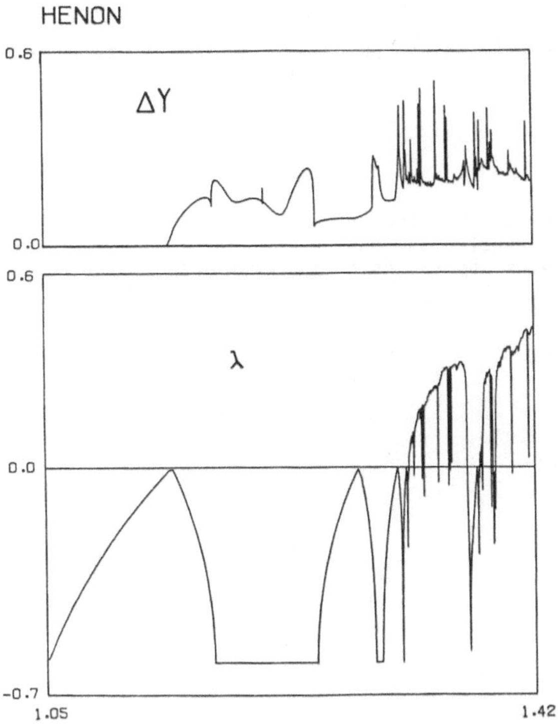

Fig.11: Largest Lyapunov exponent λ (lower part) and corresponding NUF ΔY (upper part) of the Hénon-system vs. parameter a for b=0.3

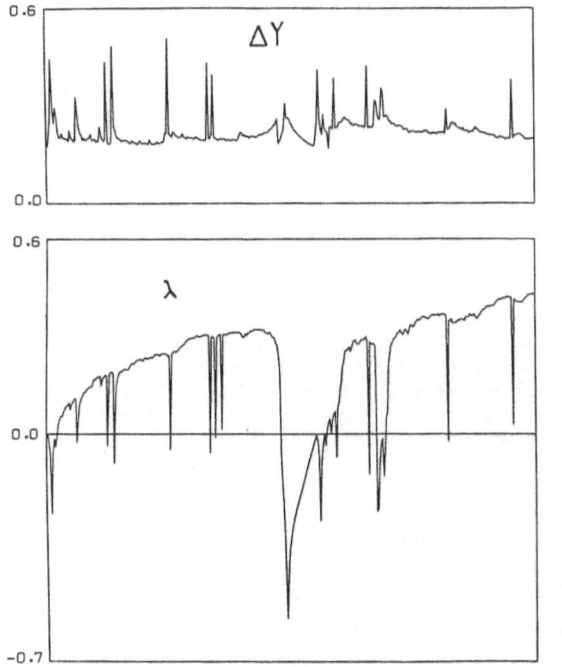

Fig.12: Magnification of the chaotic parameter range of fig. 11

c) Rössler-system

The Rössler-system is one of the simplest examples for chaotic 3-dimensional flows [25,15]:

$$X = -(Y + Z)$$
$$Y = X + aY$$
$$z = b + Z(X - c)$$

We choose a=b=0.2 and varied c from 2.6 to 6.0 . The results of the computations are shown in fig. 13. Again the upper curve denotes the variance of the local divergence rate (NUF) while the lower curve indicates the values of the mean value (Lyapunov-exponent).

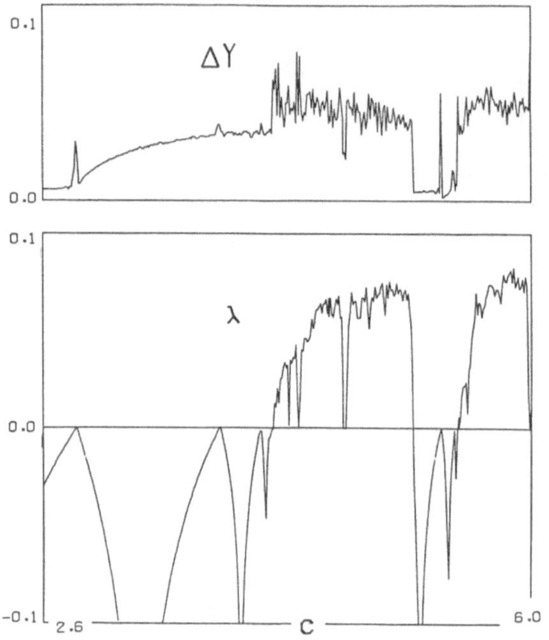

Fig.13: Largest non-zero Lyapunov exponent λ (lower part) and NUF Δ of the largest Lyapunov-exp. (upper part) of the Rössler-attractor vs. parameter c for a=b=0.2

Acknowledgement

The authors thank Professor H.Haken for his interest and support which made the present work possible. One of us (J.S.N.) expresses also his gratitude to Prof. Haken for his generous hospitality in his institute when J.S.N. served as a Gastprofessor for the academic year 1982-83. The authors are also indepted to Miss Elisabeth Straub for the careful typing of the manuscript.

References

1 H. Haken, Synergetics - An introduction, 2nd ed., Springer, Berlin, Heidelberg, New York, 1978
2 G. Nicolis and I. Prigogine, Self-organization in non-equilibrium structures, Wiley, New York [u.a.] 1977
3 D. Noton, L. Stark, Sci. Am., 224 (1971), June, p. 34
4 D. Noton, IEEE Trans. on Systems Science and Cybernetics,SSC-6 (1970), 4, p. 349
5 D. Noton, L. Stark, Science, 171 (1971), p. 308

6 W. Feller, An introduction to probability theory and applications, Third ed., Vol. 1, Wiley, New York [u.a.] 1970
7 J.M. Berger, B. Mandelbrot, IBM Journal, (1963), July, p. 224
8 E.N. Gilbert, The Bell Syst. Techn. J., (1960), Sept., p. 1253
9 J.E. Hirsch, B.A. Huberman, D.J. Scalapino, Phys. Rev. A, 25 (1982),p. 519
10 J.S. Nicolis, The role of chaos in reliable information processing, J. of the Franklin Inst., In press 1983, Also in: Synergetics, (Ed.: H. Haken), Proceedings of the Elmau Conference, 2-8 may 1983
11 J.S. Nicolis, Kybernetes, 11 (1982), Oct., p. 269 - 274
12 J.S. Nicolis, Kybernetes, 11 (1982), April, p. 123 - 132
13 J.S. Nicolis, Dynamics of hierarchical systems, To app. in Springer-Verl. 1983
14 H. Haken, Advanced synergetics, To appear in Springer-Verl.
15 A.J. Lichtenberg, M.A. Liebermann, Regular and stochastic motion, Springer, Berlin,Heidelberg,New York 1983
16 I. Shimada, T. Nagashima, Progr. Theor. Phys., 61 (1979), 6
17 G. Benettin, L. Galgani, A. Giorgilli, J.M. Strelcyn, C. R. Acad. Sci. Paris, 286 (1978) A-431
18 R.S. Shaw, Z. Naturforsch., A 36 (1981), p. 80
19 J.P. Crutchfield, N. Packard, Int. J. Teor. Phys., 21 (1982), p. 433
20 G. Mayer-Kress, H. Haken, Phys. Lett., 82 A (1981), p. 151
21 J.P. Eckmann, L. Thomas, P. Wittwer, J. Phys., A 14 (1982), p. 3153
22 G. Mayer-Kress, H. Haken, J. Stat. Phys., 26 (1981), p. 149
23 D. Farmer, J. Crutchfield, B. Huberman, Phys. Rep., 92 (1982) p. 45
24 M. Henon, Commun. Math. Phys., 50 (1976), p. 69
25 O.E. Rössler, Phys. Lett., 57 A (1976), p. 397
26 H. Haken, Phys. Lett., 94 A (1983), p. 2

Part IV

Stability and Instability in Dynamical Networks

Generalized Modes and Nonlinear Dynamical Systems

P.E. Phillipson
Department of Physics, University of Colorado, Boulder, CO 80309, USA

Dynamical systems described by either two or three coupled nonlinear differential equations can be understood in terms of generalized coordinates, or modes, constructed on the basis of dynamical considerations. The goals are to give a picture of mechanism, provide a diagnostic tool for analysis and where feasible generate approximate analytic solution. A general formalism is developed and examples are given which include the Oregonator model of the Belousov reaction, the May-Leonard equations for three competing populations and the Lorenz equations for hydrodynamic turbulence.

1. Introduction

The time evolution of many nonlinear processes are encompassed by a small number of coupled differential equations [1]. Three dimensional systems, for example, provide the simplest realistic cases of either sustained periodic limit cycle oscillation or chaotic behaviour produceable by quadratic nonlinearities. If a step in the dynamics of a process in genuinely three dimensions can be considered to be infinitely fast, however, the dynamics can be effectively reduced to two dimensions. The aim of the present work is to outline a general approach for analysis of two and three dimensional systems in terms of interaction between generalized coordinates, or modes, whose structure can serve to highlight dynamical features inherent to a given system. The goals of this program are to provide a picture of dynamical mechanism to supplement the quantitative results of computer study, provide a diagnostic tool for analysis, and finally, when possible effect approximate analytic solution. The idea, which in principle is applicable to any dimension, is to transform the differential equations originally given in chemical or physical coordinates to coordinates which bear the same relation to the former that curvilinear coordinates bear to Cartesian coordinates. Since these curvilinear coordinates find their origin more in dynamical than in geometric structure they will be referred to alternatively as generalized modes. In three dimensions the generalized modes resemble cylindrical coordinates. By formally suppressing the third cylindrical coordinate the dynamics contracts to two dimensions in terms of modes which resemble plane polar coordinates. Transformation to these modes and their rationale will be given below followed by some representative applications to demonstrate the type of information that can be obtained using a generalized modes approach.

2. Transformation Theory

Consider three coupled nonlinear autonomous differential equations

$$\frac{dX_n}{dt} = P_n(X_1,X_2,X_3), \quad n=1,2,3 \tag{1}$$

for which there exist fixed points $X^{(o)}=[X_1^{(o)},X_2^{(o)},X_3^{(o)}]$ defined by $P_n[X^{(o)}]=0$, n=1,2,3. We introduce coordinates relative to the fixed point, $x_n=X_n-X_n^{(o)}$, n=1,2,3, and expand P_n around the fixed point, with the result that (1) becomes

$$\frac{dx_n}{dt} = \sum_{m=1}^{3} a_{nm}x_m+F_n(x_1,x_2,x_3), \quad n=1,2,3 \tag{2}$$

where $a_{nm}=\left.\frac{\partial P_n}{\partial X_m}\right|_{X^{(o)}}$ so the F_n are free of linear terms. The coordinates (x_1,x_2,x_3) will now be regarded as Cartesian coordinated expressible in terms of generalized cylindrical coordinates (R,σ,Z) by the following transformation

$$\begin{aligned} x_1 &= \alpha_{11}R\cosh\sigma+\alpha_{12}R\sinh\sigma + \alpha_{13}Z \\ x_2 &= \alpha_{21}R\cosh\sigma+\alpha_{22}R\sinh\sigma + \alpha_{23}Z \\ x_3 &= \alpha_{31}R\cosh\sigma+\alpha_{32}R\sinh\sigma + \alpha_{33}Z \end{aligned} \tag{3a}$$

and its inverse

$$\begin{aligned} R\cosh\sigma &= (\alpha^{-1})_{11}x_1+(\alpha^{-1})_{12}x_2+(\alpha^{-1})_{13}x_3 \\ R\sinh\sigma &= (\alpha^{-1})_{21}x_1+(\alpha^{-1})_{22}x_2+(\alpha^{-1})_{23}x_3 \\ Z &= (\alpha^{-1})_{31}x_1+(\alpha^{-1})_{32}x_2+(\alpha^{-1})_{33}x_3 \end{aligned} \tag{3b}$$

The transformation coefficients (α) will be so chosen that the cylindrical coordinates are eigenfunctions of relevant <u>portions</u> of the eigenvalues which result from normal mode analysis of (2) in the linear limit that $F_n=0$, n=1,2,3. The rationale for this construction is that when the complete equations (2) are expressed in these generalized coordinates features of the dynamics of time evolution become highlighted in a way not always obvious from direct numerical solution of equations (2). To achieve this construction consider first the normal modes q_k and the associated eigenvalues p_k of the linearized equations

$$\frac{dx_n}{dt} = \sum_{m=1}^{3} a_{nm}x_m, \quad x_n = \sum_{k=1}^{3} C_{nk}q_k, \quad n=1,2,3 \tag{4a}$$

$$\frac{dq_k}{dt} = p_k q_k, \quad \sum_{m=1}^{3} a_{nm}C_{mk} = p_k C_{nk}, \quad k=1,2,3 \tag{4b}$$

where C_{nk} are the transformation coefficients to the normal modes and the second of (4b) gives the secular equation for the eigenvalues as the determinant of the coefficients set equal to zero.

$$\begin{vmatrix} p-a_{11} & -a_{12} & -a_{13} \\ -a_{21} & p-a_{22} & -a_{23} \\ -a_{31} & -a_{32} & p-a_{33} \end{vmatrix} = p^3-Ap^2+Bp-C=0 \tag{5a}$$

where

$$\begin{aligned} A &= p_1+p_2+p_3 = a_{11}+a_{22}+a_{33} \\ B &= p_1p_2+p_1p_3+p_2p_3 = \left[(a_{11}a_{22}-a_{12}a_{21})+(a_{11}a_{33}-a_{13}a_{31})+\right. \\ &\qquad \left.+(a_{22}a_{33}-a_{23}a_{32})\right] \\ C &= p_1p_2p_3 = \det(a) \end{aligned} \tag{5b}$$

The eigenvalues have the general structure

$$\begin{aligned} p_1 &= \lambda+\Omega & A &= 2\lambda+p_o \\ p_2 &= \lambda-\Omega & B &= 2\lambda p_o+\lambda^2-\Omega^2 \\ p_3 &= p_o & C &= p_o(\lambda^2-\Omega^2) \end{aligned} \tag{5c}$$

and the aim is to develop R, σ and Z as eigenfunctions of the eigenvalue components λ, Ω, and p_o respectively. This is achieved by the transformation

$$q_1 = R\exp(\sigma), \quad q_2 = R\exp-(\sigma), \quad q_3 = Z \tag{6}$$

From (4b) and the definitions in (5c)

$$\frac{dq_1}{dt} + \frac{dq_2}{dt} = 2\left[\cos\sigma\frac{dR}{dt} + \sinh\sigma R\frac{d\sigma}{dt}\right] = p_1q_1+p_2q_2 = 2\left[\lambda R\cosh\sigma+\Omega R\sinh\sigma\right] \tag{7}$$

so that by equating coefficients of $\cosh\sigma$ and $\sinh\sigma$

$$\begin{aligned} \frac{dR}{dt} &= \lambda R, \quad R = R(o)\exp(\lambda t) \\ \frac{d\sigma}{dt} &= \Omega, \quad \sigma = \Omega t+\sigma(o) \\ \frac{dZ}{dt} &= p_o Z, \quad Z = Z(o)\exp(p_o t) \end{aligned} \tag{8}$$

the third relation being just a redefinition of the q_3 mode. We will regard R as the radial mode, σ its conjugate angular mode and Z as the cylindrical mode. If λ is negative R is a "descending radial mode", if λ is positive R is an "ascending radial mode" and if λ is zero R is an "indifferent radial mode". Similar classification is appropriate for the cylindrical mode according as p_o is negative, positive or zero respectively. In general Ω can be either positive or purely imaginary. In the latter case $\Omega = i\omega$ and σ is imaginary, so that the transformation equations (3a,b) are modified to $\sigma = i\Theta$ and

$$\begin{aligned} x_1 &= \beta_{11}R\cos\Theta + \beta_{12}R\sin\Theta + \beta_{13}Z \\ x_2 &= \beta_{21}R\cos\Theta + \beta_{22}R\sin\Theta + \beta_{23}Z \\ x_3 &= \beta_{31}R\cos\Theta + \beta_{32}R\sin\Theta + \beta_{33}Z \end{aligned} \tag{3c}$$

and

$$\begin{aligned} R\cos\Theta &= (\beta^{-1})_{11}x_1 + (\beta^{-1})_{12}x_2 + (\beta^{-1})_{13}x_3 \\ R\sin\Theta &= (\beta^{-1})_{21}x_1 + (\beta^{-1})_{22}x_2 + (\beta^{-1})_{23}x_3 \\ Z &= (\beta^{-1})_{31}x_1 + (\beta^{-1})_{32}x_2 + (\beta^{-1})_{33}x_3 \end{aligned} \tag{3d}$$

where

$$\beta_{nm} = \alpha_{nm} \text{ if } m \neq 2, \quad \beta_{n2} = i\alpha_{n2} \tag{9}$$

$$(\beta^{-1})_{nm} = (\alpha^{-1})_{nm} \text{ if } n \neq 2, \quad (\beta^{-1})_{2m} = -i(\alpha^{-1})_{2m}$$

The characterization of these modes, dependent as they are on the eigenvalues, must be determined ultimately by the constants a_{nm} through (5b,c). The latter in turn are dependent on the nonlinear coupling constants and on the choice of singular point which arises in the transition from (1) to (2). The expansion coefficients are fixed uniquely thereby in terms of the eigenvalues and the kinetic constants a_{nm}. The explicit relations are tabulated in the Appendix A.

The final formal step is to express the complete equations (2) in terms of the generalized modes. The matrix A defined by

$$A = \begin{vmatrix} [\alpha_{11}\cosh\sigma + \alpha_{12}\sinh\sigma] & [\alpha_{11}\sinh\sigma + \alpha_{12}\cosh\sigma] & \alpha_{13} \\ [\alpha_{21}\cosh\sigma + \alpha_{22}\sinh\sigma] & [\alpha_{21}\sinh\sigma + \alpha_{22}\cosh\sigma] & \alpha_{23} \\ [\alpha_{31}\cosh\sigma + \alpha_{32}\sinh\sigma] & [\alpha_{31}\sinh\sigma + \alpha_{32}\cosh\sigma] & \alpha_{33} \end{vmatrix} \tag{10a}$$

serves to link derivatives in the Cartesian and generalized coordinates upon differentiation of equation (3a)

$$\begin{pmatrix} \dot{x}_1 \\ \dot{x}_2 \\ \dot{x}_3 \end{pmatrix} = A \begin{pmatrix} \dot{R} \\ R\dot{\sigma} \\ \dot{Z} \end{pmatrix}, \quad \begin{pmatrix} \dot{R} \\ R\dot{\sigma} \\ \dot{Z} \end{pmatrix} = A^{-1} \begin{pmatrix} \dot{x}_1 \\ \dot{x}_2 \\ \dot{x}_3 \end{pmatrix} \tag{11}$$

where the dot denotes differentiation with respect to time. Substitution of (2) gives, with (4) and (6)

$$\frac{dR}{dt} = \lambda R + [A^{-1}_{11}F_1+A^{-1}_{12}F_2+A^{-1}_{13}F_3] \tag{12a}$$

$$R\frac{d\sigma}{dt} = \Omega R + [A^{-1}_{21}F_1+A^{-1}_{22}F_2+A^{-1}_{23}F_3]$$

$$\frac{dZ}{dt} = p_0 Z+[A^{-1}_{31}F_1+A^{-1}_{32}F_2+A^{-1}_{33}F_3]$$

where

$$\begin{aligned}
(A^{-1})_{11}&=[(\alpha^{-1})_{11}\cosh\sigma-(\alpha^{-1})_{21}\sinh\sigma],\\
(A^{-1})_{12}&=[(\alpha^{-1})_{12}\cosh\sigma-(\alpha^{-1})_{22}\sinh\sigma],\\
(A^{-1})_{13}&=[(\alpha^{-1})_{13}\cosh\sigma-(\alpha^{-1})_{23}\sinh\sigma],\\
(A^{-1})_{21}&=[(\alpha^{-1})_{21}\cosh\sigma-(\alpha^{-1})_{11}\sinh\sigma],\\
(A^{-1})_{22}&=[(\alpha^{-1})_{22}\cosh\sigma-(\alpha^{-1})_{12}\sinh\sigma],\\
(A^{-1})_{23}&=[(\alpha^{-1})_{23}\cosh\sigma-(\alpha^{-1})_{13}\sinh\sigma],\\
(A^{-1})_{31}&=(\alpha^{-1})_{31},\quad (A^{-1})_{32}=(\alpha^{-1})_{32},\quad (A^{-1})_{33}=(\alpha^{-1})_{33},
\end{aligned} \tag{10b}$$

and the functions F_n are expressed in terms of the generalized modes with the use of the transformation equations (3a). If Ω is imaginary ($\Omega=i\omega$) the transformed equations have a similar structure in terms of a matrix B^{-1} which results from putting $\sigma=i\Theta$ into the A^{-1} matrix and expressing the α coefficients as β coefficients according to (9). The inverse matrix has elements

$$\begin{aligned}
(B^{-1})_{11}&=[(\beta^{-1})_{11}\cos\Theta+(\beta^{-1})_{21}\sin\Theta],\\
(B^{-1})_{12}&=[(\beta^{-1})_{12}\cos\Theta+(\beta^{-1})_{22}\sin\Theta],\\
(B^{-1})_{13}&=[(\beta^{-1})_{13}\cos\Theta+(\beta^{-1})_{23}\sin\Theta],\\
(B^{-1})_{21}&=[(\beta^{-1})_{21}\cos\Theta-(\beta^{-1})_{11}\sin\Theta],\\
(B^{-1})_{22}&=[(\beta^{-1})_{22}\cos\Theta-(\beta^{-1})_{12}\sin\Theta],\\
(B^{-1})_{23}&=[(\beta^{-1})_{23}\cos\Theta-(\beta^{-1})_{13}\sin\Theta],\\
(B^{-1})_{31}&= (\beta^{-1})_{31},\quad (B^{-1})_{32}=(\beta^{-1})_{32},\quad (B^{-1})_{33}=(\beta^{-1})_{33}
\end{aligned} \tag{10c}$$

so that

$$\frac{dR}{dt} = \lambda R+[(B^{-1})_{11}F_1+(B^{-1})_{12}F_2+(B^{-1})_{13}F_3] \tag{12b}$$

$$R\frac{d\Theta}{dt} = \omega R+[(B^{-1})_{21}F_1+(B^{-1})_{22}F_2+(B^{-1})_{23}F_3]$$

$$\frac{dZ}{dt} = p_0 Z+[(B^{-1})_{31}F_1+(B^{-1})_{32}F_2+(B^{-1})_{33}F_3]$$

where the functions F_n are expressed in terms of the generalized modes with the use of transformation equations (3c).

The two dimensional case represents formally suppression of the x_3 Cartesian coordinate and corresponding suppression of the cylindrical Z-mode so that (3c) reduces to a generalization of phase plane analysis useful in studies of nonlinear oscillating systems [2]. In that case there is typically the transformation $x_1=R\cos\Theta$, $x_2=R\sin\Theta$ while here each Cartesian coordinate depends upon both sine and cosine in a prescribed way. The present approach as applied to two dimensions - given in Appendix B - gives added flexibility to simplify the term linear in the radial coordinate not achieved in the simpler phase plane coordinate representation. An example of a two dimensional application of the formalism has been made to Michaelis-Menten kinetics for which the radial mode is descending. This mechanism, which describes the irreversible conversion of substrate to product through formation of an enzyme-substrate complex, could be traced analytically free of the traditional steady state approximation [3]. Approximate analytic solution to equation (1) appropriate to the description of a four membered catalytic positive feedback loop, or hypercycle, showed that the system is quasi-stable in that regression to the fixed point occurs at the asymptotically slow rate of $t^{-1/2}$ due primarily to the fact that an indifferent radial mode is reluctantly dragged down by a descending cylindrical mode [4]. In general, 2n-dimensional systems can always be cast in terms of n radial modes and n conjugate angular modes while systems of odd dimensionality include a cylindrical mode corresponding to a necessarily real eigenvalue.

3. Systems with Quadratic Nonlinearities

The development to this point is general in so far as there is no restriction on the forms of the nonlinear functions F_n of (2). However, many dynamical systems include only quadratic nonlinearities. For such cases in three dimensions the most general form of (1) is

$$\frac{dX_n}{dt} = c_n + \sum_{i=1}^{3} L_{ni} X_i + \sum_{\substack{i=1 \\ (i \leq j)}}^{3} \sum_{j=1}^{3} K_{ij}^{(n)} X_i X_j, \quad n=1,2,3 \tag{13}$$

where the thirty constants (c_n, L_{ni}, $K_{ij}^{(n)}$; i,j,n=1,2,3) are real and $K_{ij}^{(n)} = K_{ji}^{(n)}$. In practical situations most of the constants are zero and many of them are equal. The fixed point is given by solution to

$$c_n + \sum_{i=1}^{3} L_{ni} X_i^{(o)} + \sum_{i=1}^{3} \sum_{j=1}^{3} K_{ij}^{(n)} X_i^{(o)} X_j^{(o)} = 0, \quad n=1,2,3 \tag{14}$$

and (2) assumes the realization

$$a_{nm} = L_{nm} + K_{mm}^{(n)} X_m^{(o)} + \sum_{i=1}^{3} K_{im} X_i^{(o)} \tag{15a}$$

$$F_n = \sum_{\substack{i=1 \\ (i \leq j)}}^{3} \sum_{j=1}^{3} K_{ij}^{(n)} x_i x_j \tag{15b}$$

An example of a diagnostic application is the model of FIELD and NOYES for the Belousov-Zhabotinski reaction [5]. In dimensionaless units the system equations are [6]

$$\frac{dX_1}{dt} = \frac{X_1}{\varepsilon} + \frac{X_2}{\varepsilon} - \frac{q}{\varepsilon} X_1^2 - \frac{X_1 X_2}{\varepsilon} \qquad L_{11}=L_{12}=\frac{1}{\varepsilon},\ K_{11}^{(1)}=-\frac{q}{\varepsilon},\ K_{12}^{(1)}=-\frac{1}{\varepsilon}$$

$$\frac{dX_2}{dt} = -X_2+2fX_3-X_1X_2 \quad \left[L_{22}=-1,\ L_{23}=2f,\ K_{12}^{(2)}=-1\right] \tag{16}$$

$$\frac{dX_3}{dt} = \frac{X_1}{p} - \frac{X_3}{p} \qquad \left[L_{31}=\frac{1}{p},\ L_{33}=-\frac{1}{p}\right]$$

where the bracketed terms are an identification with the structure of (13). The parameters q,ε are typically small compared to unity while the parameter p is typically large. Assuming f=1/2 the fixed point is given by

$$\left[X_1^{(o)},X_2^{(o)},X_3^{(o)}\right]=\left[\left[\left(\frac{1}{4}+\frac{2}{q}\right)^{1/2}-\frac{1}{2}\right],\ \frac{X_1^{(o)}}{1+X_1^{(o)}},\ X_1^{(o)}\right]$$

$$\approx [U,1,U], \quad U=\left[\frac{2}{q}\right]^{1/2} \tag{17}$$

The coefficients a_{nm} deriveable from (15a) in combination with (5a,b) show that the eigenvalues are given to the same approximation as that of (17) by

$$\lambda = -\frac{1}{2}\left[\frac{3}{\varepsilon U}+U\right], \quad \Omega=\left[\frac{1}{\varepsilon}(U-4)+\frac{U^2}{4}\right]^{1/2}, \quad p_o = \frac{4}{pU} \tag{18}$$

For example, a representative set of parameter values [6] (q,ε,p)= ($8.4\text{x}10^{-6}$, $2\text{x}10^{-4}$, 310) give exact eigenvalues (λ,Ω,p_o)=(-259.58, 1574.7,$2.66\text{x}10^{-5}$) while the approximate analytical relations of (18) give(-259.35,1574.6,$2.64\text{x}10^{-5}$). In the context of the present formulation the system features a descending radial mode and a relatively weak ascending cylindrical mode. Dynamical mechanism for oscillation has been investigated on the basis that a finite closed surface must exist such that any solution of (16) within at any instant must remain within [6]. Here oscillation is due to the existence of two regions in species space which we may denote as the "Field region" (F) and the "Noyes region" (N). These regions arise because the radial mode can and does go imaginary. This can be seen from the structure of (3b)

$$R = \left[A^2-B^2\right]^{1/2}$$

$$A = \left[(\alpha^{-1})_{11}x_1+(\alpha^{-1})_{12}x_2+(\alpha^{-1})_{13}x_3\right] \tag{19}$$

$$B = \left[(\alpha^{-1})_{21}x_1+(\alpha^{-1})_{22}x_2+(\alpha^{-1})_{23}x_3\right]$$

For the particular choice of expansion coefficients $(\alpha_{11},\alpha_{12},\alpha_{13})=$ $=(2,0,1)$, [c.f.App.A] R^2 is negative most of the time ($|A|<|B|$ which defines the F region) for periods between oscillations. Just prior to an oscillation R^2 goes positive ($|A|>|B|$ which defines the N region) passing through a zero when the two bracketed terms in (19) just cancel ($|A|=|B|$). In the above approximation of (18) for the eigenvalues the coefficients in (19) assume simple analytical form so that critical values (c) for the three species such that R=0 - the common boundary of the F and N regions - which heralds the onset of oscillation is given by

$$\left[x_1^{(c)}-x_3^{(c)}\right]\sim \Omega x_2^{(c)} \tag{20}$$

Species X_1 goes through very rapid oscillation and decay during which R^2 remains positive, again goes through a zero characterized by (20) and then goes negative during the relatively slow descent of X_2 and X_3. Parallel to the behaviour of the latter, descent in the F region is characterized by strong slaving of the cylindrical mode by the radial mode. Figure (1a) shows computer solution for (16) indicating the N and F regions, Fig.(1b) shows the corresponding time dependences of the $|R|$ and $|Z|$ modes (absolute values) which illustrate how they track each other, and Fig.(1c) is a phase plot of $|Z|$ vs $|R|$ for which the virtually straight line accounts for most of the F region. That $|Z|$ and $|R|$ are strongly coupled most of the time allows an approximate contraction of the system to two dimensions to facilitate approximate solution in these generalized coordinates free of the steady state approximation $\varepsilon\to 0$ which has been invoked to analytically study this system [7].

The F and N regions correspond to the set of species values (x_1,x_2,x_3) such that the radial mode is imaginary or real respectively. For a different free choice of coefficients R^2 could be positive most of the time until just prior to oscillation and the roles of the F and N regions become reversed. The absolute statement is that to pass from one region to the other yet have solutions remain real requires the radial mode pass through a zero and the angular mode behave such that $R\to -iR'$, $\sigma\to\sigma'+i\pi/2$ and back again. In general, limit cycle oscillation associated with real eigenvalues and at least one mode ascending can be understood in terms of the periodic passage between two regions separated by a zero of the radial mode.

While development of approximate analytic solution of the Field-Noyes model on the basis of the above considerations will be given

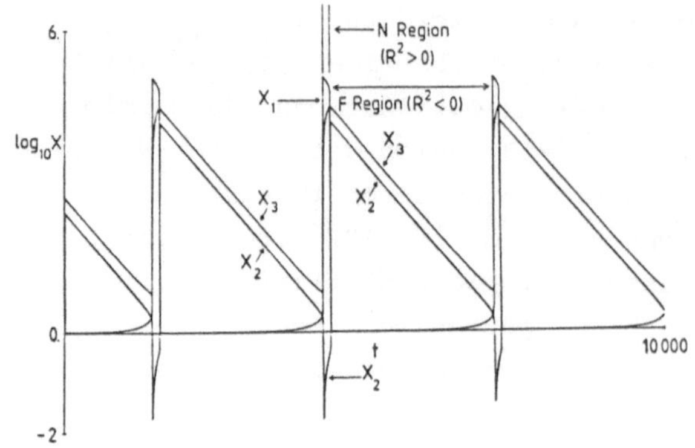

<u>Figure 1a.</u> Oscillations of Oregonator model (16) for parameter values given in text and initial conditions $X_1(0)=X_3(0)=500$, $X_2(0)=1$

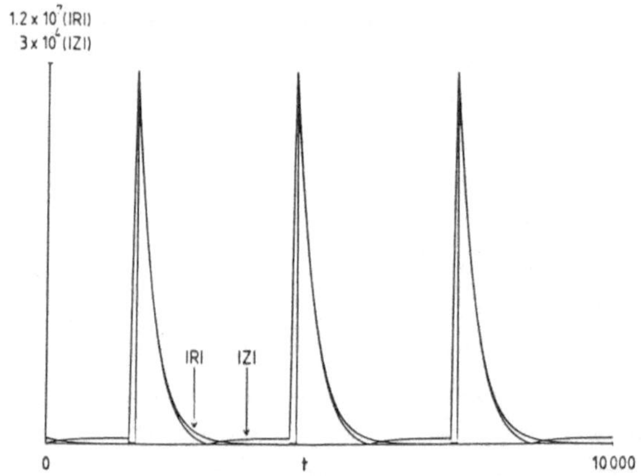

<u>Figure 1b.</u> Absolute values of the radial mode defined by (19) and cylindrical mode defined by (3b) for the Oregonator corresponding to the same time scale as Fig.1a

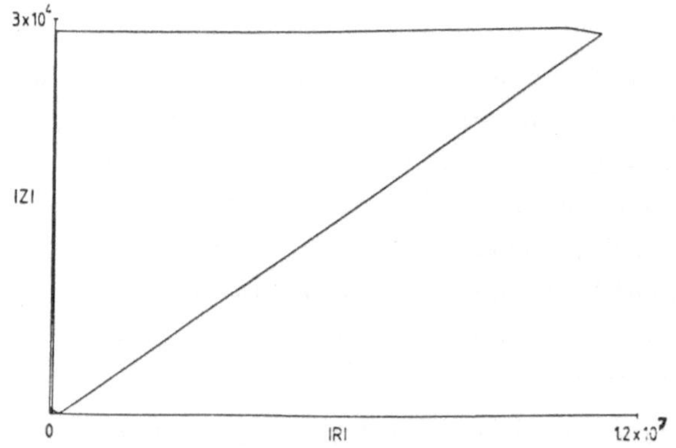

<u>Figure 1c.</u> $|Z|$ vs. $|R|$ of Fig.1b for the Oregonator

elsewhere, we will illustrate procedure by considering a special case of a model by May and Leonard [8] for competition between three populations according to the following Lotka-Volterra scheme

$$\frac{dX_1}{dt} = X_1 \left[1-X_1-\alpha X_2-\beta X_3\right], \qquad L_{11}=L_{22}=L_{33}=1$$

$$\frac{dX_2}{dt} = X_2 \left[1-\beta X_1-X_2-\alpha X_3\right], \qquad K^{(n)}_{nn}=-1 \tag{21}$$

$$K^{(1)}_{12}=K^{(2)}_{23}=K^{(3)}_{31}=-\alpha$$

$$\frac{dX_3}{dt} = X_3 \left[1-\alpha X_1-\beta X_2-X_3\right], \qquad K^{(1)}_{31}=K^{(2)}_{12}=K^{(3)}_{23}=-\beta$$

where X_k is the number of individuals of the k-th population, intrinsic growth rates are fixed at unity and competition is symmetric. The fixed point is given by $X_1^{(o)}=X_2^{(o)}=X_3^{(o)}=(1+\alpha+\beta)^{-1}$, the transformation equations are

$$\begin{aligned} x_1 &= 2R\cos\Theta+Z \\ x_2 &= -R\cos\Theta-\sqrt{3}R\sin\Theta+Z \\ x_3 &= -R\cos\Theta+\sqrt{3}R\sin\Theta+Z \end{aligned} \tag{22}$$

and for the special case $\alpha+\beta=2$ (12b) assumes the form

$$\frac{dR}{dt} = -\ 3\omega R^2\sin3\Theta-\ 3RZ \tag{23a}$$

$$\frac{d\Theta}{dt} = \omega\left[1-3R\cos3\Theta\right]+3\omega Z \tag{23b}$$

$$\frac{dZ}{dt} = -Z-3Z^2 \tag{23c}$$

This is a situation of an indifferent radial mode and a descending cylindrical mode with eigenvalues given by

$$\lambda = \frac{1}{\sigma}\left[\frac{\alpha+\beta}{2} - 1\right]=0, \quad \omega = \frac{\sqrt{3}}{2\sigma}\left[\beta-\alpha\right], \quad p_o=-1 \tag{24}$$

where $\sigma=1+\alpha+\beta=3$. In the more general situation when $\alpha+\beta>2$, $\lambda>0$, the system has an ascending radial mode which couples with the descending cylindrical mode. The net effect is that the period of the system limit cycle increases in time as it asymptotically approaches the corners of a triangular population simplex defined by (1,0,0), (0,1,0), (0,0,1). As pointed out by MAY and LEONARD, when $\alpha+\beta=2$ since there is no mechanism for evolution or decay of the system, it, as the famous two dimensional predator-prey model, exhibits limit cycle oscillation parameterized by initial conditions. The solution for (23c) goes to zero as exp(-t) but the integral of Z(t) enters into solution for R and Θ. As a consequence the initial condition Z(0) persists in their solution. We will assume (1) that Z(0)=0 in which case Z is zero for all time by (23c) and the last terms in (23a,b) are consequently zero, (2) that Θ(0)=0 since its value does not affect the dynamics of time evolution, and (3) that R has some initial value R_o which is held fixed. Then integration of (23b) with Z=0 gives

$$\tan\frac{3\Theta}{2} = \left[\frac{1-3R_o}{1+3R_o}\right]^{\frac{1}{2}} \tan\omega' t, \quad \omega' = \frac{3\omega}{2}\left[1-9R_o^2\right]^{\frac{1}{2}} \qquad (25a)$$

Substitution of this result into (23a) with Z=0 gives the time dependence of the radial mode as

$$R(t) = \frac{R_o}{1+M(t)}$$

$$M(t) = 3\omega\int_0^t dt' \sin 3\Theta(t') = -\frac{1}{3}\ln\left[\frac{1+3R_o\cos 2\omega' t}{1+3R_o}\right] \qquad (25b)$$

The integral M reflects the accumulation of memory [3,4], or the propagation of the system forward in time through past times t' up to instant t, the integrand serving as the propagator. In this formalism solution of the rate equations at any instant of time depends explicitly upon the behaviour of the generalized modes for its entire previous history. M(t) of (25b) reflects a repetitive pattern of frequency ω' which itself depends upon initial conditions through R_o. This initial condition also dictates the size of the phase plane found by elimination of time between equations (25a,b) to give

$$R(\Theta) = \frac{R_o}{1+\frac{1}{3}\ln\left[\frac{1-3R_o\cos\Theta}{1-3R_o}\right]} \qquad (25c)$$

Comparison of these simple analytical results with computer solution of (23) in Figs.(2) show that for R_o=1/6, relatively far from the theoretically allowed limit of 1/3 (which corresponds to the corners of the simplex), agreement is quite good. For large orbits (R_o=.32) close to that value the difficulty of the system point to traverse regions near the simplex corners limits the validity of the present result which in principle could be improved by further iteration. However, this limitation plus the initial restriction that Z(0)=0 find their justification in a study to be reported of the more general case λ>0 which shows that for long times Z=2λ/3. If λ is small but finite the solution generated here for R_o=1/6 describes to close

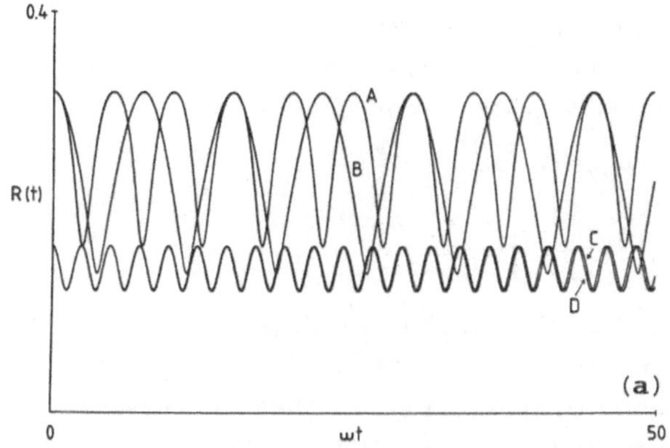

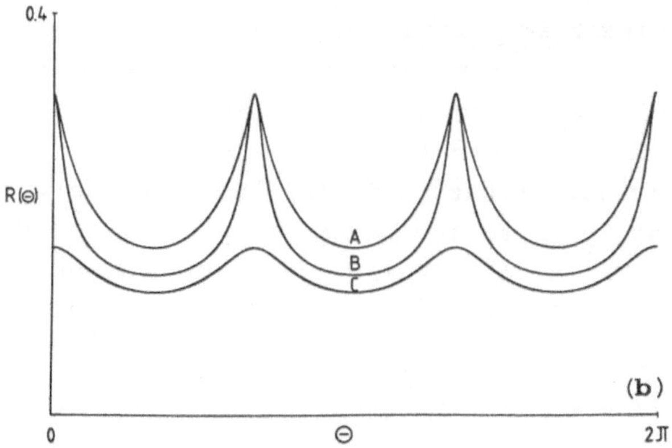

Figure 2a. Time evolution of the radial mode for May-Leonard equations (23) for $\alpha+\beta=2$ and $Z(0)=0$. A. Computer solution of (23) for $R_o=.32$ B. Equation (25b) for $R_o=.32$. C. Computer solution of (23) for $R_o=1/6$. D. Equation (25b) for $R_o=1/6$

Figure 2b. $R(\Theta)$ for May-Leonard equations (23) for $\alpha+\beta=2$ and $Z(0)=0$. A. Computer solution of (23) for $R_o=.32$. B. Equation (25c) for $R_o=.32$. C. Computer solution from (23) and analytical expression (25c) for $R_o=1/6$

approximation analytically the dynamics of the system between simplex corners for which the transit time is found to be $4\pi/3\sqrt{3}\omega$.

The purpose of this simple example is to demonstrate how it is possible to develop analytic solution through integral equations in the generalized coordinates, as exemplified by (25b), provided one can make a reasonable guess at a tractable trial solution. Even in the absence of such solutions computer solutions in terms of the generalized modes can furnish information. An example is the Lorenz equations of hydrodynamic turbulence, given by [9]

$$\frac{dX_1}{dt} = -\sigma X_1 + \sigma X_2 \qquad \left[L_{11} = -\sigma,\ L_{12} = \sigma\right] \tag{26}$$

$$\frac{dX_2}{dt} = rX_1 - X_2 - X_1X_3 \quad \left[L_{21} = r,\ L_{22} = -1,\ K^{(2)}_{13} = -1\right]$$

$$\frac{dX_3}{dt} = -bX_3 + X_1X_2 \qquad \left[L_{33} = -b,\ K^{(3)}_{12} = 1\right]$$

for which the fixed point is $\left[X_1^{(o)},\ X_2^{(o)},\ X_3^{(o)}\right] = \left[-\sqrt{b(r-1)},\ X_1^{(o)},\ r-1\right]$. With the parameter values originally used by LORENZ $(\sigma, b, r) = (10, 8/3, 28)$ the eigenvalues are $\lambda, \omega, p_o = (.094, 10.19, -13.85)$ so that the system features an ascending radial mode and a descending cylindrical mode.

Figure 3a reproduces the coherent amplification of oscillation from an early peak until about t=16.5 after which the system executes irregular chaotic motion. The corresponding behaviours of the generalized modes are shown in Figs.(3b,c,d). Parallel to the X-coordinates, the radial mode shows an orderly amplification during the relatively quiescent period from approximately .5 to 16.5. The cylindrical mode is quite flat during this same period, displaying oscillation around

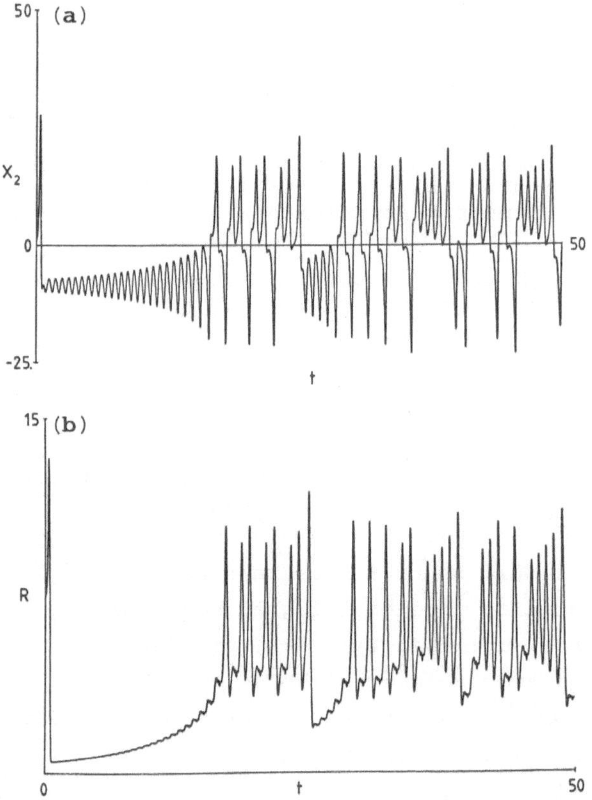

Fig.3a,b. Caption see opposite page

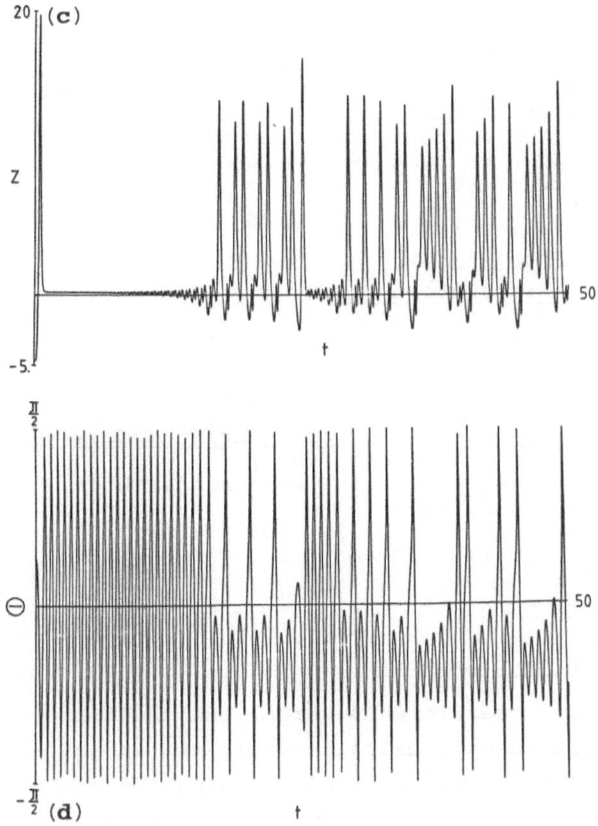

Figure 3a. $X_2(t)$ of Lorenz equations (26) for initial condition $X_1(0)=X_3(0)=0$, $X_2(0)=1$, and parameter values given in text.

Figure 3b. Corresponding radial mode for Lorenz equations (26)

Figure 3c. Corresponding cylindrical mode for Lorenz equations (26)

Figure 3d. Corresponding angular mode for Lorenz equations (26)

a median of essentially zero slope. The geometry implicit in the transformation to the generalized modes for Lorenz equations coupled with their dynamical behaviour result in the angular mode being bounded as shown between $\pm\pi/2$. In the turbulent region after 16.5 the modes mimic the irregular behaviour of Fig.2a. Due to the sensitivity of chaotic systems upon initial conditions, the absence of an ideal computer implies a unique trajectory can never be known. There are, however, locally regular patterns, or what might be called "complexions". The chaotic aspect is inability to predict how long a complexion will persist or how long the next one will be. The present formulation suggests a local approximate analysis of complexions such as these for three dimensional continuous systems which display deterministic chaos [10].

Acknowledgements

A portion of this work was done during a stay at the Institut für Theoretische Chemie und Strahlenchemie of the University of Vienna, and the author expresses his sincere appreciation to Professor Peter Schuster for his interest and support. Computer plots run by Dr.F.Kemler and the supply of computer time by the Interfakultäres EDV-Zentrum is gratefully acknowledged.

References

1 O.Gurel and O.Rössler, eds.: Bifurcation Theory and Applications in Scientific Disciplines, Ann.N.Y.Acad.Sci.,Vol.316 (1978)
2 N.Minorsky: Nonlinear Oscillations (Kreiger, New York ,1974)
3 P.E.Phillipson: Biophys.Chem. 16, 173 (1982)
4 P.E.Phillipson and P.Schuster, J.Chem.Phys. (in press)
5 R.J.Field and R.M.Noyes: J.Chem.Phys. 60, 1877 (1974)
6 J.D.Murray: Lectures on Nonlinear-Differential-Equation Models in Biology, pp.159-178 (Clarendon, Oxford, 1977)
7 J.J.Tyson: The Belousov-Zhabotinski Reaction, Lectures in Biomathematics, Vol.10 (Springer, 1976)
8 R.M.May and W.J.Leonard: SIAM J.Appl.Math. 29, 243 (1975)
9 E.N.Lorenz: J.Atmos.Sci. 21, 130 (1963)
10 O.E.Rössler, ref. 1 ,p.376

Appendix A. The Transformation Coefficients

Six of the nine transformation coefficients which link the Cartesian coordinates to the generalized cylindrical coordinates in equations (3) are uniquely specified in terms of the remaining three coefficients, which can be chosen arbitrarily, and the eigenvalues p_1, p_2, p_3. It is convenient to define the following quantities,

$$D_1^{(k)}=\left[a_{23}(p_k-a_{11})+a_{21}a_{13}\right],\quad D_2^{(k)}=\left[(p_k-a_{11})(p_k-a_{22})-a_{12}a_{21}\right],$$
$$D_3^{(k)}=\left[(p_k-a_{22})(p_k-a_{33})-a_{23}a_{32}\right],\quad D_4^{(k)}=\left[a_{12}(p_k-a_{33})+a_{13}a_{32}\right], \tag{A.1}$$
$$D_5^{(k)}=\left[a_{13}(p_k-a_{22})+a_{12}a_{23}\right],\quad k=1,2,3\text{ , where}$$

a_{nm} are the coefficients of the linear terms of equations (2). Then with the choice $\alpha_{11}=2$, $\alpha_{12}=0$, $\alpha_{13}=1$ the coefficients of equations (3a) are given by

$$\alpha_{11} = 2,\qquad \alpha_{12} = 0,\qquad \alpha_{13} = 1$$
$$\alpha_{21}=\left[\frac{D_1^{(1)}}{D_5^{(1)}}+\frac{D_1^{(2)}}{D_5^{(2)}}\right],\ \alpha_{22}=\left[\frac{D_1^{(1)}}{D_5^{(1)}}-\frac{D_1^{(2)}}{D_5^{(2)}}\right],\ \alpha_{23}=\frac{D_1^{(3)}}{D_5^{(3)}} \tag{A.2}$$
$$\alpha_{32}=\left[\frac{D_2^{(1)}}{D_5^{(1)}}+\frac{D_2^{(2)}}{D_5^{(2)}}\right],\ \alpha_{32}=\left[\frac{D_2^{(1)}}{D_5^{(1)}}-\frac{D_2^{(2)}}{D_5^{(2)}}\right],\ \alpha_{33}=\frac{D_2^{(3)}}{D_5^{(3)}}$$

and the inverse coefficients of equations (3b) are given by

$$(\alpha^{-1})_{11}=\frac{1}{2P}\left[(p_3-p_2)D_3^{(1)}+(p_1-p_3)D_3^{(2)}\right],\ (\alpha^{-1})_{12}=\frac{1}{2P}\left[(p_3-p_2)D_4^{(1)}+(p_1-p_3)D_4^{(2)}\right]$$

$$(\alpha^{-1})_{13}=\frac{1}{2P}\left[(p_3-p_2)D_5^{(1)}+(p_1-p_3)D_5^{(2)}\right],\ (\alpha^{-1})_{21}=\frac{1}{2P}\left[(p_3-p_2)D_3^{(1)}-(p_1-p_3)D_3^{(2)}\right]$$

$$(\alpha^{-1})_{22}=\frac{1}{2P}\left[(p_3-p_2)D_4^{(1)}-(p_1-p_3)D_4^{(2)}\right],\ (\alpha^{-1})_{23}=\frac{1}{2P}\left[(p_3-p_2)D_5^{(1)}-(p_1-p_3)D_5^{(2)}\right] \tag{A.3}$$

$$(\alpha^{-1})_{31}=\frac{(p_2-p_1)D_3^{(3)}}{P},\quad (\alpha^{-1})_{32}=\frac{(p_2-p_1)D_4^{(3)}}{P},\quad (\alpha^{-1})_{33}=\frac{(p_2-p_1)D_5^{(3)}}{P}$$

$$\left[P=(p_1-p_2)(p_2-p_3)(p_3-p_1);\ p_1=\lambda+\Omega,\quad p_2=\lambda-\Omega,\ p_3=p_o\right]$$

Sometimes it is more convenient to have a different choice for α_{11}, α_{12}, α_{13}. Then denoting by primed quantities the coefficients as given in equations (A.2,3), the coefficients for any choice provided $\alpha_{11}\neq\alpha_{12}$ are given by

$$\alpha_{11}\qquad\qquad \alpha_{12}\qquad\qquad \alpha_{13}$$

$$\alpha_{21}=\tfrac{1}{2}[\alpha_{11}\alpha'_{21}+\alpha_{12}\alpha'_{22}],\ \alpha_{22}=\tfrac{1}{2}[\alpha_{11}\alpha'_{22}+\alpha_{12}\alpha'_{21}],\ \alpha_{23}=\alpha_{13}\alpha'_{23} \tag{A.4}$$

$$\alpha_{31}=\tfrac{1}{2}[\alpha_{11}\alpha'_{31}+\alpha_{12}\alpha'_{32}],\ \alpha_{32}=\tfrac{1}{2}[\alpha_{11}\alpha'_{32}+\alpha_{12}\alpha'_{31}],\ \alpha_{33}=\alpha_{13}\alpha'_{33}$$

$$(\alpha^{-1})_{11}=\gamma[\alpha_{11}(\alpha'^{-1})_{11}-\alpha_{12}(\alpha'^{-1})_{21}],\ (\alpha^{-1})_{12}=\gamma[\alpha_{11}(\alpha'^{-1})_{12}-\alpha_{12}(\alpha'^{-1})_{22}]$$

$$(\alpha^{-1})_{13}=\gamma[\alpha_{11}(\alpha'^{-1})_{13}-\alpha_{12}(\alpha'^{-1})_{23}],\ (\alpha^{-1})_{21}=\gamma[\alpha_{11}(\alpha'^{-1})_{21}-\alpha_{12}(\alpha'^{-1})_{11}],$$

$$(\alpha^{-1})_{22}=\gamma[\alpha_{11}(\alpha'^{-1})_{22}-\alpha_{12}(\alpha'^{-1})_{12}],\ (\alpha^{-1})_{23}=\gamma[\alpha_{11}(\alpha'^{-1})_{23}-\alpha_{12}(\alpha'^{-1})_{13}]$$

$$(\alpha^{-1})_{31}=\frac{(\alpha'^{-1})_{31}}{\alpha_{13}},\quad (\alpha^{-1})_{32}=\frac{(\alpha'^{-1})_{32}}{\alpha_{13}},\quad (\alpha^{-1})_{33}=\frac{(\alpha'^{-1})_{33}}{\alpha_{13}}$$

$$\left[\gamma=\frac{2}{|\alpha_{11}^2-\alpha_{12}^2|}\right]$$

If $\Omega=i\omega$, then $p_1=\lambda+i\omega$, $p_2=\lambda-i\omega$, $p_3=p_o$ and the coefficients β_{mn}, $(\beta^{-1})_{mn}$ of equations (3c,d) follow from equation (9).

Appendix B. Two Dimensions

The two dimensional case is given by

$$\frac{dx_1}{dt} = a_{11}x_1+a_{12}x_2+F_1(x_1,x_2)$$
$$\frac{dx_2}{dt} = a_{21}x_1+a_{22}x_2+F_2(x_1,x_2) \tag{B.1}$$

for which the transformation and its inverse are given by

$$x_1=\alpha_{11}R\cosh\sigma+\alpha_{12}R\sinh\sigma, \quad x_2=\alpha_{21}R\cosh\sigma+\alpha_{22}R\sinh\sigma$$
$$R\cosh\sigma = (\alpha^{-1})_{11}x_1+(\alpha^{-1})_{12}x_2, \quad R\sinh\sigma=(\alpha^{-1})_{21}x_1+(\alpha^{-1})_{22}x_2 \tag{B.2}$$

where R and σ are eigenfunctions in the linear approximation with eigenvalues λ and Ω respectively. Following a procedure identical to that for three dimensions, this implies $p_1=\lambda+\Omega$, $p_2=\lambda-\Omega$ for which the secular equation is

$$\begin{vmatrix} p-a_{11} & -a_{12} \\ -a_{21} & p-a_{22} \end{vmatrix} = (p-p_1)(p-p_2)=0 \tag{B.3}$$

$$p_1+p_2=2\lambda=a_{11}+a_{22}, \quad p_1p_2=\lambda^2-\Omega^2=a_{11}a_{22}-a_{12}a_{21}$$

In this case λ and Ω are given explicitly by

$$\lambda = \frac{1}{2}(a_{11} + a_{22}) \tag{B.4}$$
$$\Omega = \frac{1}{2}[(a_{11}+a_{22})^2-4(a_{11}a_{22}-a_{12}a_{21})]^{\frac{1}{2}} \quad \text{so that}$$

$$\frac{dR}{dt} = \lambda R+(A^{-1})_{11}F_1+(A^{-1})_{12}F_2$$
$$R\frac{d\sigma}{dt} = \Omega R+(A^{-1})_{21}F_1+(A^{-1})_{22}F_2 \tag{B.5}$$

$$(A^{-1})_{11}=\left[(\alpha^{-1})_{11}\cosh\sigma-(\alpha^{-1})_{21}\sinh\sigma\right],$$
$$(A^{-1})_{12}=\left[(\alpha^{-1})_{12}\cosh\sigma-(\alpha^{-1})_{22}\sinh\sigma\right],$$
$$(A^{-1})_{21}=\left[(\alpha^{-1})_{21}\cosh\sigma-(\alpha^{-1})_{11}\sinh\sigma\right],$$
$$(A^{-1})_{22}=\left[(\alpha^{-1})_{22}\cosh\sigma-(\alpha^{-1})_{12}\sinh\sigma\right],$$

If $4(a_{11}a_{22}-a_{12}a_{21})>(a_{11}+a_{22})^2$ then $\Omega=i\omega$ for which the appropriate equations are

$$x_1=\beta_{11}R\cos\Theta+\beta_{12}R\sin\Theta, \quad x_2=\beta_{21}R\cos\Theta+\beta_{22}R\sin\Theta$$
$$R\cos\Theta=(\beta^{-1})_{11}x_1+(\beta^{-1})_{12}x_2, \quad R\sin\Theta=(\beta^{-1})_{21}x_1+(\beta^{-1})_{22}x_2 \tag{B.6}$$ and

$$\frac{dR}{dt} = \lambda R + (B^{-1})_{11}F_1 + (B^{-1})_{12}F_2$$

$$R\frac{d\Theta}{dt} = \omega R + (B^{-1})_{21}F_1 + (B^{-1})_{22}F_2 \tag{B.7}$$

$$(B^{-1})_{11} = \left[(\beta^{-1})_{11}\cos\Theta + (\beta^{-1})_{21}\sin\Theta\right],$$

$$(B^{-1})_{12} = \left[(\beta^{-1})_{12}\cos\Theta + (\beta^{-1})_{22}\sin\Theta\right],$$

$$(B^{-1})_{21} = \left[(\beta^{-1})_{21}\cos\Theta - (\beta^{-1})_{11}\sin\Theta\right],$$

$$(B^{-1})_{22} = \left[(\beta^{-1})_{22}\cos\Theta - (\beta^{-1})_{12}\sin\Theta\right]$$

The coefficients of these transformations are given, finally, by α_{11} and α_{12} arbitrary $(\alpha_{11} \neq \alpha_{12})$

$$\alpha_{21} = \left[\left(\frac{a_{22}-a_{11}}{2a_{12}}\right)\alpha_{11} + \left(\frac{\Omega}{a_{12}}\right)\alpha_{12}\right], \quad \alpha_{22} = \left[\left(\frac{\Omega}{a_{12}}\right)\alpha_{11} + \left(\frac{a_{22}-a_{11}}{2a_{12}}\right)\alpha_{12}\right]$$

$$(\alpha^{-1})_{11} = \gamma\left[\alpha_{11} + \left(\frac{a_{22}-a_{11}}{2\Omega}\right)\alpha_{12}\right], \quad (\alpha^{-1})_{12} = -\gamma\left[\left(\frac{a_{12}}{\Omega}\right)\alpha_{12}\right] \tag{B.8}$$

$$(\alpha^{-1})_{21} = -\gamma\left[\left(\frac{a_{22}-a_{11}}{2\Omega}\right)\alpha_{11} + \alpha_{12}\right], \quad (\alpha^{-1})_{22} = \gamma\left[\frac{a_{12}}{\Omega}\ \alpha_{11}\right]$$

$$\left[\gamma = \frac{1}{(\alpha_{11}^2 - \alpha_{12}^2)}\right]$$

and β_{11} and β_{12} arbitrary

$$\beta_{21} = \left[\left(\frac{a_{22}-a_{11}}{2a_{12}}\right)\beta_{11} + \left(\frac{\omega}{a_{12}}\right)\beta_{12}\right], \quad \beta_{22} = \left[-\left(\frac{\omega}{a_{12}}\right)\beta_{11} + \left(\frac{a_{22}-a_{11}}{2a_{12}}\right)\beta_{12}\right]$$

$$(\beta^{-1})_{11} = \gamma\left[\beta_{11} - \left(\frac{a_{22}-a_{11}}{2\omega}\right)\beta_{12}\right], \quad (\beta^{-1})_{12} = \gamma\left[\left(\frac{a_{12}}{\omega}\right)\ \beta_{12}\right] \tag{B.9}$$

$$(\beta^{-1})_{21} = \gamma\left[\left(\frac{a_{22}-a_{11}}{2\omega}\right)\beta_{11} + \beta_{12}\right], \quad (\beta^{-1})_{22} = -\gamma\left[\left(\frac{a_{12}}{\omega}\right)\beta_{11}\right]$$

$$\left[\gamma = \frac{1}{(\beta_{11}^2 + \beta_{12}^2)}\right]$$

Sometimes the parameters a_{nm} can assume such values that Ω vanishes identically $4(a_{11}a_{22} - a_{12}a_{21}) = (a_{11}+a_{22})^2$ in which case the eigenvalues $p_{1,2}$ become degenerate. In such a case the coordinate conjugate to the radial coordinate is defined in terms of the limit

$$\lim_{\substack{\sigma \to 0 \\ \Omega \to 0}} \left(\frac{\sigma}{\Omega}\right)$$

An example of this situation is afforded by the van der Pol equation. A similar limiting process would be required in higher dimensional systems if eigenvalues become degenerate.

Dynamics of Linear and Nonlinear Autocatalysis and Competition

J. Hofbauer and P. Schuster

Institut für Mathematik, Universität Wien, Strudlhofgasse 4
A-1090 Wien, Austria and

Institut für Theoretische Chemie und Strahlenchemie, Universität Wien
Währinger Straße 17, A-1090 Wien, Austria

Abstract

The aim of this contribution is to analyse some conventional kinetic equations of reaction mechanisms which include autocatalytic steps. In case of first-order or linear autocatalysis represented by the elementary step

$$A + I \underset{f'}{\overset{f}{\rightleftarrows}} 2I \tag{1}$$

the dynamics is very simple: we find a uniquely defined asymptotically stable steady state under almost all conditions. Higher order autocatalytic processes, e.g. the trimolecular elementary step

$$A + 2I \underset{f'}{\overset{f}{\rightleftarrows}} 3I \tag{2}$$

which has been used in several investigations of chemical non-equilibrium phenomena (SCHLÖGL [1,2]; PRIGOGINE and LEFEVER [3]) lead to much richer dynamics. In higher dimensional systems we encounter limit cycles and chaotic attractors.

1. Introduction

Competition, selection and permanence [4] are particularly important notions in the theory of biological evolution. Recently, experimental investigations were undertaken [5] which allow to study these phenomena in molecular systems far from equilibrium. Polynucleotides, in particular RNA and enzymes of simple bacteriophages are used in these experiments. Polynucleotides have an intrinsic capability to act as autocatalysts built into their molecular structures. It is of particular interest, therefore, to find model systems which allow to control the distance from the equilibrium state and which are simple enough for extensive qualitative analysis. A combination of an autocatalytic reaction step, a degradation reaction and a recycling process was found to represent an appropriate mechanism for this purpose. For reasons which will become clear soon we distinguish between linear (1) and second-order (2) autocatalysis. Although trimolecular reactions like (2) are commonly excluded in chemistry as acceptable elementary steps, they may appear in overall kinetics and particularly in biochemical reactions involving biological macromolecules.

2. Linear Autocatalysis

The single-step reactions (1) and (2) are well understood and easy to analyze - they lead essentially to the logistic or to the hypologistic equation respectively. As indicated above we embed the two elementary steps in many-step mechanisms which keep the autocatalytic reaction off equilibrium in a controllable manner. The autocatalytic reaction is coupled to a first-order degradation process and a recycling reaction driven by external energy. We consider the open system

$$A + I \underset{f'}{\overset{f}{\rightleftharpoons}} 2I \tag{3a}$$

$$I \underset{d'}{\overset{d}{\rightleftharpoons}} B \tag{3b}$$

$$B \xrightarrow{g(E)} A\,. \tag{3c}$$

As indicated in (3c) the rate constant of the recycling reaction g(E) is determined by an external energy source E. An example of such a process is a photochemical reaction using a light source.

The dynamics of the mechanism (3) is described by the differential equations

$$\dot{a} = gb + f'x^2 - fax \tag{4a}$$

$$\dot{b} = dx - (d'+g)b \tag{4b}$$

$$\dot{x} = fax+d'b-f'x^2-dx \tag{4c}$$

wherein we denote the concentrations of A, B and I by a, b and x respectively. Obviously, the total concentration of all substances is constant:

$$a + b + x = c_o = \text{const.} \tag{5}$$

Hence, we are left with a two-dimensional system, defined on the state space

$$S = \{(a,b,x) \quad R^3_+ : a+b+x = c_o\}$$

which is positively invariant under the flow of (4). The fixed points of (4) are the solutions of

$$d\bar{x} - (d'+g)\bar{b} = 0 \tag{6a}$$

$$f(c_o-\bar{b}-\bar{x})\bar{x}+d'\bar{b}-f'\bar{x}^2-d\bar{x} = 0\,. \tag{6b}$$

They are given by

$$P_o = (\bar{b}_o,\bar{x}_o) = (0,0) \tag{7a}$$

and $$P_1 = (\bar{b}_1,\bar{x}_1) = \frac{fc_o(d'+g)-dg}{(f+f')(d'+g)+fd}\left(\frac{d}{d'+g},1\right)\,. \tag{7b}$$

If $c_o<c_{crit}=\frac{dg}{f(d'+g)}$, P_1 is outside of the physically relevant state space S; at $c_o=c_{crit}$, P_1 coincides with P_o and enters S for $c_o>c_{crit}$. Local stability analysis performed by computing the linearization of (4) at P_o and P_1 shows that P_o is a sink for $c_o<c_{crit}$, and P_1 is asymptotically stable whereas P_o is a saddle for $c_o>c_{crit}$.

In order to globalize these results we use the Bendixson-Dulac test to exclude the existence of limit cycles (see ANDRONOV et al. [6]: we apply the Dulac-function x^{-1} to the vector field (4) and obtain

$$(\dot{b}) = d-(d'+g)\,\frac{b}{x} \tag{8a}$$

and $$(\dot{x}) = f(c_o-b-x) + d'\frac{b}{x} - f'x - d \tag{8b}$$.

Equation (8) obviously has a strictly negative divergence on S. Therefore the flow (8) is area contracting, and periodic orbits are impossible. Then, the Poincaré-Bendixson theorem implies that the stable stationary solutions P_o or P_1 respectively are indeed globally stable: every solution starting in S converges to the stable fixed point (see figs.1a and 1b).

For d'=0, global stability of P, can be proved also by means of the Lyapunov function

$$V = f(b-\bar{b}_1)^2+2d(x-\bar{x}_1\log x),$$ where

$(\bar{b}_1,\bar{x}_1)$ are the coordinates of P_1 given by (7b)

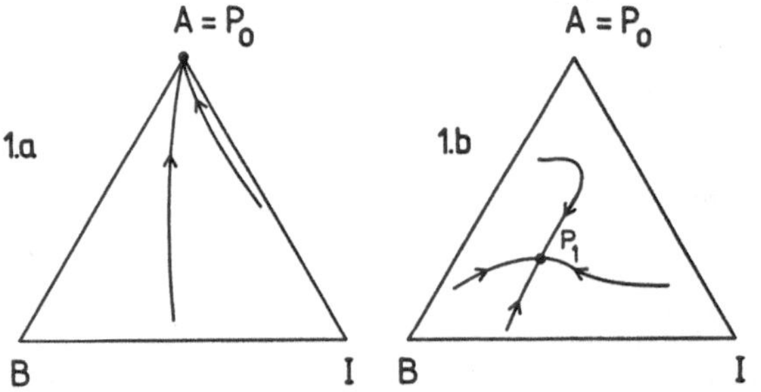

Figure 1. Phase portrait of equation (4) (a: $c_o < c_{crit}$ and b: $c_o > c_{crit}$).

The reaction mechanism (3) thus admits only two qualitatively different types of dynamical behaviour. The plane of external parameters (g, c_0) is split into two regions corresponding to the stability of either P_o or P_1 (see Fig.2).

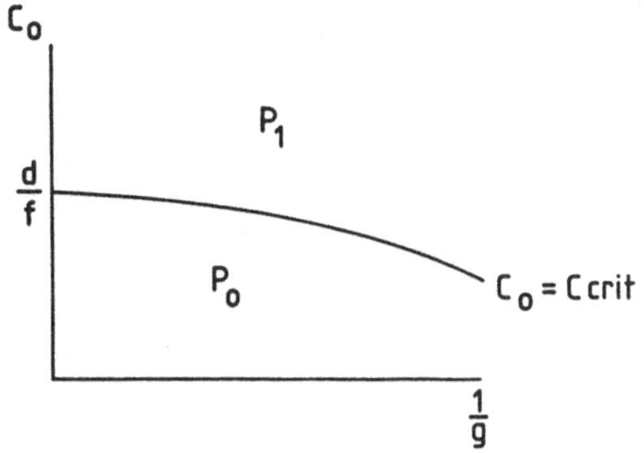

Figure 2. The plane of external parameters (g, c_0) of equation (4) is split into two regions corresponding to the stability of either P_o or P_1. The two regions are separated by the curve $c_0 = c_{crit}$.

If all reactions are irreversible ($f' = d' = 0$), a condition which is approximately true for most biochemical and biological systems, the concentration of A at the stationary solution P_1 and the critical value c_{crit} coincide and are independent of g: $\bar{a} = c_{crit} = \frac{d}{f}$. The irreversible autocatalytic processes consume the reactant A to the ultimate limit: P_1 would become unstable if the concentration of A went below c_{crit}.

3. Competition between First-Order Autocatalytic Reactions

Now, we consider a network of 2n+1 reactions: n first-order autocatalytic reaction steps followed by n first-order degradation processes which are coupled to a common irreversible recycling reaction which again is controlled from outside and which allows to drive the system far off equilibrium.

$$A + I_i \underset{f_i'}{\overset{f_i}{\rightleftarrows}} 2I_i; \quad i=1,2,\ldots,n \tag{9a}$$

$$I_i \overset{d_i}{\rightleftarrows} B; \quad i=1,2,\ldots,n \tag{9b}$$

$$B \xrightarrow{g(E)} A\,. \tag{9c}$$

This reaction scheme may be considered as competition between n autocatalysts for the common source of material A for synthesis. In the case of reversibility of reactions (9a) and (9b) there is a thermodynamic restriction of the choice of rate constants because of the uniqueness of the thermodynamic equilibrium (by $\bar{b}$ and $\bar{a}$ we denote equilibrium concentrations):

$$\frac{\bar{b}}{\bar{a}} = \frac{f_i d_i}{f_i' d_i'}; \quad i=1,2,\dots,n .$$

The dynamics of mechanism (9) is described by the differential equation (concentrations of I_i are denoted by x_i):

$$\dot{a} = gb + \sum_{i=1}^{n} f_i' x_i^2 - \sum_{i=1}^{n} f_i a x_i \tag{10a}$$

$$\dot{b} = \sum_{i=1}^{n} d_i x_i - (g + \sum_{i=1}^{n} d_i')b \tag{10b}$$

$$\dot{x}_i = x_i(f_i a - f_i' x_i - d_i) + d_i' b . \tag{10c}$$

Equations (10) imply the conservation law

$$a + b + \sum_{i=1}^{n} x_i = c_o = \text{const.}$$

Since the analysis is rather involved we consider first the case of irreversible degradation ($d_i'=0$). Then, the fixed points are readily computed. There are 2^n fixed points of equation (10). The most convenient notation uses sets of indices which correspond to the non-vanishing components: P_i is the fixed point with $\bar{x}_i \neq 0$ and $\bar{x}_k=0$ for $k \neq i$; for P_{ij} we have $\bar{x}_i \neq 0$, $\bar{x}_j \neq 0$ and $\bar{x}_k=0$ for $k \neq i,j$, etc.

Without losing generality we arrange indices such that

$$\frac{d_1}{f_1} < \frac{d_2}{f_2} < \dots\dots < \frac{d_n}{f_n} .$$

Then, it is straightforward to show from (10) that the presence of species I_i at a stable stationary point implies also the presence of species I_j with $j<i$. Therefore all fixed points different from P_o, P_1, P_{12}, $P_{123}, \dots, P_{12\dots n}$ are unstable irrespective of the values of the external parameters g and c_o. In fact only one of the fixed points listed above is stable depending on the values of g and c_o: if $f_i'>0$, we obtain a sequence of double point bifurcations of the type $P_{12\dots i} \longrightarrow P_{12\dots i,i+1}$ with increasing c_o. At the point of coincidence a change in stability takes place (Fig.3) and a new species is introduced into the system at the stable stationary state. The regions of stability in the (g,c_o) plane are shown in Fig.4. The larger the environment, sustaining a larger total concentration c_o, the more

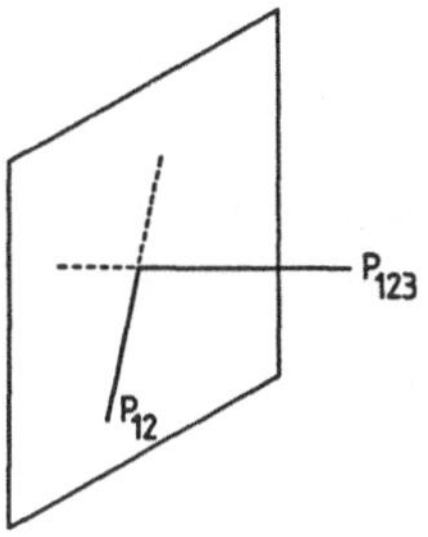

Figure 3. Double point bifurcation introducing a new species into the stable solutions of equations (10).

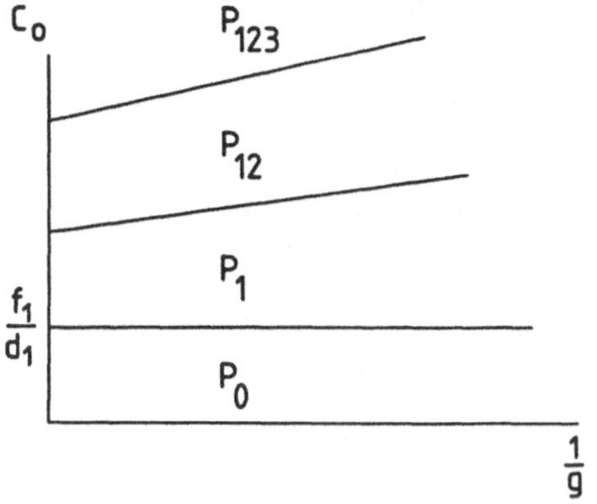

Figure 4. The (g,c_o) plane is split into n+1 regions of stability of the fixed points $P_o, P_1, P_{12}, \ldots, P_{12\ldots n}$ of equations (10) with $f_i' > 0$ $(i = 1,2,\ldots,n)$.

species can coexist. We mention that a similar phenomenon has been found for the competition of self-replicating macromolecules the reproduction of which is based on Michaelis-Menten type kinetics [7,8] and in Lotka-Volterra food chains [9,10].

In case the autocatalytic reactions are irreversible ($f_i'=0$), a condition usually met by ecological systems, and if the values of the quotients d_i/f_i are distinct, only P_o and the one species fixed points, $P_1, P_2, \ldots, P_n$, exist at finite values of c_o, and P_1, the steady state corresponding to the lowest value of d_i/f_i, is stable for $c_o > d_1/f_1$. In this case we observe selection of the fittest species, this is the one which is characterized by the smallest quotient of degradation rate over replication rate.

The following theorems give some information on the global behaviour of the dynamical systems (10) with $d_i'=0$, $i=1,2,\ldots,n$.

Theorem 1: If there is no stationary point in the interior of the state space, i.e. $P_{12\ldots n}$ does not exist, then at least one species dies out.

Proof: If the system of linear equations

$$\frac{\dot{x}_i}{x_i} = f_i a_i - f_i' x_i - d_i = 0; \; i=1,2,\ldots,n \tag{11a}$$

$$\dot{b} = \sum_{i=1}^{n} d_i x_i - gb = 0 \tag{11b}$$

has no solution $(\bar{a},\bar{b},\bar{x}_1,\ldots,\bar{x}_n)$ with positive entries then a well known convexity argument ensures the existence of real numbers q_i $(i=0,1,\ldots,n)$ such that

$$q_o\left(\sum_{i=1}^{n} d_i x_i - gb\right) + \sum_{i=1}^{n} q_i (f_i a - f_i x_i^! - d_i) > 0 \quad \text{for all } a,b,x_i>0,\ i=1,2,\ldots,n.$$

But then the function $V = q_o b + \sum_{i=1}^{n} q_i \log x_i$, defined for $x_i>0$, satisfies

$$\dot{V} = q_o \dot{b} + \sum_{i=1}^{n} q_i \frac{\dot{x}_i}{x_i} > 0$$

and therefore is a Lyapunov function. Hence, all orbits converge to points where at least one of the concentrations x_i vanishes.

Conversely, if there is an interior fixed point $P_{12\ldots n}$, then it is stable. We were not able to prove its global stability. But we can provide a proof for the somewhat weaker statement.

Theorem 2: If the interior fixed point $P_{12\ldots n}$ exists then the system is permanent [4]. The time averages of every solution starting with positive concentrations $x_i>0$, $i=1,2,\ldots,n$, converge to the stationary solution $P_{12\ldots n}$.

Proof: Firstly, it is easy to derive that if $P_{12\ldots n}$ exists, then all 2^n fixed points P_J (with J any subset of $\{1,2,\ldots,n\}$) exist too and furthermore that the relation

$$\bar{a}(P_J) > \bar{a}(P_{12\ldots n}) > \frac{d_i}{f_i} \tag{12}$$

holds for all i.

Therefore, we may assume (by induction on the number n of competitors) that on every boundary face $\{x_i=0 \text{ for } i\notin J\}$ the restricted system (10) is permanent and time averages converge to P_J. But then we can verify the condition of the theorem derived in section 4 of [4] with

$$P = \prod_{i=1}^{n} x_i$$

as average Lyapunov function:

$$\frac{1}{T}\int_0^T \psi(x(t))\,dt = \frac{1}{T}\int_0^T \sum_{i=1}^{n} \frac{\dot{x}_i}{x_i}\,dt =$$

$$= \sum_{i=1}^{n} \frac{1}{T}\int_0^T \left[f_i a(t) - f_i^! x_i(t) - d_i\right] dt \longrightarrow \sum_{i=1}^{n} \left[f_i \bar{a}(P_J) - f_i^! \bar{x}_i(P_J) - d_i\right]$$

for $T \to +\infty$, and

$$\sum_{i=1}^{n}\left[f_i\bar{a}(P_J)-f_i'\bar{x}_i(P_J)-d_i\right] = \sum_{i\in J}\left[f_i\bar{a}(P_J)-d_i\right]>0$$

from (12). Hence (10) is permanent. But then the convergence of the time averages is obtained immediately by integration of (11) using similar arguments as in [11].

A final remark is to be made on the reversible case, $d_i'>0$. Here, only the first double point bifurcation $P_0 \rightarrow P_1$ occurs and at P_1 all competitors I_i, $i=1,2,\ldots,n$, are present with positive concentrations. The further bifurcations do not take place since their branches do not intersect (the points $P_{12},\ldots,P_{12\ldots n}$ are always outside the state space). At small values of d_i' the branches, nevertheless, may come very close and we are dealing with "imperfect" bifurcations (see [12]).

4. Second-Order Autocatalysis

In order to provide some feeling for the enormous richness of the dynamics of higher order catalytic systems we study now the trimolecular reaction (2) coupled to degradation and recycling in analogy to the reaction scheme (3):

$$A + 2I \underset{f'}{\overset{f}{\rightleftharpoons}} 3I \tag{13a}$$

$$I \underset{d'}{\overset{d}{\rightleftharpoons}} B \tag{13b}$$

$$B \xrightarrow{g(E)} A. \tag{13c}$$

The dynamics of this mechanism is described by the differential equation

$$\dot{a} = gb + f'x^3 - fax^2 \tag{14a}$$

$$\dot{b} = dx - (d'+g)b \tag{14b}$$

$$\dot{x} = fax^2 + d'b - f'x^3 - dx. \tag{14c}$$

Again the total concentration is conserved:

$$a+b+x = c_0 = \text{const.}$$

To simplify the calculations we consider only the case $d'=0$ here, although the results obtained are valid also for systems with $d'>0$. The first basic difference compared to equation (4) one observes in analyzing (14) is the fact that the fixed point $P_0(\bar{a}=c_0,\ \bar{b}=\bar{x}=0)$ is always stable. The positions of the other two fixed points P and Q are given by

$$\bar{x} = \frac{c_o f \pm \left[c_o^2 f^2 - 4d(f+f' + \frac{fd}{g})\right]^{1/2}}{2(f+f'+\frac{fd}{g})} .$$

Let $p>q$ denote the two values of x at the corresponding interior fixed points P and Q. Note that p grows and q shrinks with increasing c_o. It is then easy to derive that Q is always a saddle and P either a sink or a source.

So for $c_o<c_{SN} = \frac{1}{f}\left[4d(f+f'+f\frac{d}{g})\right]^{1/2}$, P_o is the only fixed point and as c_o crosses c_{SN}, by a saddle-node bifurcation P and Q emerge into the interior of the state space S.

In order to determine the stability of P we evaluate the trace of the Jacobian (A) at P:

$$\text{trace } A(x=p) = d-g-p^2(f+f') \tag{16}$$

which is a <u>decreasing</u> function of c_o.

If expression (16) is negative at the birth of P ($c_o=c_{SN}$) then P is stable for all $c_o>c_{SN}$. This corresponds to a direct transition from region I into region V in fig.5 and it happens in particular always if $d\leq g$. In this special case a similar argument as applied in section 2, using now x^{-2} as a Dulac function, excludes the existence of periodic orbits so that all solutions have to converge to one of the fixed points P_o, P or eventually Q.

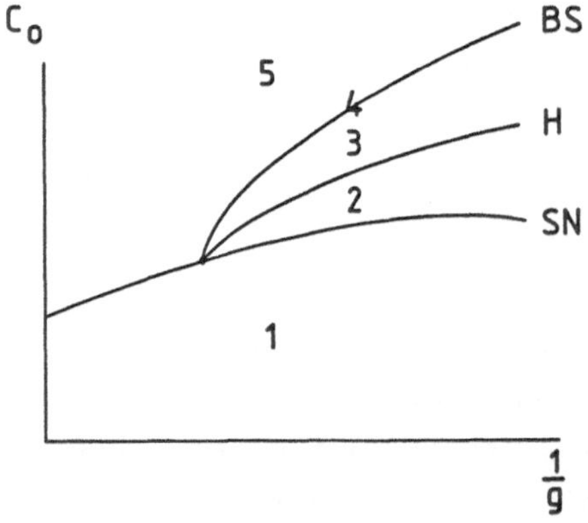

Figure 5. The regions of different dynamical behaviour of equation (14) in the (g,c_o) plane. We observe three types of bifurcations: SN = saddle-node bifurcation, H = Hopf bifurcation and BS = blue sky bifurcation. The regions denoted by 1,2,...,5 refer to different phase portraits shown in Fig.6.

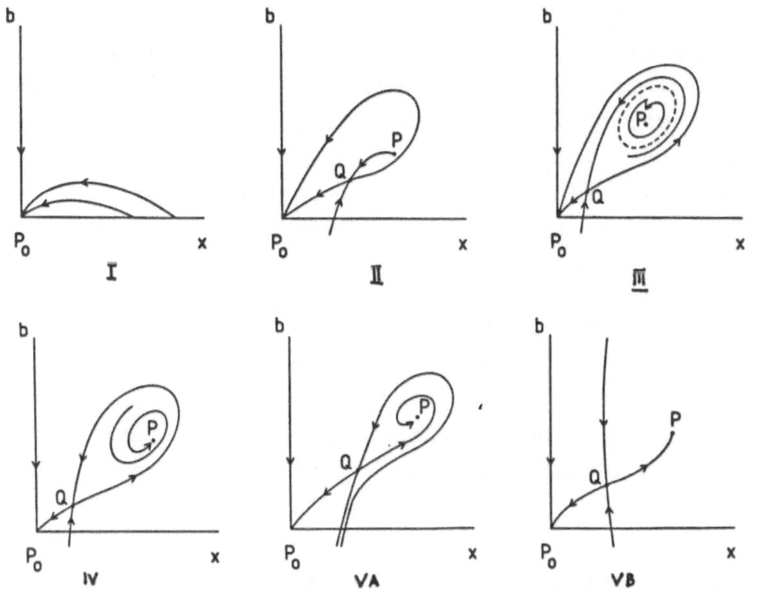

Figure 6. Phase portraits of the regions of different dynamical behaviour of equation (14) according to Fig.5. I: $c_o < c_{SN}$; P_o is the only fixed point in S. II: $c_{SN} < c_o < c_H$; Q is a saddle, P a source. III: $c_H < c_o < c_{BS}$; P is a sink whose basin of attraction is bounded by an unstable periodic orbit. IV: $c_o = c_{BS}$; a homoclinic orbit including Q. V: A: $c_o \gtrsim c_{BS}$ the basin of P contains only a small strip of the x axis. B: $c_o \gg c_{BS}$; P has a large basin of attraction.

If, however, the numerical value of (16) at $c_o = c_{SN}$ is positive, the fixed point P is first an unstable node (fig.6,II). With increasing c_o the eigenvalues become complex conjugate. Since (16) is decreasing the eigenvalues have to cross the imaginary axis at a certain value $c_o = c_H$ at which a Hopf bifurcation occurs. The value of c_H may be obtained from (16) by setting trace A(x=p)=0. A lengthy calculation based on the treatment given by MARSDEN and MC CRACKEN [13] shows that this Hopf bifurcation is always subcritical. This means that the bifurcating periodic orbit is always unstable and occurs for $c_o > c_H$ when P is stable. In fact, it separates the basins of attraction for the two stable fixed points P_o and P (fig.6,III). Further increase of c_o leads to growth of the periodic orbit until it includes the saddle Q at the next critical value $c_o = c_{BS}$. Accordingly, the periodic orbit changes into a homoclinic orbit (fig.6,IV) and disappearsfor $c_o > c_{BS}$. This phenomenon is sometimes called "blue sky bifurcation" [14]. The numerical value of c_o at which the blue sky bifurcation occurs (c_{BS}) can be computed approximately only.

At total concentrations $c_o > c_{BS}$ also orbits starting from the boundary of S have a chance to converge to P. At values close to c_{BS} the admissible range of initial concentrations x(o) for final survival of the autocatalyst is very small (fig.6,VA): if x(o) is too small or if it is too large the orbit will converge to P_o and the autocatalyst will die out. For large values of c_o, of course, the basin of attraction of P becomes almost the entire state space S, since the saddle Q tends towards P_o for $c_o \to \infty$ (fig.6,VB).

Finally, we mention that the existence of two stable solutions on S as found in regions III, IV and V (fig.5,6) is a common and well-known feature of higher order autocatalytic reactions. One of the most striking properties of reaction networks including higher order autocatalytic processes is the enormous sensitivity on initial conditions as illustrated in fig. 6 (VA).

5. Competition Between Higher Order Autocatalytic Reactions

We conclude this contribution with some remarks on the second order analogue of mechanism (9). By second order we refer to the trimolecular autocatalytic reaction (2):

$$A + 2I_i \underset{f_i'}{\overset{f_i}{\rightleftharpoons}} 3I_i; \quad i=1,2,\ldots,n \tag{17a}$$

$$I_i \underset{d_i'}{\overset{d_i}{\rightleftharpoons}} B; \quad i=1,2,\ldots,n \tag{17b}$$

$$B \xrightarrow{g(E)} A. \tag{17c}$$

For the sake of simplicity we restrict ourselves to two competitors (n=2) and irreversible degradation ($d_i'=0$). Then, the dynamics is described by the differential equation

$$\dot{b} = d_1x_1 + d_2x_2 - gb \tag{18a}$$

$$\dot{x}_1 = x_1(f_1ax_1 - f_1'x_1^2 - d_1) \tag{18b}$$

$$\dot{x}_2 = x_2(f_2ax_2 - f_2'x_2^2 - d_2) \tag{18c}$$

with $a = c_o - x_1 - x_2 - b$.

Again the fixed point $P_o(\bar{x}_1=\bar{x}_2=\bar{b}=0)$ is always stable. Moreover, it is obvious that any x_i which is very small is characterized by $\dot{x}_i<0$ and hence will tend towards zero. In contrast to the competition between "linear" autocatalytic processes (9) all invariant surfaces of S, coordinate planes where some x_i are zero, are attracting. On both planes, $x_1=0$ and $x_2=0$, we observe all the dynamical behaviour described in the previous section. In particular, we obtain two stable fixed points P_1 ($\bar{x}_1>0$, $\bar{x}_2=0$) and $P_2(\bar{x}_1=0, \bar{x}_2>0)$ after having passed through the regions of transition shown in fig.6.

Increasing c_0 further in the case $f_i'>0$ we observe creation of two pairs of fixed points in the interior of the state space. If c_0 is large enough one of the four fixed points, $P_{12}(\bar{x}_1>0, \bar{x}_2>0)$, will be stable and finally, we end up with four stable and five unstable fixed points. The various bifurcations occurring with increasing c_0 are difficult to analyze and one might expect some chaotic behaviour near the Hopf and homoclinic bifurcations. The methods developed by Coullet [15] will be applied to this problem in the near future.

In the case $f_i'=0$ only one pair of fixed points emerges in the interior of S. These are always a source and a saddle. There is no stable fixed point in the interior not even for large values of c_0. This indicates that in the irreversible case all orbits converge to one of the planes $x_1=0$ or $x_2=0$. At least one of the competitors has to die out. As we found also for the "linear" case (9) there is no region of coexistence of the two competitors when the autocatalytic processes are irreversible.

6. Conclusion

In this contribution we tried to present an impressive illustration for the difference between linear and higher order autocatalytic processes. In case of linear autocatalysis we were able to derive global results, i.e. results for a given set of external parameters which are valid in the entire phase space S. This is not the case for higher order autocatalytic processes: in general, we find more than one attractor and hence, only results of local validity can be obtained. No global fitness functions or global section criteria exist in these systems.

Acknowledgements: The work reported here has been supported financially by the"Austrian Fonds zur Förderung der Wissenschaftlichen Forschung" projects no 3502 and 4506. Technical assistance in preparing the manuscript by Mrs.J.Jakubetz and Mr.J.König is gratefully acknowledged.

References:

1 F.Schlögl: Z.Physik 248, 446 (1971)
2 F.Schlögl: Z.Physik 253, 147 (1972)
3 I.Prigogine, R.Lefever: J.Chem.Physics 48, 1695 (1968)
4 K.Sigmund, P.Schuster: Permanence and Uninvadability for Deterministic Population models (this volume)
5 C.K.Biebricher: Darwinian Selection of Self-replicating RNA-Molecules, Evolutionary Biology 16, pp.1-52 (Plenum Press, New York 1983)
6 A.Andronov, E.Leontovich, I.Gordon, A.Maier: Qualitative Theory of Second Order Dynamic Systems (Halsted Press, New York, 1973)
7 I.Epstein: J.theor.Biol. 78, 271 (1979)
8 J.Hofbauer, P.Schuster, K.Sigmund: J.Math.Biol. 11, 155 (1981)
9 J.So: J.Theor.Biol. 80, 185 (1979)
10 T.Gard, T.Hallam: Bull.Math.Biol. 41, 877 (1979)
11 P.Schuster, K.Sigmund,R.Wolff: J.Math.Anal.Appl. 78, 88 (1980)
12 G.Iooss, D.D.Joseph: Elementary Stability and Bifurcation Theory (UTM Springer 1980)
13 J.E.Marsden, M.McCracken: The Hopf Bifurcation and its Applications (Applied Math.Sciences 19, Springer, New York 1976)
14 R.Abraham, J.E.Marsden: Foundations of Mechanics (2nd Ed.,Benjamin/ Cummings, Reading-Mass., 1987)
15 P.Coullet: Complex Behaviour in Macrosystems near Polycritical Points (this volume)

Permanence and Uninvadability for Deterministic Population Models

K. Sigmund and P. Schuster

Institut für Mathematik, Universität Wien, Strudlhofgasse 4
A-1090 Wien, Austria and

Institut für Theoretische Chemie und Strahlenchemie, Universität Wien
Währinger Straße 17, A-1090 Wien, Austria

The notion of permanence is used to deal with population dynamical systems which are too complicated to allow a detailed analysis of their asymptotic behaviour. This paper offers an exposition of some general mathematical results, illustrated by applications from population genetics, ecology, sociobiology and - somewhat more detailed - from prebiotic evolution.

1. Introduction

Looking for robust features in population dynamical models has often been confined to a search for stable equilibria. The nonlinear features of such models will often, however, lead to cyclic, quasiperiodic or chaotic behaviour. A complete mathematical analysis may be impossible in many cases.

Proceeding along several lines, the systematic exploration of strange attractors aims at an understanding of their basic typology (see PEITGEN [1] COULLET [2] and ROESSLER et al. [3] in this volume). It cannot be expected, however, that every chaotic behaviour occurring in population models will be elucidated completely. In particular, a given model in ecology, say, or in game dynamics may simply not be worth the effort; its interest could lie in certain traits which are considerably less sophisticated than the full description of its asymptotic behaviour. In such a case, a satisfactory understanding of the model would consist in an analytical proof of the relevant traits, supplemented by a few illustrative samples of numerical simulations.

One of the basic questions for any model of interacting populations concerns the number of species persisting under fluctuations. In this paper, we shall discuss it with the help of the notions of permanence and uninvadability. The first part deals with the definitions and the few general results presently known. The second part offers a brief overview on some applications in population genetics, ecology, sociobiology and macromolecular evolution.

2. Equations for Population Dynamics

We shall be concerned with the simplest deterministic dynamics of populations, given by differential equations in the continuous case and by difference equations in the case of discrete generations. (Differential equations with time delays and integro-differential equations could be discussed along the same lines). The state space will be either the positive orthant

$$R_+^n = \{\vec{x} = (x_1,\dots,x_n): x_i>0 \text{ for all } i\}$$

or, if we are interested in relative frequencies, the simplex

$$S_n = \{\vec{x} \in R_+^n : \Sigma x_i = 1\}.$$

The population will consist of replicators (organisms, genes or polynucleotides). We shall neglect any migration phenomena or spatial inhomogeneities. Thus, we may assume that the boundaries of the state space - i.e. the subsets where $x_i=0$ for some indices i - are invariant. Furthermore, we shall always assume that no population explodes, i.e. that there is a bounded subset of the phase space capturing all positive half-orbits. (These assumptions are unnecessarily restrictive in some cases).

It is well known that dynamical systems of this type may exhibit very complicated behaviour. In particular, even the simplest discrete models for one population, like $x \rightarrow rx(1-x)$ or $x \rightarrow x \exp r(1-\frac{x}{K})$, will display a chaotic regime for some parameters (see MAY [4]). Differential equations can admit limit cycles for two, and strange attractors for three interacting populations: in particular, SMALE [5] has shown that any attractor may occur in the framework of competitive population equations, and ARNEODO [6] found strange attractors in the three-dimensional Lotka-Volterra equations. Obviously, linearization around steady states cannot capture such complicated behaviour, and important robust features may escape fixed point analysis.

3. Permanence and Uninvadability

We shall call a population dynamical system permanent if all populations survive (provided they are initially present). More precisely, a system will be called permanent if there exists some level $k>0$ such that if $x_i(0)>0$ for $i=1,\dots,n$ then $x_i(t)>k$ for all t larger than some time T which depends on the $x_i(0)$.

This property was called cooperativity in SCHUSTER et al. [7] and several subsequent papers, e.g. HOFBAUER [8]. For several reasons, however, this name is not very appropriate in general situations. It was used in the context of molecular evolution, where polynucleotides effectively helped each other through catalytic interaction. But the notion applies equally well in ecology, for example, and it is rather awkward to speak of cooperative predator-prey communities. Furthermore, HIRSCH [9] defined a differential equation $\dot{\vec{x}} = \vec{f}(\vec{x})$ as cooperative if $\partial f_i/\partial x_j > 0$ for all $j \neq i$, which seems a more appropriate use of this term.

An equivalent way of defining a permanent system would be to postulate the existence of a compact set M in the interior of the state space such that all orbits in the interior end up in M. Since the distance from M to the boundary of the state space is strictly positive, this implies that the system is proof against fluctuations, provided these are sufficiently small and rare: indeed, the effect of a small fluctuation upon a state in M would not be enough to send it all the way to the boundary and would be compensated by the dynamics which would lead the orbit of the perturbed state back into M. Whatever happens in M is of no relevance in this context.

One can also define a permanent system as one admitting a compact set F in the interior of the state space such that F is a globally stable attractor. F is called stable if for any neighbourhood V of F there exists a neighbourhood V or F such that no positive half-orbit issued from V ever leaves U. If furthermore all orbits issued in V converge to F, then F is said to be an attractor; and if, in addition, every orbit in the interior of the state space converges to F, then F is a globally stable attractor.

We mention that permanent systems can also be characterized as having a boundary which is strictly repulsive (GREGORIUS [10]) or, in other words, which is an unreachable repellor (SIEVEKING [11]). Since this notion corresponds, under time reversal, to that of a stable attractor, we shall not elaborate it further. We shall also not discuss here systems which are conditionally permanent in the sense that the coexistence of the populations depends on the initial conditions.

Roughly speaking, a community of populations is permanent if internal strife cannot destroy it; it is uninvadable, on the other hand, if it is proof against disturbance from without. Of course, uninvadability is not a property per se: it has always to be made clear against which invaders the community is protected.

Thus let $x_1,...,x_m$ be the frequencies (or densities) of a community of "established" populations and $x_{m+1},...,x_n$ the frequencies of potential"invaders". Then the "established" community is uninvadable if (a) it is permanent and (b) small "invading" populations will persist. Mathematically, this means that the $(x_1,...,x_m)$ community is uninvadable in the $(x_1,...,x_n)$ space if there exists a compact set F in the interior of the $(x_1,...,x_m)$ subspace which is a stable attractor in the $(x_1,...,x_n)$ space and even a globally stable attractor in the $(x_1,...,x_m)$ subspace.

Obviously, any proper subset of a permanent community can be invaded by the complementary subset.

4. Necessary and Sufficient Conditions for Permanence

There are few general results on permanence. GREGORIUS [10] proved that a necessary and sufficient condition is the existence of a Ljapunov function. This is a function V from the state space into R_+ such that (a) $V(\vec{x}) = 0$ iff $\vec{x}$ is on the boundary; (b) V is continuous on some neighbourhood N of the boundary and (c) if $\vec{y}$ is on the positive half-orbit of $\vec{x}$, with both $\vec{x}$ and $\vec{y}$ in N, then $V(\vec{x})>0$ implies $V(\vec{x})<V(\vec{y})$. There is no general method, however, for finding such a Ljapunov function.

An important necessary condition for permanence is the existence of an equilibrium in the interior of the state space. This is essentially a consequence of the Brouwer fixed point theorem, and has been proved, for discrete systems, by HUTSON and MORAN [12], and for continuous systems by SIEVEKING [11].

An obvious sufficient condition for the permanence of a continuous time model is that the coordinates x_i should increase, whenever they are small. This condition is far from being necessary, however. A small population could drop further, for a while, before picking up and compensating the previous loss. Besides, the condition does not hold for discrete systems, where orbits may approach the boundary by a sequence of increasing, oscillatory jumps.

A very useful condition for the permanence of continuous systems of the type $\dot{x}_i=x_iF_i(\vec{x})$ was given by HOFBAUER [8]. If there exists a function P on the state space which vanishes on the boundary and is strictly positive in the interior, and if the time derivative of P along the orbits satisfies $\dot{P}=P\psi$, where ψ is a continuous function with the property

(+) for any $\vec{x}$ on the boundary there is $T>1$ with

$$\frac{1}{T}\int_0^T \psi(\vec{x}(t))dt > 0$$

then the system is permanent.

Condition (+) looks rather technical. If ψ , however, is strictly positive on the boundary, then (+) is always fulfilled and P is just a Ljapunov function. Condition (+), then, means that P is a sort of "average Ljapunov function".

HUTSON and MORAN [12] showed that a discrete system $\vec{x} \to T\vec{x}$ given by $(T\vec{x})_i = x_i F_i(\vec{x})$ is permanent if there exists a nonnegative function P on the state space which vanishes exactly on the boundary and satisfies

$$\sup_{k>0} \liminf_{\substack{\vec{y} \longrightarrow \vec{x} \\ \vec{y} \text{ interior}}} \frac{P(T^k \vec{y})}{P(\vec{y})} > 1$$

for all $\vec{x}$ on the boundary.

5. Protected Polymorphism

In population genetics, permanence problems are usually treated in terms of "protected polymorphism". We shall only mention a few papers dealing with one-locus models in this context. If x_i denotes the frequency of allele A_i, then populations for which all x_i are strictly positive are called polymorphic. Many authors discuss the aspect of the initial progress of new genes (often in the context of models with only two alleles). This is, of course, a very restricted aspect of permanence and invadability problems.

In the classical theory of viability selection, the fitness parameters of the genotypes are constants, and the fundamental theorem, which states that the average fitness increases, reduces the problem to a discussion of equilibria. For temporally or spatially varying environments, however, i.e. for fitness parameters subject to change, the dynamics can be considerably more involved. In many cases, there is little known about the long term behaviour. The existence of a protected polymorphism, then, is one of the few questions which can be successfully attacked. We refer for temporally varying environments with changes in the direction of selection to HALDANE and JAYAKAR [13], for spatially varying environments (multiple niche theory) to LEVENE [14], and for more information on both fields to ROUGHGARDEN [15]. In the theory of fecundity selection, where fitness depends on the mating pairs, a recent paper by GREGORIUS and ZIEHE [16] gives some necessary and sufficient conditions for protectedness of alleles in terms of extinction in finite time and initial increase.

6. Coexistence of Species

Questions of persistence and resilience have played a prominent role in the investigation of multispecies communities. We mention, as example, the so-called exclusion principle which states (roughly) that n species cannot coexist on m resources, if $m<n$. If the dependence of the growth rates on the resources is linear, this can be easily proved. But for more general types of dependence, this principle is not valid. There are no point attractors in the interior of the state space, but other attractors, for example periodic ones, may occur (cf. ARMSTRONG and McGEHEE [17] and NITECKI [18]).

Many results concern conditionally permanent communities. But recently, HUTSON and VICKERS [19] have obtained necessary and sufficient conditions for the permanence of communities consisting of one predator and two preys, within the framework of differential equations of Lotka-Volterra type. Their results cover all but some marginal cases, and include situations where permanence holds but no asymptotically stable equilibrium exists.

Furthermore, HUTSON and MORAN [12] have studied discrete predator-prey equations $\vec{x} \to T\vec{x}$ given by an analogue of the Lotka-Volterra model, namely

$$(T\vec{x})_1 = x_1 \exp(b_1 - a_{11}x_1 - a_{12}x_2)$$

$$(T\vec{x})_2 = x_2 \exp(-b_2 + a_{21}x_1 - a_{22}x_2)$$

with all b_i and $a_{ij}>0$). They showed that the system is permanent if and only if it admits a fixed point in the interior of the state space. This result is similar to the continuous Lotka-Volterra case. But while for the differential equation, permanence implies global stability, the discrete case allows for considerably more complicated asymptotic behaviour.

7. Evolutionarily Stable Strategies

MAYNARD SMITH [20] used game theoretical models in the context of innerspecific conflicts. Strategies, here, correspond to behavioural phenotypes and the payoff is the change in Darwinian fitness. If x_i

denotes the frequency of strategy E_i $(i=1,\ldots,n)$ and if the payoff for strategy E_i against strategy E_j is given by the matrix $A=(a_{ij})$, then the average payoff for a member of a population described by $\vec{x}$ against an opponent from a population described by $\vec{y}$ is given by the inner product $\vec{x}.A\vec{y}$.

The state $\vec{p}$ is called evolutionarily stable if it is proof against invasion by (initially rare) mutants. This means that if $\vec{p}$ is the established state and $\vec{x}$ an alternative ("mutant") state, then $\vec{p}$ fares better than $\vec{x}$ within a population which consists of mostly $\vec{p}$ and some small amount of $\vec{x}$. Thus $\vec{p}$ is an evolutionarily stable state iff, for all $\vec{x}\neq\vec{p}$,

$$\vec{p}.A(\varepsilon\vec{x}+(1-\varepsilon)\vec{p}) > \vec{x}.A(\varepsilon\vec{x}+(1-\varepsilon)\vec{p})$$

provided ε is sufficiently small.

This formulation within the static framework of game theory can be discussed in terms of a dynamical evolution (see TAYLOR and JONKER [21]). Assuming that the rate of increase $\dot{x}_i/x_i$ of strategy E_i is given by the difference between the payoff for E_i and the average payoff, one obtains the game dynamical equation

$$\dot{x}_i = x_i((A\vec{x})_i - \vec{x}.A\vec{x})$$

for S_n. It is easy to see that if $\vec{p}$ is an evolutionarily stable state, then it is a stable attractor; if $\vec{p}$ is in the interior of S_n, then it is even globally stable (see HOFBAUER et al. [22] and ZEEMAN [23]). The existence of an evolutionary stable state $\vec{p}$ with all $p_i>0$, then, implies permanence. If $\vec{p}$, on the other hand, lies on the boundary (say $p_i>0$ for $1\leq i\leq m$ and $p_i=0$ for $m<i\leq n$), then the strategies E_1 to E_m are uninvadable by the other ones. However, the static approach via evolutionarily stable states yields only a fraction of permanence and uninvadability properties. There may be equilibria which are stable attractors, or which correspond to time averages, but which are not evolutionarily stable. And there may be attractors of the game dynamical equation which are not equilibria at all.

8. Catalytic Cooperation and the Hypercycle

In their theory of prebiotic evolution, EIGEN and SCHUSTER [24] consider the interactions of several species $M_1,\ldots,M_n$ of self-replicating polynucleotides, with concentrations $x_1,\ldots,x_n$, in an evolution reactor where a dilution flow controlled from the exterior keeps the total concentration constant (say, equal to 1). With $G_i(\vec{x})$ as the growth rate of x_i as determined by the chemical reaction, one obtains

$$\dot{x}_i = x_i(G_i-\phi) \text{ with } \phi=\sum_j x_j G_j$$

as the kinetic equation describing the evolution of the state $\vec{x}=(x_1,\ldots,x_n)$ in the concentration space S_n.

In general, the competition of the molecular species for the same type of energy-rich material (GTP, ATP, CTP and UTP) will lead to exclusion and to a corresponding loss of the information stored in the polynucleotides. In order to guarantee permanence, the reactor network must somehow implement cooperation between the molecular species. Eigen and Schuster proposed the hypercycle as one such mechanism: a closed loop of catalytic growth enhancement leading from M_1 to M_2, to M_3, etc. and finally from M_n back to M_1. Through such a "ganging up" each molecular species would have an "interest" in the survival of the others.

The correponding chemical growth term G_i will accordingly be of the form $x_{i-1}F_i(\vec{x})$ (with indices counted mod n), where $F_i(\vec{x})$ is a function strictly positive on S_n. An (unrealistically) simple approach, which can be viewed as a limiting case for low concentrations (cf. EIGEN et al. [25]) would have $F_i(\vec{x})$ equal to some constant $k_i>0$. In this case the unique equilibrium in the interior of S_n is globally stable if and only if $n\leq 4$. For $n\geq 5$ a periodic orbit is the attractor, as shown by numerical simulations (see EIGEN and SCHUSTER [24]). More realistic models of the hypercycle, using intermediate enzymes, lead to considerably more complicated functions F_i, which sometimes can be obtained only in implicit form (see HOFBAUER et al. [26]).

In any case, the hypercycle turns out to be permanent. This is its "raison d'être" and causes no surprise. The proof, however, is not straightforward. In general, the concentration x_i can decline even if it is very small, and a handy Ljapunov function does not seem to be known. But Hofbauer's criterion (see section 4) leads to the required result.

Indeed it is enough to show that the function $P=x_1x_2\ldots x_n$, which vanishes on the boundary of S_n and is strictly positive in the interior, is an "average Ljapunov function". One has $\dot{P}=P\psi$ with

$$\psi = \int \Sigma x_{i-1}F_i - n\phi.$$

Assume that condition (+) does not hold. Then there is an x on the boundary of S_n such that

$$T^{-1}\int_0^T \psi(x(t))\,dt\leq 0$$

for all T>0, or

$$n\,T^{-1}\int_0^T \phi(\vec{x}(t)) \geqslant T^{-1}\int_0^T \Sigma x_{i-1}F_i \;.$$

Since there is a k>0 with $F_1 \geqslant k$ for all i, this implies $T^{-1}\int_0^T \phi > \frac{k}{n}$. We shall show by induction that for all i, $x_i(t)$ converges to 0 for $t\to+\infty$, and thus obtain a contradiction to $\Sigma\, x_i=1$. The proof uses induction. There certainly is some i with $x_i\to 0$, indeed even with $x_i\equiv 0$, since $\vec{x}$ is on the boundary. Now if $x_i\to 0$, then $x_{i+1}\to 0$, as can be seen by taking the average (from 0 to T) of

$$(\log x_{i+1})^{\cdot} = \frac{\dot{x}_{i+1}}{x_{i+1}} = x_i F_{i+1} - \phi \;.$$

This yields

$$T^{-1}[\log x_{i+1}(T)-\log x_{i+1}(0)] = T^{-1}\int_0^T x_i F_{i+1} - T^{-1}\int_0^T \phi < \frac{k}{2n} - \frac{k}{n}$$

(since $x_i\to 0$). Thus

$$x_{i+1}(T) \leqslant x_{i+1}(0)\ \exp\left(-\frac{kT}{2n}\right)$$

and so $x_{i+1}\to 0$.

Having shown the permanence of hypercycles, one can deduce their uninvadability. More precisely, a hypercycle H_1 cannot be invaded by a disjoint hypercycle H_2 (see Fig.1).

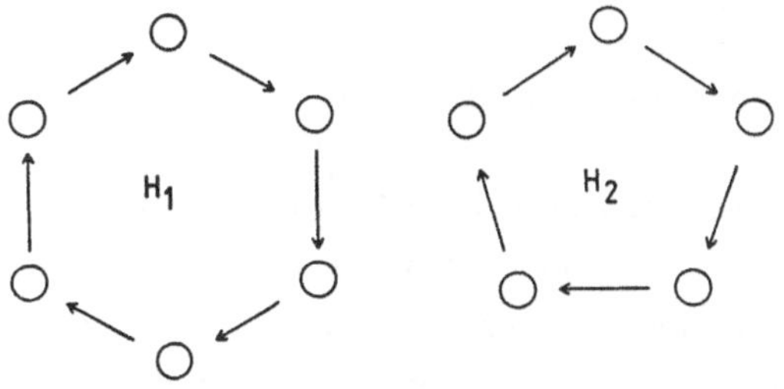

Figure 1. The competition of two disjoint hypercycles H_1 and H_2

Indeed, let l_1 and l_2 be the lengths of H_1 and H_2, and set

$$m = \min_{i\in H_1} F_i(\vec{x}) \qquad M = \max_{j\in H_2} F_j(\vec{x}) \;.$$

If H_1 is established, there is some $\varepsilon>0$ such that $x_i>\varepsilon$ holds for all $i\in H_1$, as shown by the previous argument. Therefore, the dilution flow is larger than $l_1\varepsilon m$. If H_2 is not complete, i.e. if $x_j=0$ for some j H_2, then x_{j+1} and all successive concentrations of H_2 converge to 0. Even if a small mutation introduces the hitherto missing species of H_2, their concentrations will initially be very low, so that $x_j < m\varepsilon l_1/M$ will hold for all $j\in H_2$. But this implies that $\dot{x}_j = x_j(x_{j-1}F_j - \phi)$ is negative, and hence that the "intruding" hypercycle will vanish.

This sheds some light on the competition of hypercycles in the evolution reactor. The first one which is established cannot be threatened by a disjoint one: an eventual successor must necessarily inherit some components from his predecessor.

HOFBAUER [27] has shown that "discrete hypercycles" given by difference equations of the type $\vec{x} \to T\vec{x}$ with

$$(T\vec{x})_i = x_i \frac{C + k_i x_{i-1}}{\Sigma x_j (C + k_j x_{j-1})}$$

on S_n, are permanent if and only if $C>0$. Another result due to HOFBAUER [8] is that the "inhomogeneous hypercycle", given by a differential equation of the type

$$\dot{x}_i = x_i(b_i + k_i x_{i-1} - \phi)$$

is permanent if and only if it admits a fixed point in the interior of S_n. This can be shown by using the function $P = \pi_i x_i^{1/k_i}$ as "average Ljapunov function".

Some results on permanence have been obtained for "autocatalytic systems" described by equations of the type

$$\dot{x}_i = x_i(\Sigma a_{ij} x_j - \phi)$$

with $a_{ij} \geqslant 0$. With each such equation, one may associate an interaction graph whose vertices correspond to the molecular species M_1 to M_n and where an arrow from M_j to M_i indicates that M_j effectively catalyses the reproduction of M_i, i.e. that a_{ij} is strictly positive. Some sufficient conditions for permanence are illustrated in Fig.2, where the ratios of weak and strong interactions have to satisfy certain relations.

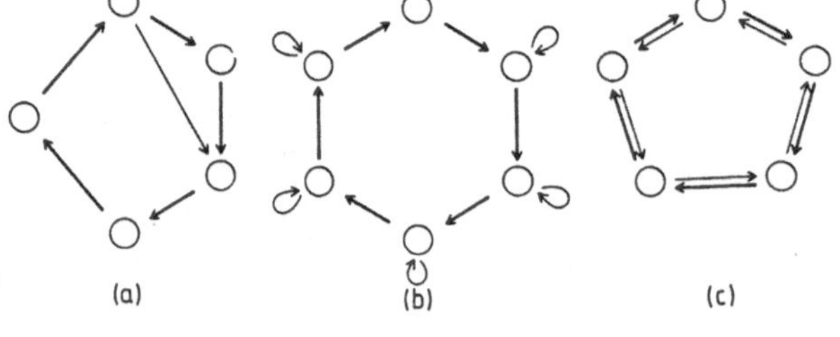

Figure 2. The interaction graphs of some permanent systems. If two arrows issue from one vertex, the bolder one corresponds to the larger catalytic effect

It seems of considerable interest to find necessary conditions for permanence, too. In particular, one may ask whether a permanent network must have a Hamiltonian graph (a graph with a closed loop of arrows visiting each node exactly once). This would single out the hypercycles as"minimal" permanent networks. Here, we shall show a weaker result: a permanent network is irreducible.

Indeed, suppose that the network is reducible, i.e. that there exists a proper subset D of $\{1,\ldots,n\}$ which is "closed" in the sense that $a_{ij}=0$ for all $i\in D$ and all $j\notin D$.We set

$$m = \min_{i\in D} \sum_{t\in D} a_{it} \qquad\qquad M = \max_{1\leq j\leq n} \sum_{s=1}^{n} a_{js} \; .$$

Clearly, we may assume $M>0$, as otherwise all coefficients would vanish and all points would be equilibria, a contradiction to permanence. Now if $m > 0$, then the set

$$G = \left\{\vec{x}\in S_n;\ \ \frac{x_j}{x_i} < \frac{m}{M} \text{ for all } i\in D \text{ and all } j\notin D\right\}$$

is positively invariant and all its orbits converge to the boundary, since for all $\vec{x}\in G$

$$\left(\frac{x_j}{x_i}\right)^{\bullet} = \left(\frac{x_j}{x_i}\right)\left(\sum_{s=1}^{n} a_{js}x_s - \sum_{t=1}^{n} a_{it}x_t\right)$$

$$\leq \left(\frac{x_j}{x_i}\right)\left(\sum_{s\in D} a_{js}x_s - \sum_{t\in D} a_{it}x_t\right)$$

$$\leq \left(\frac{x_j}{x_i}\right)\left(M \max_{s\in D} x_s - m \min_{t\in D} x_t\right)<0.$$

There remains the case that $m=0$. In this case, there is an $i\in D$ such that $a_{it}=0$ for all $t\in D$, and one may replace the set D by D $\{i\}$, which is again closed. Repeating the argument if necessary, one ends up either with a closed set for which $m>0$, or with a closed set $\{k\}$ consisting of a single species for which $a_{kk}=0$. This would imply that $a_{jk}=0$ for all j, i.e. that k has no effect on the growth of any species. If the system is permanent, there exists an interior fixed point, i.e. a strictly positive solution of

$$\sum_j a_{1j}x_j = \sum_j a_{2j}x_j = \qquad\qquad = \sum_j a_{nj}x_j$$

and

$$x_1+\ldots+x_n=1 \; .$$

Since x_k does not appear explicitly in the first n-1 of these linear equations, it follows that the solutions form an at least one-dimensional linear manifold of fixed points, which has to intersect the boundary of S_n. But then this boundary is no repellor, which contradicts the persistence of the system.

Acknowledgement: The work reported here has been supported financially by the Austrian "Fonds zur Förderung der wissenschaftlichen Forschung" project no 3502 and 4506. Technical assistance in preparing the manuscript by Mrs.J.Jakubetz and Mr.J.König is gratefully acknowledged.

References:

1 H.O.Peitgen: "A mechanism for spurious solutions of nonlinear boundary value problems", this volume
2 P.Coullet: "Complex behaviour in macrosystems near polycritical points", this volume
3 O.E.Rössler,J.L.Hudson,J.D.Farmer: "Noodle map chaos:a simple example", this volume
4 R.M.May: Nature 261, 459 (1976)
5 S.Smale: J.Math.Biol. 3, 5 (1976)
6 A.Arnedo,P.Coullet, C.Tresser: Physics Letters 79, 259 (1980)
7 P.Schuster,K.Sigmund,Wolff,R: J.Diff.Equs. 32, 357 (1979)
8 J.Hofbauer: Monatsh.Math. 91, 233 (1981)
9 M.Hirsch: SIAM Math.Analysis 13, 167 (1982)
10 H.Gregorius: Int.J.Systems Sci, 8, 863 (1979)
11 F.Sieveking: to appear
12 V.Hutson, W.Moran: J.Math.Biol. 15, 203 (1982)
13 J.B.S.Hadane, S.D.Jayakar: J.Genet. 58, 237 (1963)
14 H.Levene: Amer.Nat. 87,311 (1953)
15 J.Roughgarden: Theory of Population Genetics and Evolutionary Ecology: an Introduction (Macmillan, New York, 1979)
16 H.Gregorius, M.Ziehe: Math.Biosciences 61, 29 (1982)
17 R.A.Armstrong, R.McGehee: J.Theor.Biol. 56, 499 (1976)
18 Z.Nitecki: J.Math.Anal.and Appl. 72, 446 (1979)
19 V.Hutson, G.T.Vickers: Math.Biosciences 63, 253 (1983)
20 J.Maynard Smith: Evolutionary game theory. (Cambridge Univ.Press, 1982)
21 P.Taylor, L.Jonker, Math.Biosciences 40, 145 (1978)
22 J.Hofbauer, P.Schuster, K.Sigmund: J.Theor.Biol. 81, 609 (1979)
23 E.C.Zeeman: "Population dynamics from game theory" in Springer Lecture notes in Mathematics 81, (1979)
24 M.Eigen, P.Schuster: "The Hypercycle: a Principle of Natural Self-organization (Springer Berlin-Heidelberg, 1978)
25 M.Eigen, P.Schuster, K.Sigmund, R.Wolff: BioSystems 13, 1 (1980)
26 J.Hofbauer, P.Schuster, K.Sigmund: J.Math.Biol. 11, 155 (1981)
27 J.Hofbauer: SIAM J.Appl.Math., to appear.

Part V

Stochasticity in Complex Systems

Random Selection and the Neutral Theory - Sources of Stochasticity in Replication

P. Schuster and K. Sigmund

Institut für Theoretische Chemie und Strahlenchemie, Universität Wien
Währinger Straße 17, A-1090 Wien, Austria and

Institut für Mathematik, Universität Wien, Strudlhofgasse 4
A-1090 Wien, Austria

Three internal sources of stochasticity in polynucleotide replication are discussed together with a survey of experimental data. Replication is visualized as a multi-type branching process. In case of kinetic degeneracy - when the kinetic parameters are the same for all replicating elements - the replicating system can be analysed as a stochastic process which consists of simultaneous, but independent linear birth and death processes. Then, we were able to derive analytical expressions for the occurence of random selection in mathematical model systems. Some models are discrete, some continuous in time. Random selection may be described in terms of "sequential extinction times" which represent a special class of first passage times. Simple expressions are obtained also for the probabilities of gene fixation in asexually multiplying populations.

1. Sources of Stochasticity in Replicating Systems

Polynucleotides, DNA and RNA, represent the only presently known classes of molecules which carry the capability of replication in their molecular structure. Indeed, it is the formation of a double helix which is basic to the molecular copying mechanism. Experimental studies on polynucleotide replication, in particular those on template induced RNA-synthesis reached such a high degree of precision that detailed kinetic analysis became possible [1-3]. Although the full mechanism of RNA replication is highly complicated and consists of many elementary steps, the over-all process can be described well by a single-step autocatalytic reaction provided the experimental boundary conditions meet some general requirements [4]: (1) activated monomers, GTP, ATP, CTP and UTP, are present in large excess, (2) the replicating enzyme usually a virus specific RNA replicase is present in excess to the total RNA concentration and (3) all external parameters, pressure, temperature, ionic strength etc., are kept constant. Then, the total concentration of RNA (denoted by c_N^o in Fig.1) grows exponentially.

In case condition (2) is not fulfilled and the total concentration of RNA (c_N^o) exceeds that of the enzyme ($c_E^o < c_N^o$) we observe a range of linear growth (Fig.1). At still higher concentrations of RNA the rate of RNA synthesis decreases further as a result of product inhibition [1].

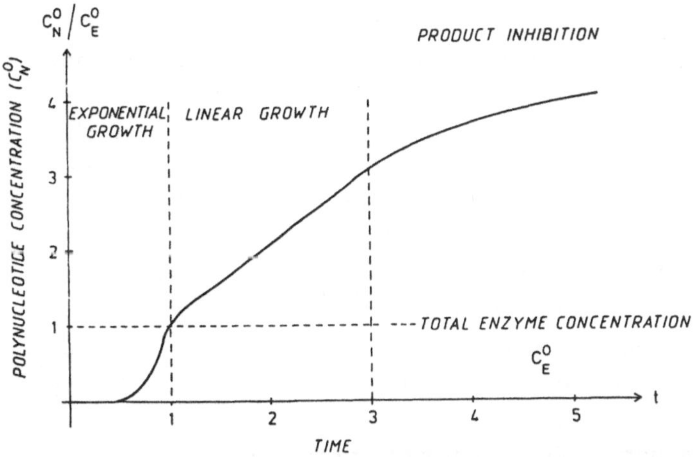

Figure 1. RNA synthesis in the test tube. Template induced polymerization is carried out by means of a specific RNA polymerase. The enzyme is present at constant total concentration (C_E^O), the activated monomers GTP, ATP, CTP and UTP, in large excess. A characteristic experimentally recorded curve shows three phases of growth: exponential growth at low template concentration, linear growth above saturation of the enzyme by polynucleotides and further levelling off at still higher concentrations of template where enzyme reactivation becomes the rate determining step.

In this section we shall discuss sources of randomness in "replicating systems". A replicating system consists of an ensemble of "replicators"[1)] under conditions which allow self-organization. Hence, replicating systems are open systems far off equilibrium. We shall be mainly concerned with polynucleotide replication, the mathematical treatment, however, is of more general nature and the results to be derived apply also to other replicators in biology [6]. In particular, the stochastic processes discrete in time are models of populations with defined generations as found with higher organisms.

Stochasticity in the evolution of a replicating system can originate from external sources, e.g. from a fluctuating environment, or from internal sources. These internal sources can be traced back to the molecular details of the replication process. Three internal factors of randomness are of primary importance:

(1) *Variability*. The number of individuals in any natural population is by far smaller than the number of possible polynucleotide sequences [7]. Using the words of JACOB [8] we might sag: the "actual" is a tiny fraction of the "possible". Rare mutants are present in a few copies only, most of them occur only once or not at all in a given population. Thus, the appearance and the extinction of these mutants are random events.

---

1) The notion of a "replicator" was used first by DAWKINS [5] as a synonym for the long-winded expression replicating element.

(2) *Chaotic dynamics*. In replicating systems with autocatalysis of second or higher order[2)] strange attractors may occur for dimensions $n \geq 3$ [9,10]. Then, the complex dynamics of the system introduces another source of stochasticity.
(3) *Kinetic degeneracy*. In case different replicators have the same kinetic parameters the deterministic description of the replicating system breaks down completely. The numbers of individual polynucleotide sequences are subjected to random drift. Experimental data obtained from laboratory studies on polynucleotides [2,11], from microbiology and from investigations of higher organisms [12] stress the importance of such mutations which are selectively neutral for the evolutionary process.

Here we shall not discuss chaotic dynamics which is treated extensively in other contributions. The other two sources of randomness may be studied by means of stochastic processes.

2. Replication as a birth and death process

A great variety of stochastic processes in a single variable, say $X(t)$ fall into the class of birth and death processes. Such a birth and death process is characterized by a set of transition probabilities from a given state x of the stochastic variable X into the two neighbouring states $x \pm 1$. In particular, we assign the probability

λ_x *to the birth process* $x \longrightarrow x + 1$

and the probability

μ_x *to the death process* $x \longrightarrow x - 1$

A birth and death process is completely defined by the set of λ_x's and μ_x's and the initial conditions. Four eventually occuring states are of special importance, upper and lower absorbing and reflecting states, since they confine the range which is accessible to the variable of the stochastic process (Table 1).

The replicating system and its time development are characterized by probability densities

$$P_x(t) = \mathit{Prob}\{X(t)=x\} \qquad (1)$$

$P_x(t)$ obviously depends on initial conditions. We shall be dealing here with sharp initial densities exclusively:

---

2) By the order of a catalytic process we denote the number of catalytically active molecules[4]. A simple autocatalytic reaction $A+I \rightarrow 2I$ is of first order: I acts as catalytically active template. The reaction $A+I+E \rightarrow 2I+E$ is of second order: I acts as template, E as catalyst.

Table 1. Characterization of upper and lower, absorbing and reflecting states

	Absorbing state	Reflecting state
Upper special state "u"	$\lambda_u=0,\ \mu_u=0$ $P_x=0\ \forall x>u$, or $X = \dots,u-2,u-1,u$	$\lambda_u=0,\ \mu_u=0$ $P_x=0\ \forall x>u$, or $X = \dots,n-2,n-1,u$
Lower special state "l"	$\lambda_l=0,\ \mu_l=0$ $P_x=0\ \forall x<l$, or $X = l,l+1,l+2,\dots$	$\lambda_l=0,\ \mu_l=0$ $P_x=0\ \forall x<l$, or $X = l,l+1,l+2,\dots$

$$P_x(0) = \delta_{xm},$$

at time t=0 we had exactly m replicators. These initial conditions are incorporated into (1) by means of the conditional probabilities

$$P_{xm}(t) = \mathit{Prob}\{X(t) = x|X(0) = m\} \tag{2}$$

For reviews of the results derived for birth and death processes we refer to the literature [13,14].

Applying mass action kinetics to the transition probabilities λ_x and μ_x we obtain polynomials in x:

$$\lambda_x = \lambda^{(0)}+\lambda^{(1)}x+\lambda^{(2)}x^2+\dots \tag{3a}$$

$$\mu_x = \mu^{(0)}+\mu^{(1)}x+\mu^{(2)}x^2+\dots \tag{3b}$$

In this contribution we restrict ourselves to examples which contain the first two terms of the expansion exclusively. Depending on the molecular mechanism of the process under consideration the higher order terms may be more involved. Often the kinetics of second and higher order reactions cannot be described by birth and death processes.

Processes for which only the constant terms $\lambda^{(0)}$ and $\mu^{(0)}$ are different from zero fall into the class of random walks. Those in which only $\lambda^{(1)}$ and $\mu^{(1)}$ are considered are commonly called linear birth and death processes. Note, that for a pure linear birth and death process the zero state (X=0) is an absorbing state $\lambda_o=\mu_o=0$. Particle numbers are necessarily positive, hence it is a lower absorbing state.

Two quantities of birth and death process are of particular interest: the probability of absorption into the state X(t)=0 from the initial state X(0)=m, commonly called the probability of extinction, $P_{om}(t)$, or its limit for long times $P_{om}(\infty)$ respectively and the mean time of extinction $E\{T_o\}$. Here, T_o is a stochastic variable, a so-called first passage time, namely the time at which X(t) becomes zero. T_o depends on the initial state and we may account for this dependence by a second index T_{om}. KARLIN [13] gives some general expressions:

$$\lim_{t\to\infty} P_{om}(t) = \begin{cases} \dfrac{\sum_{i=m}^{\infty} \left(\prod_{j=1}^{i} \frac{\mu_j}{\lambda_j}\right)}{1+\sum_{i=1}^{\infty} \left(\prod_{j=1}^{i} \frac{\mu_j}{\lambda_j}\right)} & \text{if } \sum_{i=1}^{\infty} \left(\prod_{j=1}^{i} \frac{\mu_j}{\lambda_j}\right) < \infty \\ 1 & \text{if } \sum_{i=1}^{\infty} \left(\prod_{j=1}^{i} \frac{\mu_j}{\lambda_j}\right) = \infty \end{cases} \qquad (4)$$

$$E\{T_{om}\} = \begin{cases} \sum_{i=1}^{\infty} \rho_i + \sum_{r=1}^{m-1} \left(\prod_{k=1}^{r} \frac{\mu_k}{\lambda_k}\right) \sum_{j=r+1}^{\infty} \rho_j & \text{if } \sum_{i=1}^{\infty} \rho_i < \infty \\ \infty & \text{if } \sum_{i=1}^{\infty} \rho_i = \infty \end{cases} \qquad (5)$$

where $\rho_i = (\lambda_1 \lambda_2 \cdot \ldots \cdot \lambda_{i-1})/(\mu_1 \mu_2 \cdot \ldots \cdot \mu_i)$. In case of a pure linear birth and death process one derives easily the useful expression for the mean time of absorption from state X(0)=1 $(\mu^{(1)} > \lambda^{(1)})$:

$$E\{T_{01}\} = -\frac{\mu^{(1)}}{\lambda^{(1)}} \ln\left(1 - \frac{\lambda^{(1)}}{\mu^{(1)}}\right).$$

Several examples of birth and death processes as stochastic models of chemical reactions have been studied in the past (see e.g. [15]). Autocatalytic reactions in chemistry and multiplication in biology can be modelled as birth and death processes with a lower absorbing barrier. This is the natural barrier state X=0 in case of linear birth and death processes ($\lambda_o = 0 \cdot \lambda$, $\mu_o = 0 \cdot \mu$) but it has to be artificially introduced into physical models for the process with constant transition probabilities. Thus, X(t) is confined to the range $0 \leq X(t) < \infty$. We distinguish between models which are discrete or continuous in time. Chemical reactions are commonly described by continuous processes. Phase relations are readily dissipated in ordinary chemical systems. In mathematical biology we find both classes of stochastic models. The existence of distinct generations in many populations justifies the assumption of discrete time. A primary requirement for the maintenance of synchronism in replication is a property of the mechanism: the mean time of replication has to be long compared to its fluctuations.

Then, synchronized populations lose phase only after very long time and external pacemakers can easily readjust the replicating system. Replication of polynucleotide templates takes an intermediate position: synchronization of the polymerization process breaks down after a few replication rounds [3].

In this section we compare some simple discrete and continuous mathematical models for replication in order to study their common features and the differences between them.

2.1 Processes with constant transition probabilities

A birth and death process with constant transition probabilities may be characterized as a random walk on a linear lattice the elements of which are the numbers the random variable X(t) can take. The discrete process is shown schematically in Fig.2. The stochastic variable X changes by ±1 per generation. The probability of increase is denoted by $\lambda_i = \lambda^{(0)} = \lambda$, i=1,2,..., and that of decrease by $\mu_i = \mu^{(0)} = \mu$, i= 1,2,...; otherwise we have $\lambda_i = \mu_i = 0$, in particular $\lambda_0 = \mu_0 = 0$. By conservation of probability we have $\lambda + \mu = 1$.

For reasons which will become clear later on we are particularly interested in the case $\lambda = \mu = 1/2$. Then, the stochastic process is an example of coin tossing [13]. Applying the initial condition X(0)=1

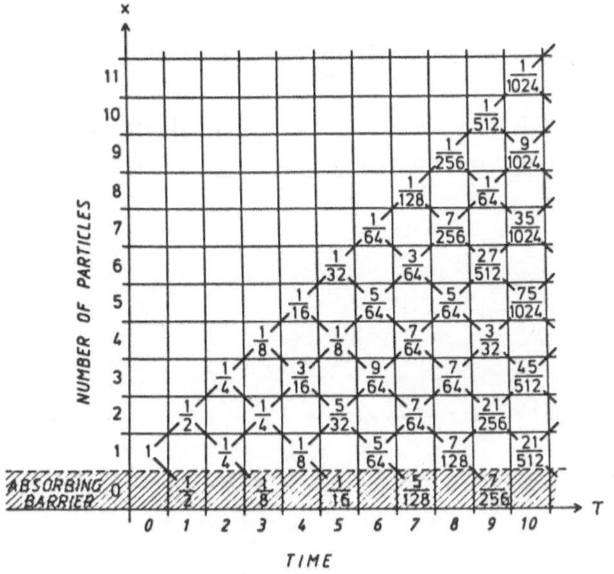

Figure 2. RNA replication in the range of linear growth modelled as a random walk with the zero state as an absorbing barrier [16]. The initial condition X(0)=1 and the transition probabilities $\lambda = \mu = 1/2$.

as in Fig.2 we obtain for the probability of extinction[3]:

$$P_{01}^{(2k)} = 1- \frac{\binom{2k}{k}}{2^{2k}}; \quad P_{01}^{(2k-1)} = P_{01}^{(2k)}; \quad k=0,1,2,... \qquad (6)$$

A numerical example is shown in Fig.3.

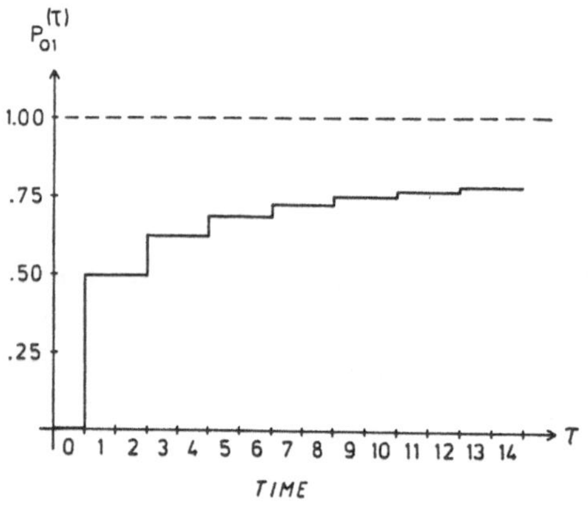

Figure 3. The probability of extinction for the process shown in Fig.2 according to (6)

Note, that the zero state (X=0) can be reached every second step only and hence P_{01} changes only in odd numbered generations. The asymptotic behaviour of (6) at long times can be derived easily by means of Stirling's formula

$$P_{01}^{(2k)} \propto 1 - \frac{1}{\sqrt{\pi}} k^{-1/2} \text{ (for large k)} \qquad (7)$$

It is interesting that the process with constant transition probabilities approaches the zero state slowly with a $t^{-1/2}$ dependence.

The continuous random walk with the zero state as an absorbing barrier is described by the Master equation

$$\frac{dP_{om}}{dt} = \mu P_{1m} \qquad (8a)$$

$$\frac{dP_{1m}}{dt} = \mu P_{2m} - (\lambda+\mu) P_{1m} \qquad (8b)$$

$$\frac{dP_{km}}{dt} = \lambda P_{k-1,m} + \mu P_{k+1,m} - (\lambda+\mu) P_{km}, \quad k=2,3,... \qquad (8c)$$

Analytical solutions of (8) are given by COX and MILLER [17]:

---

[3] We want to express the difference between discrete and continuous processes in the notation already: time is written as an argument in the continuous process, e.g. $P(t)$, whereas discrete time is given as superscript, e.g. $P^{(\tau)}$.

$$P_{nm}(t) = \left(\frac{\lambda}{\mu}\right)^{(n-m)/2} \exp\{-(\lambda+\mu)t\}\left[I_{n-m}(t')-I_{n+m}(t')\right] \quad n=1,2,3.. \quad (9)$$

They involve modified Bessel-functions I_ν of the argument $t'=2\sqrt{\lambda\mu t}$. The probability of extinction $P_{om}(t)$ cannot be calculated directly from (9). It is obtained from (8a) by integration

$$P_{om} = \int_0^t \mu P_{1m}\, dt \quad (10)$$

As in the discrete case we are interested in the asymptotic behaviour of the probability of extinction. For the sake of simplicity we assume $X(0)=1$ and $\lambda=\mu=1/2$. The long time dependence of $P_{11}(t)$ can be calculated from the asymptotic expressions for Bessel-functions of large arguments:

$$P_{11}(t) \simeq \frac{1}{\sqrt{2\pi t}}\left(\frac{2}{t} - \frac{3}{4t^2} + \ldots\right) \quad (11)$$

We partition the integral for $P_{01}(t)$ into a time independent constant A and the time dependent part f(t)

$$P_{01}(t) = \frac{1}{2}\int_0^t P_{11}(\tau)d\tau = \frac{1}{2}\int_0^{t_1} P_{11}(\tau)d\tau + \frac{1}{2}\int_{t_1}^t P_{11}(\tau)d\tau = A+f(t) \quad (12)$$

The time t_1 is chosen such that the asymptotic expression is fulfilled with sufficient accuracy by the first term of (11). Then, we find

$$P_{01}(t) \simeq A - \sqrt{\frac{2}{\pi}}\, t^{-1/2} \quad (13).$$

Again, we observe a $t^{-1/2}$ dependence of the probability of extinction as with the discrete process. Indeed, both expressions (7) and (13) become identical in case we substitute $\tau=2k$ in (7).

The random walk is an appropriate model for enzymatic polynucleotide replication and degradation under the condition of excess polynucleotides. Then, all enzyme molecules are occupied by templates and the rates are independent of polynucleotide concentrations. We may visualize replication under these conditions as an example of a queueing process: polynucleotides are the customers and enzyme molecules the servers which are in short supply. In absence of a degradation process ($\mu^{(o)}=0$) the concentrations of polynucleotides grow linearly as observed experimentally [1]. The condition under which no net growth is observed in the deterministic equations, $\lambda^{(0)}=\mu^{(0)}$, has been studied here. According to (4) extinction is certain, $P_{om}(\infty)=1$ but the expectation value for the time of extinction given in (5) diverges $E\{T_{om}\}=\infty$. The replicating system approaches the zero state with a very slow $t^{-1/2}$ dependence which has the consequence that selection is also a very slow process under these conditions.

2.2 Replication as a linear birth and death process

Replication and degradation of polynucleotides in the exponential phase can be modelled as a linear birth and death process [16,18,19]. As in the previous section we consider the discrete process first: every individual in the population gives rise to a birth with probability $\lambda^{(1)}=f$ or dies with probability $\mu^{(1)}=d$ in a given time interval (Fig.4). The corresponding stochastic process falls into the general class of branching processes [13]. We describe it appropriately by means of a generating function

$$F_m^{(\tau)}(s) = \sum_{x=0}^{\infty} P_{xm}^{(\tau)} s^x \qquad (14).$$

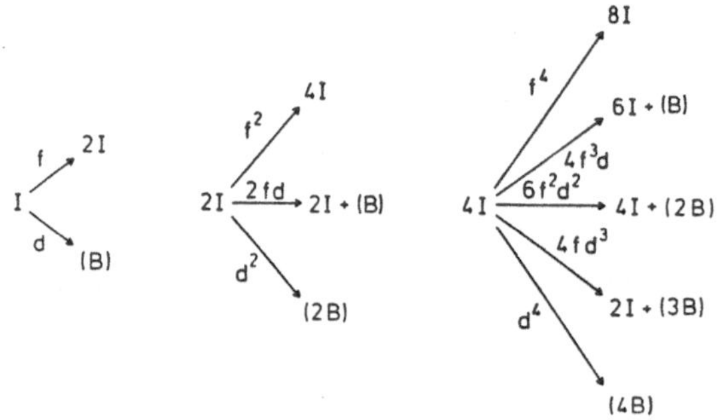

Figure 4. RNA replication in the range of exponential growth modelled as a linear birth and death process: $f=\lambda^{(1)}$ and $d=\mu^{(1)}$ according to (3a,b). This birth and death process falls into the general class of branching process

Starting from m molecules I at time $\tau=0$, which corresponds to $F_m^{(o)}=s^m$, we obtain for the probability generating function in the first generation

$$F_m^{(1)}(s) = (d+fs^2)^m \qquad (15).$$

The generating function for all further generations can be calculated from the recursion formula

$$F_m^{(k+1)}(s) = d+f\{F_m^{(k)}(s)\}^2;\ k=1,2,... \qquad (16).$$

In this context we are mainly interested in the probability of extinction

$$P_{om}^{(\tau)} = F_m^{(\tau)}(0) \qquad (17).$$

Although $P_{om}^{(\tau)}$ can be readily computed sequentially from the generating functions it is very difficult to derive an analytical formula. In

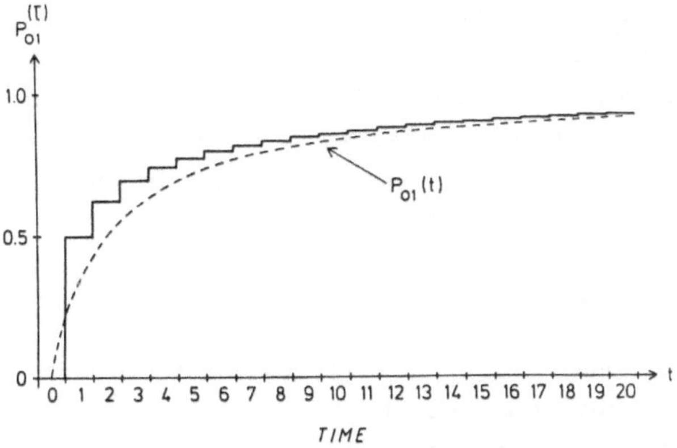

Figure 5. The probability of extinction for the discrete process shown in Fig. 4 ($P_{01}^{(\tau)}$) together with the corresponding quantity ($P_{01}(t)$) for the continuous process as derived from the master equation (18). Initial conditions and parameters were: X(o)=1; f=d=½

Fig.5 we present a characteristic numerical example: X(0)=1 and $f=d=\frac{1}{2}$. After a short initial transient the time dependence of $P_{01}^{(\tau)}$ is very close to t^{-1}.

The continuous process which is analogous to the discrete model shown in Fig.4 is described by the master equation

$$\frac{dP_{xm}}{dt} = (x-1)fP_{x-1,m}+(x+1)dP_{x+1,m}-x(f+d)P_{xm} \qquad (18).$$

Solutions of (18) have been derived by BARTHOLOMAY [20]. Again we apply the technique of probability generating functions

$$F_m(s,t) = \sum_{x=0}^{\infty} P_{xm}(t)s^x \qquad (19).$$

After some straight forward computations we find:

$$F_m(s,f) = \{\frac{d\alpha-d-(d\alpha-f)s}{f\alpha-d-(f\alpha-f)s}\}^m \qquad (20a)$$

with $\alpha = \exp\{(f-d)t\}$ for $f \neq d$ and

$$F_m(s,t) = \{\frac{ft-(ft-1)s}{ft+1-fts}\}^m \qquad (20b)$$

for d=f.

From (20b) we derive the probability of extinction for the case f=d:

$$P_{om}(t) = \{\frac{ft}{1+ft}\}^m = \{1-\frac{1}{1+ft}\}^m \qquad (21)$$

For long times this expression can be approximated by

$$P_{om} \simeq 1-\frac{m}{f}t^{-1} \qquad (22).$$

A characteristic curve for $f=d=\frac{1}{2}$ is shown in Fig.5. For long times the probabilities of extinction for the discrete and the continuous process approach each other closely. In the limit $t\to\infty$ both become unity which means certain extinction. According to (5) the expectation value of the time of extinction T_{om}, $E\{T_{om}\}$, diverges. We note the different time behaviour of the replication process in the linear and in the exponential phase of growth: $t^{-1/2}$ and t^{-1} dependence respectively.

2.3 Discrete and continuous processes

In the two cases treated in sections 2.1 and 2.2 we compared the discrete and the continuous process and found convergence in the limit of long times. In order to work out the differences between them we consider as an example the two linear birth and death processes[4].

Discretization of the continuous process can be performed by a commonly used technique: the generating function of the discrete process is obtained from that of the continuous process by setting t=1:

$$F_m^{(1)}(s) = F_m(s,1) \qquad (23).$$

For the linear birth and death process with d=f we obtain

$$F_m^{(1)}(s) = F_m(s,1) = \left\{\frac{f+(1-f)s}{1+f-fs}\right\}^m \qquad (24).$$

Assume that we started from a single molecule: X(O)=m=1. We find easily the probability density of the discrete process which has the form of a geometric progression (Fig.6):

$$P_{x1}^{(1)} = b \cdot p^{x-1};\ x=1,2,\ldots \qquad (25a)$$

with $b=(1+f)^{-2}$ and $p=f(1+f)^{-1}$ and

$$P_{01}^{(1)} = 1 - \sum_{x=1}^{\infty} P_{x1}^{(1)} = \frac{f}{1+f} \qquad (25b)$$

From (24) and (16) we get the definition of the complete stochastic process. Hence, the discretized version of the continuous linear birth and death process is a branching process but not a discrete birth and death process. This basic difference in the nature of the two stochastic processes leads to different behaviour in the initial phase. After a sufficiently large number of individual steps this difference becomes negligibly small and the solution curves of both processes become finally identical.

---

[4] The other case, discrete and continuous random walk would serve as an example equally well. For the sake of conciseness we dispense here from this redundancy.

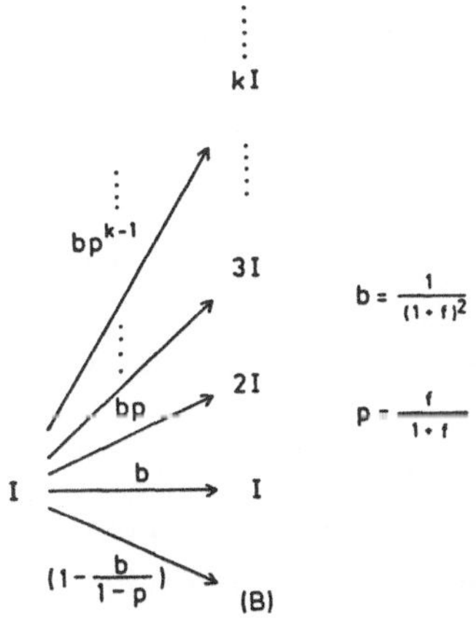

Figure 6. The discretized version of the continuous linear birth and death process as a branching process

3. Sequential extinction times

Let us consider an ensemble of n independent replicators which can be described by a chemical model of 2n parallel reactions, n autocatalytic reaction steps coupled to n decay processes:

$$(A) + I_i \xrightarrow{f_i} 2I_i; \quad i=1,2,\ldots,n \tag{26a}$$

$$I_i \xrightarrow{d_i} (B); \quad i=1,2,\ldots,n \tag{26b}$$

The basic assumptions of the model are the same as before: the concentration of A is buffered; hence, it is constant and does not enter as a variable into the equations of the replication process. Here, we shall be concerned with the quasistationary or critical case $f_i=d_i$ which cannot be properly described by the deterministic approach. The particle numbers of the individual replicators are represented by independent stochastic variables: $[I_i]= X_i$; $i=1,2,\ldots,n$. The replicating system then is described by a probability density which can be factorized:

$$\begin{aligned} P_{x_1x_2\ldots x_n}(t) &= \mathit{Prob}\,\{X_1(t)=x_1,\ X_2(t)=x_2,\ldots,X_n(t)=x_n\} = \\ &= P^{(1)}_{x_1}(t)\cdot P^{(2)}_{x_2}(t)\cdot\ \ldots\cdot P^{(n)}_{x_n}(t) \end{aligned} \tag{27}$$

Initial conditions can be incorporated by means of conditional probabilities

$$\begin{aligned} &P_{x_1m_1,x_2m_2,\ldots,x_nm_n}(t) = \\ &= \mathit{Prob}\{X_1(t)=x_1,\ X_2(t)=x_2,\ldots,X_n(t)=x_n \,|\, X_1(0)=m_1,\ X_2(0)=m_2,\ldots,X_n(0)=m_n\} = \\ &= P^{(1)}_{x_1m_1}(t)\cdot\ P^{(2)}_{x_2m_2}(t)\cdot\ \ldots\ \cdot P^{(n)}_{x_nm_n}(t) \end{aligned} \tag{28}$$

A knowledge of the linear birth and death process in one variable is sufficient to describe the system (26).

3.1 The case of maximum kinetic degeneracy

As we pointed out in section 1 the case $f_1=f_2=\ldots=f_n=f$ is of particular interest. Here, we expect random drift [21] to prevail. By means of our simple model we can derive exact analytical expressions for the time course of random selection.

Let us start from an ensemble of n different molecules[5)] each one present in a single copy: $m_1=m_2=\ldots=m_n=1$. We introduce new stochastic variables which decribe sequential extinction of replicators (Fig.7).

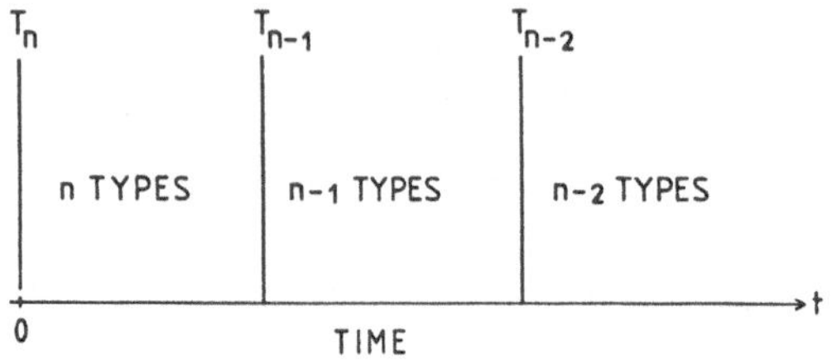

Figure 7. Sequential extinction times

By T_k we denote the time up to extinction of (n-k) types of replicators, i.e. the waiting time until k types are left. Accordingly, $T_n \equiv 0$ and T_o is the extinction time of the whole population. Since the master equation has been derived for single events, birth or death, within an infinitesimal time interval, we have

$$T_n \equiv 0 < T_{n-1} < T_{n-2} < \ldots < T_o \qquad (29).$$

The sequential extinction times are characterized by probability distributions

$$H_k(t) = \mathcal{P}rob\{T_k < t\} \qquad (30)$$

$H_o(t)$ is simply the probability that all molecular types have died out

$$H_o(t) = \sum_{i=1}^{n} P_{01}^{(i)}(t) = \{P_{01}(t)\}^n \qquad (31)$$

$H_1(t)$ is the probability to have one survivor or none.

---

5) According to the initial assumptions the molecules are indistinguishable with respect to their kinetic parameters but they differ as to other other properties, e.g. as to their polynucleotide sequences. Hence, they can be distinguished by sequence analysis.

$$H_1(t) = H_o(t) + \sum_{j=1}^{n} \{1-P_{01}^{(j)}(t)\} \prod_{i \neq j} P_{01}^{(i)}(t) =$$

$$= H_o(t) + n\,\{1-P_{01}(t)\}\{P_{01}(t)\}^{n-1} \quad (32)$$

The higher probability distributions can be derived by means of the recursion formula

$$H_k(t) = H_{k-1}(t) + \binom{n}{k}\{1-P_{01}(t)\}^k\{P_{01}(t)\}^{n-k} \quad (33).$$

Expectation values of the sequential extinction times are obtained from the probability distributions by differentiation and integration

$$E\{T_k\} = \int_0^{\infty} t\,\frac{dH_k(t)}{dt}\,dt \quad (34).$$

Recently, the expectation values were calculated from the expression for P_{01}(21) for the linear birth and death process [19]:

$$E\{T_k\} = \frac{n-k}{k} \cdot \frac{1}{f}\,; \quad k=1,2,\ldots,n \quad (35).$$

In Fig.8 we present one numerical example. Most types die out within the first two time units ($2f^{-1}$). All expectation values are finite except that of T_o. Accordingly, the number of different types will decrease until the population becomes uniform. This process is called "random selection" characteristicly. It leads to fixation of one of the initially n different types. Since they were present in equal numbers at the beginning the probability of fixation is 1/n for a given type.

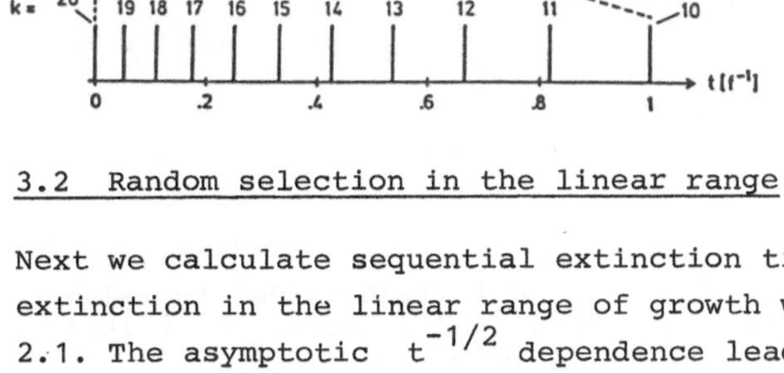

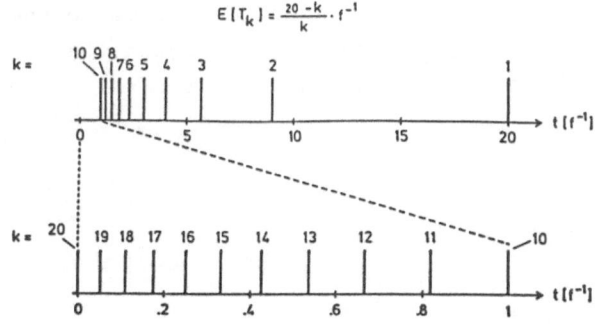

Figure 8. Sequential extinction times for the linear birth and death process with n=20

3.2 Random selection in the linear range

Next we calculate sequential extinction times for the probability of extinction in the linear range of growth which we computed in section 2.1. The asymptotic $t^{-1/2}$ dependence leads to a random selection process which is much less efficient than the corresponding process between replicators in the exponential range which we treated in the previous section. Indeed we find that the expectation values of T_o and T_1 diverge. Thus the mean value of the time of fixation of one type is infinite. Hence, we cannot expect to observe complete selection in finite times.

3.3 Generalization to different replication rates and non-uniform initial conditions

In this section we return to the linear birth and death process and drop first the assumption that the rate constants f_i are equal[6]. The expressions for $E\{T_k\}$ become considerably more complicated but the main conclusion remains uneffected: $E\{T_1\}$ is finite and $E\{T_o\}$ is not [19].

The probability that type I_i is the only survivor at time t under the condition that fixation has occurred is

$$\frac{1}{f_i t}\left(\sum_{j=1}^{n} \frac{1}{f_j t}\right)^{-1}$$

which is independent of time. Hence, the probability for eventual fixation of I_i is just the same expression namely

$$\frac{f_i^{-1}}{\sum_j f_j^{-1}} \qquad (36).$$

Secondly, we drop the assumption $X_i(0)=m_i=1$. In this case we may use the same expressions as before but we assign m_i identical terms to the type I_i. After some lengthy algebraic operations we find again that $E\{T_1\}$ is finite and that $E\{T_o\}$ diverges. The probability for the fixation of I_i is now

$$\frac{m_i f_i^{-1}}{\sum_j m_j f_j^{-1}} \qquad (37).$$

---

[6] We still retain the condition $f_i=d_i$ which is much harder to release for technical reasons. Moreover, $f_i \neq d_i$ and different f_i values lead to problems beyond the framework of neutral mutations.

4. Random Selection and the Neutral Theory

Let us now consider the spreading of a mutant gene in a population. The corresponding stochastic treatment of a population of diploid organisms goes back to ideas of FISHER [21] and was developed in detail by KIMURA (see e.g. KIMURA and OHTA [22]). Recently, the neutral theory of evolution was presented together with a collection of experimental data on molecular evolution by KIMURA [23].

The mathematical analysis centers around two stochastic variables which are shown schematically in figure 9:

(1) the time of fixation (ΔT_f) of a selectively neutral mutant allele in a given population, and

(2) the time of replacement (ΔT_r) of an allele by another neutral allele in a given population. The neutral theory provides simple expressions for the expectation values:

$$E\{\Delta T_f\} = 4n_e\tau \qquad \text{and} \qquad E\{\Delta t_r\} = \frac{1}{v}\,\tau \qquad \text{(38) and (39)}$$

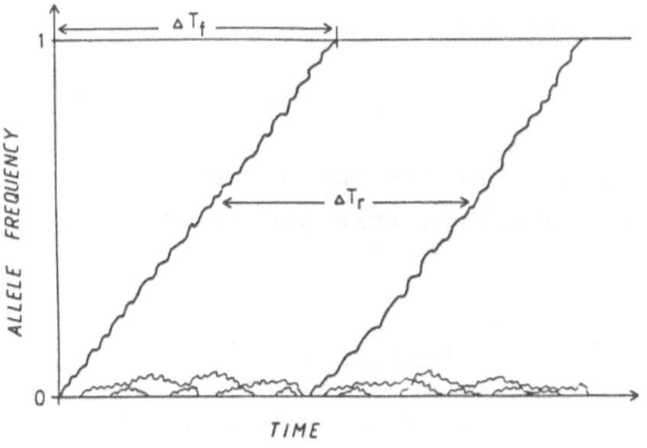

Figure 9. Time of fixation (ΔT_f) and time of replacement (ΔT_r) of a selectively neutral allele in a finite population

The unit of the time scale (τ) is chosen to be the mean time between two consecutive generations. By n_e we denote the effective population size [22] which we replace here by the actual number of replicators ($n_e \cong n$). The microscopic mutation rate per generation is v. Interestingly, this is also the rate at which new mutant alleles appear macroscopically in the population. Equation (38) shows, that fixation of neutral mutations is a phenomenon of finite populations. In an infinite population the expectation value of ΔT_f diverges.

Let us now calculate both quantities from our simple model which may serve as an example for asexually replicating systems. Apart from the mechanism of replication - sexual or asexual - our model differs from the neutral theory also with respect to the constraint imposed on the population: the expressions of the neutral theory were derived for constant population size whereas we allow for fluctuations of the total number of replicators; only its expectation value is constant. Any mutation starts from a single copy. Since replication steps are assumed to be independent we can mimic the process of mutant fixation by the initial condition of n-1 replicators of type I_1 and one of type I_2. Provided the replicator I_2 reaches fixation at all the time ΔT_f will be identical with the sequential extinction time T_1:

$$E\{\Delta T_f\} \;=\; E\{T_1\} \;=\; (n-1)\cdot \frac{1}{f} \approx \frac{n}{f} \tag{38a}$$

Indeed, (38) and (38a) are closely related. The reciprocal rate of re-plication (f^{-1}) is the analogue of the mean generation time (τ). The numerical factor 4 in (38) can be explained partly by the number of autosomal genes in a dipolid population which is twice the number of individuals, and partly by differences in the mechanism of replication and the constraints applied. The expectation value for the time of re-placement can be obtained from the mutation rate (v), the probability of fixation(1/n) and the population size (n):

$$E\{\Delta T_r\} = \frac{1}{v}\cdot\frac{1}{f}\cdot\frac{1}{n}\cdot n \;=\; \frac{1}{v}\cdot\frac{1}{f} \tag{39a}$$

We find complete analogy to (39). Note, that the factors which con-tain the population size cancel here. These factors are 1/2n and 2n respectively in the diploid population.

5. Replication as a multi-type branching process

The basic assumption of independent replication made in sections 3 and 4 is only an approximation to reality. All realistic systems are characterized by finite probabilities of mutation. Hence, the repli-cation of a sequence I_i leads to an ensemble of sequences which is characterized by a probability density:

$$(A) + I_i \longrightarrow 2I_i; \quad i=1,2,\ldots,n \tag{40}$$

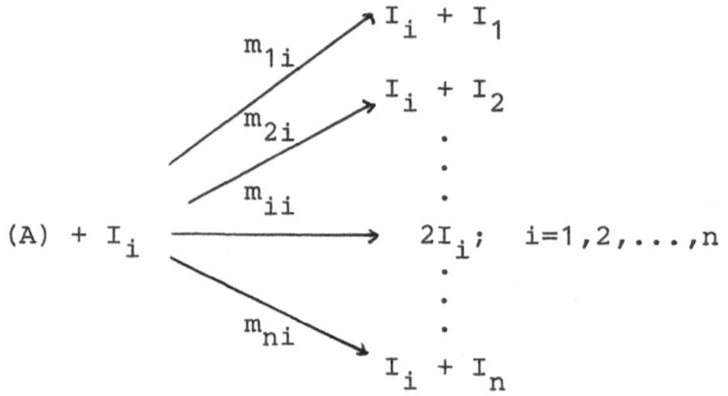

By m_{ji} we denote the average number of offspring of type I_j from an individual of type I_i. Accordingly, we have

$$\sum_j m_{ji} = m_i \qquad (41),$$

the total offspring of type I_i. If all $m_{ij}>0$ the mechanism (40) forms a fully connected network of chemical reactions. The individual stochastic variables $X_i=[I_i]$ ($i=1,\ldots,n$) are no longer independent. We combine the n variables to a stochastic state vector $\vec{X}$ of the system and the probabilities m_{ij} to a transition matrix M.

For the sake of simplicity we describe the dynamics of the reaction network by a discrete multi-type branching process [7]. The problem thus is to find solutions $\vec{X}^{(\tau)}$ ($=1,2,3\ldots$) for given initial conditions $\vec{X}^{(o)}$. Provided the transition matrix M is positive regular ($M^{(\tau)}>0$ for $\tau \geq \tau_k$) we can derive a few general results [13]. Let ρ be the dominant eigenvalue of M which is unique (i.e. non-degenerate), positive and largest in absolute value according to FROBENIUS [24]. All components of the corresponding, the dominant eigenvector are positive. We distinguish two cases:

(1) $\rho \leq 1$ all types I_i die out with probability one, and
(2) $\rho > 1$ all types I_i survive with some finite probability p ($0<p<1$).

Now we introduce expectation values for the variables of the multi-type branching process

$$y_i^{(\tau)} = E\{X_i^{(\tau)}\} \quad \text{and} \quad \vec{y}^{(\tau)} = \{y_i^{(\tau)};\ i=1,2,\ldots,n\} \qquad (42)$$

The variables $\vec{y}$ are determined by initial conditions $\vec{y}^{(0)}$ and the difference equation

$$\vec{y}^{(\tau+1)} = M\,\vec{y}^{(\tau)} \qquad (43)$$

Equation (43) is readily converted into a linear differential equation

$$\frac{d\vec{y}}{dt} = W\,\vec{y} \quad \text{with} \quad W = M-E \qquad (44)$$

where E is the unit matrix of dimension n. Matrix W has the dominant eigenvector

$$\lambda = \rho - 1 \qquad (45).$$

From the two cases distinguished above follows now:

---

[7] Here we follow an unpublished suggestion of Lloyd Demetrius (Max-Planck Institut für biophysikalische Chemie, Göttingen, B.R.D. [36]

(1) $\rho<1 \to \lambda<0$ and $\vec{y}(t) \to 0$ for large t, and

(2) $\rho>1 \to \lambda>0$ and $||\vec{y}(t)|| \to \infty$ for for large t.

Transformation to normalized variables

$$x_i = \frac{y_i}{\sum_{j=1}^{n} y_j} \quad \text{and} \quad \vec{x} = \{x_i;\ i=1,2,\dots,n\} \tag{46}.$$

corresponds to a mapping onto the unit simplex

$$S^n:\ \{\vec{x} \in \mathbb{R}^n;\quad 0 \leq x_i \leq 1;\ \Sigma x_i = 1;\quad i=1,2,\dots,n\}.$$

The differential equation (44) then is of the form

$$\frac{d\vec{x}}{dt} = W\vec{x} - \vec{x} \sum_k (W\vec{x})_k \tag{47}$$

with $(W\vec{x})_k = \sum_j w_{jk} x_j$. This equation is identical with the selection equation originally derived by EIGEN [18] and lateron studied extensively by several groups [4,25-27]. The expectation values of the multi-type branching process, thus, fulfil the deterministic selection equation. The elements of the matrix W can be specified further

$$w_{ii} = f_i Q_{ii} - d_i \tag{48a}$$

and $$w_{ij} = f_j Q_{ij} \tag{48b}.$$

Herein, f_i describes the total rate of multiplication counting correct copies and mutations together; Q_{ij} is the probability to obtain I_i as an error copy of I_j. By conservation of probability we have

$$\sum_{i=1}^{n} Q_{ij} = 1 \tag{48c}$$

The diagonal elements of the matrix Q are a measure of the accuracy of the replication process: $Q_{ii}=1$ means ultimate precision, every copy is correct.

Equation (47) has a unique stable steady state which corresponds to the dominant eigenvector of the matrix W. All orbits starting in S_n converge to this steady state.

It is of some interest to consider also cases where M is not positive regular. We mention here one example which we have studied previously [28,29]. Suppose that the backward mutations leading to one type, say I_1, are suppressed: $Q_{1i}=0$ for all $i \neq 1$, which means $m_{1i}=0$ for all $i \neq 1$. Then, we can represent the development of I_1 by a single type branching process.The role of the dominant eigenvalue is now played by the average number of offspring of type I_1 from an I_1 individual. In the previous notation this is the matrix element m_{11}. Let us assume that I_1 gives rise to a total progeny of $m_1=\Sigma$ offspring. The accuracy of the replication process is described by the quality factor $Q_{11}=Q$. The number of correct offspring is a random variable X with binomial distribution. Thus, the probability to have k descendants of type I_1 is of the form

$$Prob\{X=k\} = \binom{\Sigma}{k} Q^k (1-Q)^{\Sigma-k} \qquad (49).$$

This distribution has an expectation value of

$$E\{X\} = \Sigma.Q = m_{11} \qquad (50).$$

The condition of a non-zero probability of survival (case (2) from above) simply yields:

$$m_{11}>1 \quad \text{or} \quad Q > \Sigma^{-1} \qquad (51).$$

Obviously, there is a minimum accuracy $Q_{min}=\Sigma^{-1}$ below which I_1 goes extinct with probability one.

Let us finally turn back to polynucleotide replication where we can define an average single digit accuracy q. This single digit accuracy is the probability of incorporation of the correct digit at a given position in the sequence, 1-q is the probability of making an error at this position. Replication errors are assumed to occur independently. Thus, a sequence of ν digits will be replicated with an accuracy

$$Q = q^{\nu} \qquad (52)$$

The minimum accuracy Q_{min} corresponds for a given single digit accuracy q to a maximum chain length ν_{max} of the polymer:

$$Q_{min} = q^{\nu_{max}} \qquad (53)$$

$$\text{or} \quad \nu_{max} = -\frac{\ln\Sigma}{\ln q} \simeq \frac{\ln\Sigma}{1-q}, \text{ if } q\simeq 1 \qquad (54).$$

Equation (54) is the probabilistic analogue of the error threshold derived from the deterministic equations [4,18]. It is a stronger condition than the deterministic expression: the corresponding solution of the deterministic equation

$$y(t) = y(0)\exp\{(m_{11}-1)t\}$$

does not approach zero as long as $m_{11}>1$ is fulfilled.

Master equations corresponding to the continuous analogue of the discrete multi-type branching process (40) under special constraints were derived by several authors [30-34]. An alternative approach starts out from a Langevin-type equation [35]. For the sake of conciseness we can only mention these studies but we have to dispense from details. Although considerable progress has been made towards an understanding of the stochasticity of evolutionary processes the situation in in this field is still unsatisfactory. Analytical expressions can be derived for special cases only which are of very limited physical relevance. Numerical simulations are illustrative but they do not provide deeper insight into the stochastic process.

Acknowledgements: The work reported here has been supported financially by the Austrian "Fonds zur Förderung der wissenschaftlichen Forschung" project no 3502 and 4506. Technical assistance in preparing the manuscripts by Mrs. J.Jakubetz and Mr.J.König is gratefully acknowledged.

References

1 C.K.Biebricher, M.Eigen, R.Luce: J.Mol.Biol. 148, 369 and 391 (1981)
2 C.K.Biebricher, S.Diekmann, R.Luce: J.Mol.Biol. 154, 629 (1982)
3 C.K.Biebricher, M.Eigen, W.C.Gardiner: Biochemistry 22, 2544 (1983)
4 M.Eigen, P.Schuster: The Hypercycle - A Principle of Natural Self-Organization (Springer, Berlin 1979)
5 R.Dawkins: The Extended Phenotype (Freeman, San Francisco 1982)
6 P.Schuster, K.Sigmund: J.theor.Biol. 100, 533 (1983)
7 M.Eigen, P.Schuster: J.Mol.Evol. 19, 47 (1982)
8 F.Jacob: The Possible and the Actual (Pantheon Books, New York 1982)
9 A.Arneodo, P.Coullet, C.Tresser: Phys. Letters 79A, 259 (1980)
10 J.Hofbauer: Nonlinear Anal.5, 1003 (1980)
11 C.K.Biebricher: Evolutionary Biology 16, pp.1-52 (1983)
12 M.Kimura: Sci.Am. 241 (5), 94 (1979)
13 S.Karlin: A First Course in Stochastic Processes (Academic, New York 1966)
14 N.S.Goel, N.Richter-Dyn: Stochastic Models in Biology (Academic, New York 1974)
15 D.A.McQuarrie: J.Appl.Prob.4, 413 (1967)
16 P.Schuster: Chemie i.u.Zeit 6, 1 (1972)
17 D.R.Cox, H.D.Miller: The Theory of Stochastic Processes (Wiley, New York 1965)
18 M.Eigen: Naturwissenschaften 58, 465 (1971)
19 P.Schuster, K.Sigmund: Bull.Math.Biol. 45, in press (1983)
20 A.F.Bartholomay: Bull.Math.Biophys. 20, 97 (1958)
21 R.A.Fisher: Proc.Roy.Soc.Edinburgh 50, 205 (1930)
22 M.Kimura, T.Ohta: Theoretical Aspects of Population Genetics (Princeton Univ.Press, Princeton 1971)
23 M.Kimura, ed.: Molecular Evolution, Protein Polymorphism and the Neutral Theory (Springer, Berlin 1983)
24 G.Frobenius: Sitz.-Ber.Akad.Wiss.Phys.-math.Klasse Berlin 417 (1908) and 456 (1912)
25 C.J.Thompson, J.L.McBride: Math.Bioscience 21, 127 (1974)
26 B.L.Jones, R.H.Enns, S.S.Rangnekar: Bull.Math.Biol. 38, 15 (1976)

27 J.Swetina, P.Schuster: Biophys.Chem. 16, 329 (1982)
28 P.Schuster, K.Sigmund: "Self-organization of Biological Macromolecules and Evolutionarily Stable Strategies" in H.Haken, ed.: Dynamics of Synergetic (Springer, Berlin 1980), pp.156-169
29 P.Schuster, K.Sigmund: "From Biological Macromolecules to Protocells-The Principle of Early Evolution" in W.Hoppe, W.Lohmann, H. Markl and H.Ziegler, eds.: Biophysics, 2nd ed.(Springer, Berlin 1983)pp.874-912
30 W.Ebeling, R.Feistl: Ann.Physik 34, 81 (1977)
31 W.Ebeling, R.Mahnke: Prob.Contemp.Biophys. 4, 119 (1979)
32 B.L.Jones, H.K.Leung: Bull.Math.Biol. 43, 665 (1981)
33 W.Ebeling, R.Feistl: Physik der Selbstorganisation und Evolution (Akademie Verlag, Berlin 1982)
34 R.Heinrich, I.Sonntag: J.theor.Biol. 93, 325 (1981)
35 H.Inagaki: Bull.Math.Biol. 44, 17 (1982)
36 L.Demetrius: J.theor.Biol. 103, in press (1983)

The Dynamics of Catalytic Hypercycles - A Stochastic Simulation

A.M. Rodriguez-Vargas[1] and P. Schuster

Institut für Theoretische Physik, Universität Stuttgart, Pfaffenwaldring 57
D-7000 Stuttgart 80, Fed. Rep. of Germany

Institut für Theoretische Chemie und Strahlenchemie, Universität Wien
Währinger Straße 17, A-1090 Wien, Austria

A stochastic simulation of elementary hypercycles of dimensions n = 3, 4, 5 and 8 is presented. In populations of sizes which can be realized in nature and in laboratory experiments, hypercycles with dimensions $n \geq 5$ die out after a short time. This instability of hypercycles of higher dimensions is to be distinguished from a general instability of the model system presented: all model systems considered here are metastable in the sense that the system dies out with probability one for infinite time.

1. The model system and its deterministic analysis

In analogy to the model systems analysed by HOFBAUER and SCHUSTER [1] we consider the following network of autocatalytic chemical reactions as a model for elementary hypercycles [2]:

$$A + I_1 + I_n \xrightarrow{k_1} 2I_1 + I_n \qquad (1a)$$

$$A + I_2 + I_1 \xrightarrow{k_2} 2I_2 + I_1 \qquad (1b)$$

$$\vdots$$

$$A + I_n + I_{n-1} \xrightarrow{k_n} 2I_n + I_{n-1} \qquad (1c)$$

$$I_1 \xrightarrow{d_1} B \qquad (1d)$$

$$\vdots$$

$$I_n \xrightarrow{d_n} B \qquad (1e)$$

$$B \xrightarrow{g(E)} A\,. \qquad (1f)$$

This network consists of n second-order autocatalytic reactions

---

[1] Permanent address : Departamento de Fisica, Universidad de los Andes, Bogota, D. E., Columbia

(1a) - (1c), n degradation reactions (1d) - (1e) and a recycling reaction (1f) which is driven by an external energy source (E). The evolution of this open system is described within the frame of conventional chemical kinetics by the differential equation :

$$\dot{a} = gb - a \sum_{i=1}^{n} k_i x_i x_{i-1} \quad (2a)$$

$$\dot{b} = \sum_{i=1}^{n} d_i x_i - gb \quad (2b)$$

$$\dot{x}_i = x_i \ (k_i a x_{i-1} - d_i) \ ; \ i=1,2,. \ ,n. \quad (2c)$$

The index "i" is understood mod n : i-1 = n for i = 0, etc. Concentrations are denoted by [A] = a, [B] = b, $[I_1] = x_1, \ldots,$ $[I_n] = x_n$. We introduce a total concentration of autocatalists

$$\sum_{i=1}^{n} x_i = c . \quad (3)$$

Obviously, equation (2) fulfils a conservation law

$$a + b + c = c_o = \text{const.} \quad (4)$$

and hence one variable can be eliminated.

Equation (2) has three fixed points. The zero state

$$P_o : \bar{a}^{(o)} = c_o \ ; \ \bar{b}^{(o)} = \bar{x}_i^{(o)} = \ldots = \bar{x}_n^{(o)} = 0 \quad (5)$$

and two states Q and P which are obtained as solutions of the equations :

$$\bar{a}^{(1),(2)} = \frac{1}{2} \left\{ c_o \pm (c_o^2 - 4D^2)^{\frac{1}{2}} \right\} \quad (6)$$

$$\bar{x}_{i-1}^{(j)} = \frac{d_i}{k_i} \cdot \frac{1}{\bar{a}^{(j)}} \quad (7)$$

and

$$\bar{b}^{(j)} = \frac{1}{g} \left(\sum_{i=1}^{n} \frac{d_i \cdot d_{i+1}}{k_{i+1}} \right) \cdot \frac{1}{\bar{a}^{(j)}} \ ; \ j = 1,2 \quad (8)$$

with

$$D^2 = \frac{1}{g} \sum_{i=1}^{n} \frac{d_{i+1}}{k_{i+1}} (d_i + g) . \quad (9)$$

The two states Q and P exists only above the critical concentration

$$c_o \geq c_o^{crit} = 2D \tag{9a}$$

at which we observe a saddle node bifurcation.

One property of equation (2) will be discussed later on :

<u>Theorem:</u> There exist no stationary states of equation (2) at which some but not all concentrations of the autocatalysts I_i are zero. In other words the orbits converge either to the point $a = c_o$ or the time averages of all concentrations are strictly positive: $\langle x_i \rangle > 0 \; \forall \; i=1,\dots,n$ (see also [3]).

<u>Proof :</u> We consider the subspace of the variables $x_i \{\vec{x} \in \mathbb{R}^n ; x_i \geq 0 ; \sum x_i \leq c_o ; i = 1,2,\dots,n\}$ on the boundary of the concentration space $\{c \in \mathbb{R}^{n+2} ; c = (a,b,x_1,\dots,x_n) ; a \geq 0 , b \geq 0 , x_i \geq 0 ; a + b + \sum x_i = c_o\}$.

If $x_k = 0$ we have $\dot{x}_{k+1} = -d_{k+1}x_{k+1}$ and hence $x_{k+1} \to 0$.

This implies $\dot{x}_{k+2} = -d_{k+2}x_{k+2}$ and $x_{k+2} \to 0$. Consequently, all x_i ($i=1,2,\dots,n$) and also b converge to zero. The orbit finally approaches the point $a = c_o$.

In order to perform stability analysis of the three points P_o,Q and P we calculate the Jacobian matrix of (2) which after elimination of b is of the form (the concentration vector reads: $a,x_1,x_2,\dots,x_n$) :

$$\begin{vmatrix} -(g+\sum_{i=1}^{n} k_i x_i x_{i-1}) & -g-a(k_1x_n+k_2x_2) & -g-a(k_2x_1+k_3x_3) & \cdot & -g-a(k_nx_{n-1}+k_1x_1) \\ k_1x_1x_n & k_1ax_n-d_1 & 0 & \cdots & k_1ax_1 \\ k_2x_2x_1 & k_2ax_2 & k_2ax_1-d_2 & \cdots & 0 \\ \vdots & \vdots & \vdots & & \vdots \\ k_nx_nx_{n-1} & 0 & 0 & & k_nax_{n-1}-d_n \end{vmatrix}$$

The computation of the eigenvalues at P_o is trivial and yields:

$$w_1^{(o)} = -d_1 \;,\; w_2^{(o)} = -d_2 \;,\; \dots \;,\; w_n^{(o)} = -d_n \text{ and } w_{n+1}^{(o)} = -g . \tag{10}$$

Hence, the zerostate is always stable.

Stability analysis is somewhat more involved in case of the two states Q and P. We dispense here with details and present only the final results. It is straightforward to express the elements of the Jacobian as functions of the rate constants and the stationary concentration $\bar{a}$. Thus, the algebraic structures of the eigenvalue problem at the stationary points Q and P are the same, the only difference concerning different values of the stationary concentration $\bar{a}$.

In order to be able to solve the eigenvalue problem analytically we assume equality of the rate constants of the autocatalytic reactions

$$k_1 = k_2 = \dots = k_n = k$$

as well as of those of the degradation processes

$$d_1 = d_2 = \dots = d_n = d \,.$$ [2)]

Then, the eigenvalues of the Jacobian are solutions of the equations

$$\sum_{j=1}^{n} (w^{(l)})^{n-j} \, d^{j-1} = 0 \tag{11}$$

and $$(w^{(l)})^2 + pw^{(l)} + q = 0 \tag{12}$$

with $$p = g - d + \frac{nd^2}{k\bar{a}^2} \tag{12a}$$

and $$q = \frac{nd^2}{k\bar{a}^2}\,(\,g+d\,) - dg. \tag{12b}$$

The upper index "l" denotes state Q (l=1) or P (l=2). The solutions of (11) can be expressed in terms of the complex roots of unity :

$$w_j^{(l)} = d \exp\left(\frac{2\pi i}{n}\, j\right) \;;\; j = 1,2,\dots,n-1. \tag{11a}$$

Hence, the n-1 roots of the linearized dynamical systems around P (or Q) are identical with those of the elementary hypercycles which have been studied extensively in the past. (For details see [2,5] and [6]). The central fixed point in these n-1 dimensional dynamical systems is asymptotically stable for n=2,3 or 4. In case n≥5 it is unstable and we find a stable closed orbit.

Exponential stability of P (or Q) implies that both roots of equation (12) have negative real parts. This leads to the two inequalities

---

2) The choice of equal rate constants does not necessarily imply a loss in generality. In case of elementary hypercycles it is possible to find a nonlinear transformation which changes every hypercycle into a topologically equivalent dynamical system with equal rate constants [4].

$p > 0$ and $q > 0$ which can be transformed into inequalities imposing conditions on $\bar{a}$. For convenience we distinguish two cases :
(1) $d > g$, where the two conditions are

$$\bar{a}^2 < \frac{nd^2}{k(d-g)} \tag{13a}$$

and

$$\bar{a}^2 < \frac{nd(d+g)}{kg} \; . \tag{13b}$$

We start by an inspection of (13b) and recall the value of $\bar{a}$ at the critical concentration $c_o = c_o^{crit}$:

$$\bar{a}(c_o^{crit}) = \frac{c_o^{crit}}{2} = D = \left(\frac{nd(d+g)}{kg}\right)^{\frac{1}{2}} \; .$$

Hence, condition (13b) is fulfilled by the eigenvalues at P and is not fulfilled by the eigenvalues at Q. Consequently, Q is a saddle. Condition (13a) is more subtle. It is a stronger condition than (13b) iff

$$dg < d^2 - g^2$$

which leads to

$$d > \frac{1+\sqrt{5}}{2} g \; . \tag{13c}$$

In this case P is formed as an unstable node at the saddle-node bifurcation at c_o^{crit}. Since $\bar{a}_2$ decreased with increasing c_o, P undergoes a second bifurcation at the point

$$c_o(\bar{a}_{crit}) = (d^2-g^2+dg)\left\{\frac{n}{kg^2(d-g)}\right\}^{1/2}; \; \bar{a}_{crit} = \left\{\frac{nd^2}{k(d-g)}\right\}^{1/2} \tag{13d}$$

where it becomes stable by the omission of an unstable closed orbit. For n=2,3 and 4 the stationary state P thus is asymptotically stable beyond the HOPF bifurcation at $\bar{a}_2 = \bar{a}_{crit}$.
(2) $d < g$. In this case we have instead of (13a)

$$\bar{a}^2 > \frac{nd^2}{k(d-g)} \; . \tag{13e}$$

This inequality is trivially fulfilled since $\bar{a}^2$ is always positive. Thus, stability of both roots of equation (12) is determined exclusively by condition (13b) and hence P is asymptotically stable if all eigenvalues given by (11a) have negative real parts (n=2,3).
P is asymptotically stable also for n=4 but in this case we have very slow convergence to the stable point [5,6]. Q is always a saddle. The results are summarized schematically in Fig.1.
Although we have not yet explored the results from global qualitative analyses of equation (2) we may expect very rich dynamics

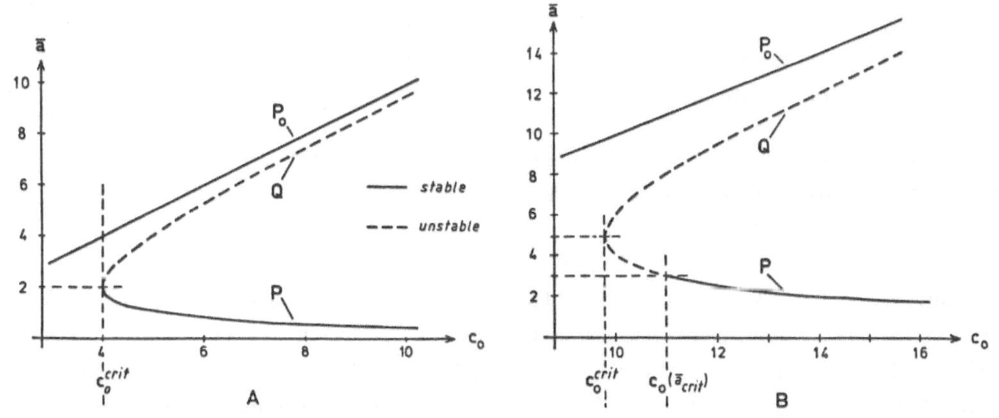

Figure 1. The three steady states of equation (2) as functions of c_0. We distinguish two qualitatively different cases. A: $d<0.5(1+\sqrt{5})g$; $n=2$, $k=d=g=1$; P is stable and Q is unstable and B: $d>0.5(1+\sqrt{5})g$; $n=2$, $k=g=1$, $d=3$; P is unstable at the saddle-node bifurcation at c_{co}^{crit} and becomes asymptotically stable only at some higher value of c_0 through a subcritical HOPF bifurcation, i.e. through omission of an unstable limit cycle. These results apply strictly also to the cases n=3 and n=4. For $n\geq5$ P remains unstable as it is the case in the elementary hypercycle.

particularly in the cases of higher dimensions $n\geqslant5$ and in the range $c_0^{crit}/2 < \bar{a}_2 < \bar{a}_{crit}$ in systems with $d \gg g$.

2. Stochastic computer simulations

The formulation of chemical reactions as stochastic processes is based on the fact that collisions of molecules in a system at thermal equilibrium occur in an essentially random manner and thus introduce a stochastic element into chemical kinetics. Finite numbers of particles as they occur in all realistic systems, additionally, introduce another source of randomness in regions of complex dynamical behaviour, e.g. near bifurcation points, near homoclinic orbits and in the basins of chaotic attractors. All these factors make a stochastic description necessary, and not only desirable, for a complete understanding of chemical reaction systems like the one given by equation (1). The time evolution of a spatially homogeneous chemical system can be described by a stochastic process continuous in time but discrete in particle numbers. The formulation of chemical reactions by master equations is now well established [7]. The basic problem concerns the search for procedures to derive solutions in general cases where the deterministic kinetics is non-linear. Much progress has been made during the last decade [8,9], but, nevertheless, one is forced to use numerical simulation in order to obtain the first information on the stochastic dynamics of the system under consideration. In this paper we present just the results of such preliminary studies on the stochastic treatment of the reaction mechanism (1).

Computer simulation of the reaction mechanism (1) was performed by means of a numerical formalism proposed by GILLESPIE [10]. Assume the system is in the state $(A,B,X_1,\dots,X_n)$ at time t. By upper case letters A(t), B(t) and $X_i(t)$, i=1,...,n we denote the stochastic variables which describe a particular run of our system. Now, we create a random pair (τ,μ) from the set of random pairs whose probability density is given by $P_t(\tau,\mu)$. $P_t(\tau,\mu)$ is the probability that given the state $(A,B,X_1,\dots,X_n)$ at time t the next reaction will occur in the time interval $(t+\tau\ ,\ t+\tau+d\tau)$ and it will be of the type μ. In our case μ has 2n+1 possible values corresponding to the individual reactions of mechanism (1). It is not necessary to describe the numerical algorithm in detail since it has been published together with a computer program in FORTRAN [11].

The mechanism (1) has an interesting property with respect to the stochastic treatment. First of all we have conservation of mass leading to a conservation law similar to equation (4):

$$A + B + \sum_i X_i = C_o . \qquad (14).$$

Secondly, the state $A=C_o$ is an absorbing state; since we have B=O and X_i=O, i=1,...,n, at this state there is no way out of it. Moreover, all other states are transients and hence all trajectories have to converge ultimately to $A=C_o$ because they are all connected. What we distinguish here is metastability in the sense that the typical trajectories converge to $A=C_o$ after very long times or in the limit $t \rightarrow \infty$ only, and short-lived transients which converge $A=C_o$ very soon.

3. Results of computer simulation

All stochastic simulations on elementary hypercycles embedded in mechanism (1) - we studied systems with n=3,4,5 and 8 - have a number of features in common :

(1) In case the total concentration is below the critical limit, $C_o < C_o^{crit}$, the system converges readily to the state of extinction, the zero state $A=C_o$, a numerical example is shown in Fig.2. The critical concentration is given by equation (9a). We use equal rate constants for the autocatalytic and the decomposition reactions and

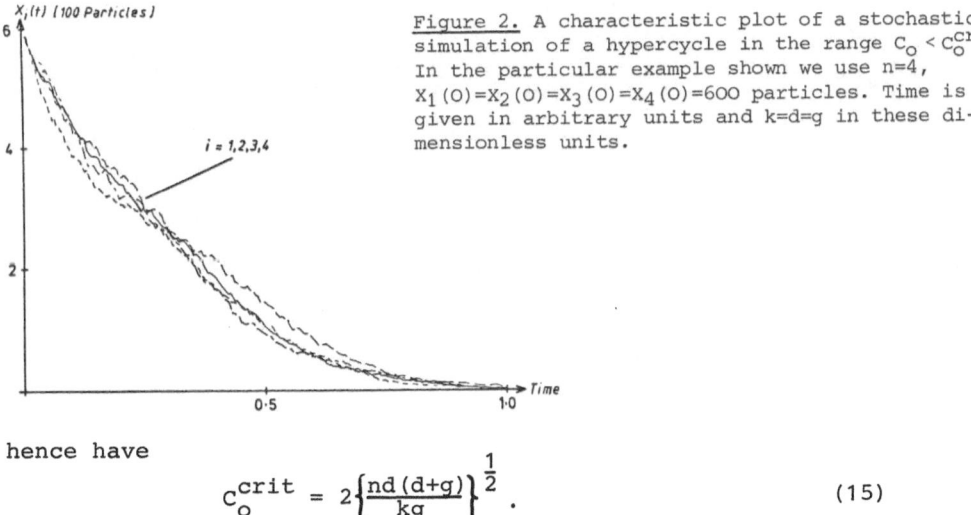

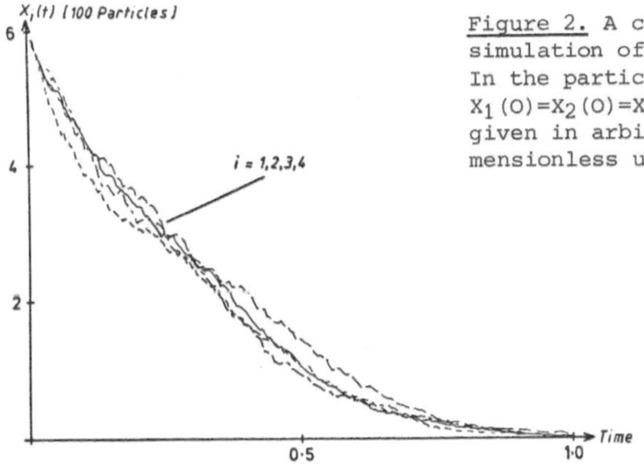

Figure 2. A characteristic plot of a stochastic simulation of a hypercycle in the range $C_o < C_o^{crit}$. In the particular example shown we use n=4, $X_1(0)=X_2(0)=X_3(0)=X_4(0)=600$ particles. Time is given in arbitrary units and k=d=g in these dimensionless units.

hence have

$$C_o^{crit} = 2\left\{\frac{nd(d+g)}{kg}\right\}^{\frac{1}{2}} . \tag{15}$$

(2) If the total concentration is above the critical limit, $C_o > C_o^{crit}$, the system usually converges to the metastable state P unless the initial value of A is around the saddle point Q or even higher. Then, depending on fluctuations the system evolves either to the zero state or to state P (Fig.3). As expected stochasticity is very important around the saddle point Q.

(3) At total concentrations around the critical value, $C_o \cong C_o^{crit}$, trajectories converge either to P_o or to P. Near the bifurcation point fluctuations become extremly large and the prediction of survival or extinction is highly uncertain (Fig.4).

(4) In general agreement with the deterministic analyses the system is metastable at high total concentrations $C_o > C_o^{crit}$ for n=3 and n=4 and converges into a limit cycle for n≥5.

(5) The oscillating systems (n ≥ 5) have a non-negligible probability to become extinct in short times. This probability decreases with increasing total concentration and it increases with increasing n. This mechanism of early extinction simply reflects the fact that some concentrations $X_i(t)$ become very small at the peaks of the oscillations and then the theorem derived in section 1 comes into operation. After one species died out by fluctuations, the others follow in sequence and the whole system converges to the zero state very fast. Oscillations, thus, bring about an additional source of instability against fluctuations which makes large hypercycles inherently unstable [12].

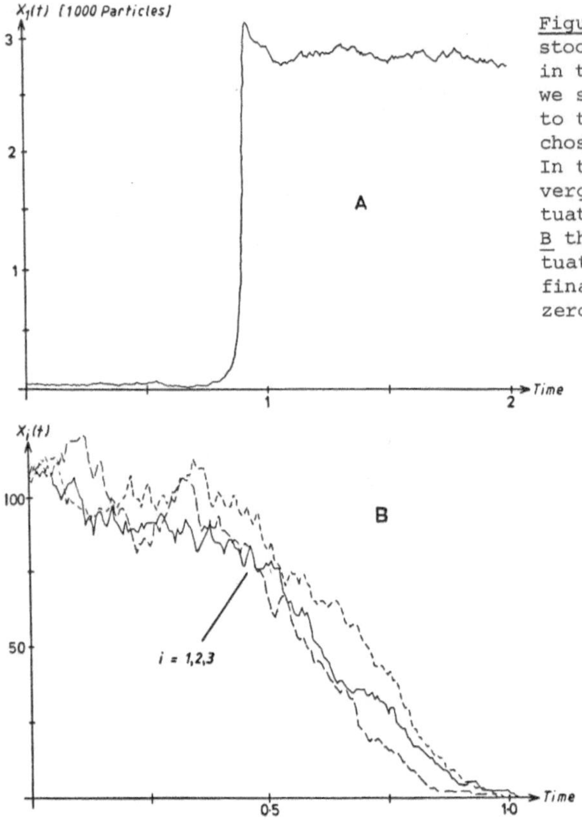

Figure 3. Two characteristic plots of stochastic simulations of hypercycles in the range $C_o > C_o^{crit}$. In both cases we start from initial conditions close to the (unstable) saddle point Q. We chose n=3 and $X_1(0)=X_2(0)=X_3(0)=108$. In the plot shown in A the system converges to the stable state P and fluctuates around it. In the plot shown in B the system is driven by initial fluctuations in the opposite direction and finally reaches the (always stable) zero state P_o.

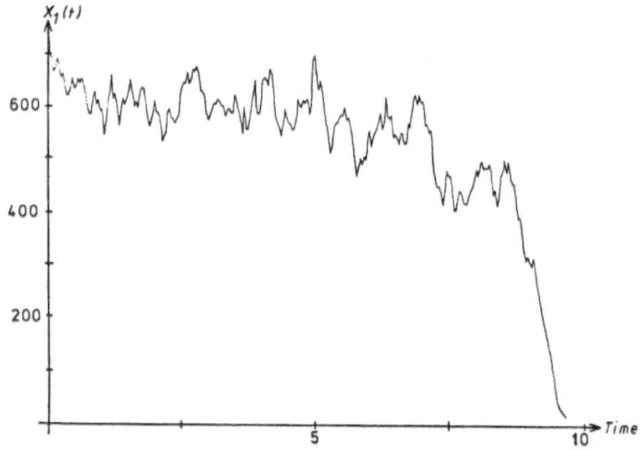

Figure 4. A characteristic plot of a stochastic simulation of a hypercycle around the critical point $C_o \simeq C_o^{crit}$. We realize particularly large fluctuations and final convergence to the zero state P_o. We chose n=3 and $X_1(0)=X_2(0)=X_3(0)=760$.

Some examples of oscillating systems with n=4 and n=5 are presented in Fig.5 and Fig.6.

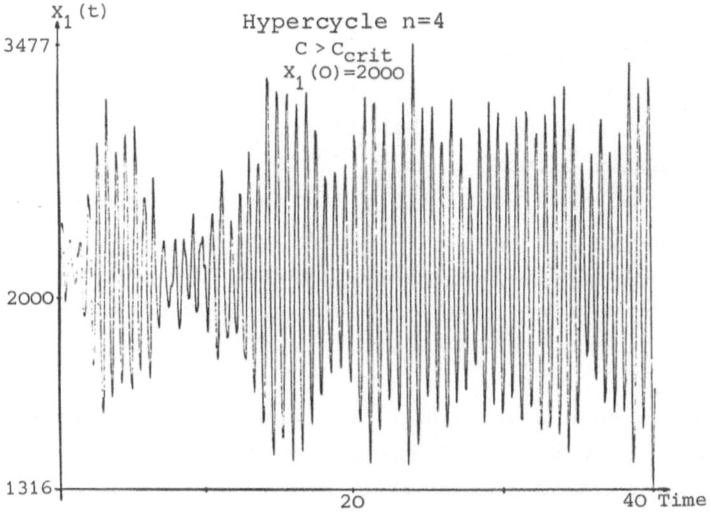

Figure 5. A characteristic plot of a stochastic simulation of a hypercycle with n=4 in the range $C_O > C_O^{crit}$. The system oscillates with an irregularly fluctuating amplitude. This behaviour is easily understood in terms of the underlying deterministic equation (2). According to equation (11a) the stationary point P is a centre in the linear approximation: $\omega_{1,3} = \pm i$. It owes its asymptotic stability to a non-linear stabilizing term [5] which becomes extremely small in the neighbourhood of the centre. Hence, the system is extremely sensitive to fluctuations in this region and oscillates with a highly variable amplitude. We chose $X_1(0)=X_2(0)=X_3(0)=X_4(0)=2000$ as initial condition

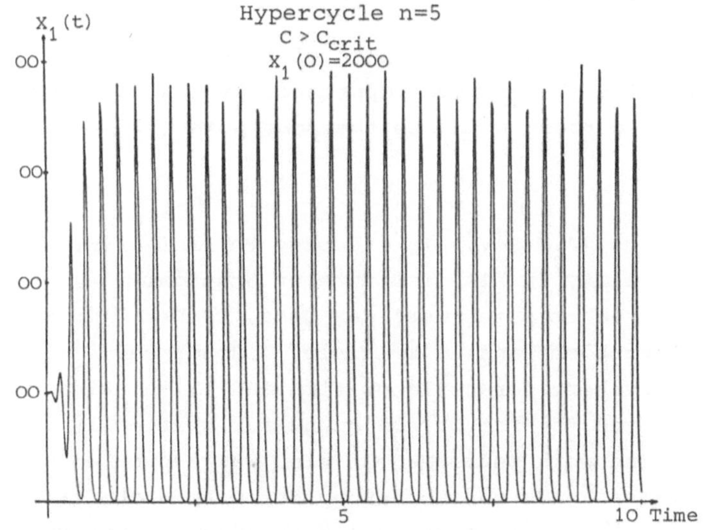

Figure 6. A characteristic plot of a stochastic simulation of a hypercycle with n=5 in the range $C_O > C_O^{crit}$. After a short initial period the system oscillates with almost regular amplitude in general agreement with the underlying deterministic equation which shows a stable limit cycle for n=5 [5,6]. Note that the individual particle numbers become extremely small at the minima. We chose $X_1(0)=X_2(0)=X_3(0)=X_4(0)=2000$ as initial conditions.

4. Concluding remarks

Qualitative analyses of the reaction system (1) corresponding to an elementary hypercycle embedded in degradation and recycling reactions indicates several sources of instability:

(1) A saddle-node bifurcation occurring at c_o^{crit}.

(2) In systems characterized by $d > 0.5(1+\sqrt{5})g$ the range of potential instability is extended since the node created at the bifurcation point is unstable and stabilizes at some higher value of c_o by means of a subcritical HOPF bifurcation. Hence, the whole range between the saddle-node bifurcation and the HOPF bifurcation is highly unstable.

(3) In systems with n=4 we have weak stability around P because the linearized deterministic equation (2) characterizes this stationary point as a centre. This result is independent of c_o and, hence, the weak stability in the neighbourhood of P remains at $c_o \gg c_o^{crit}$.

(4) In systems with $n \geq 5$ the stationary state P is always unstable. In the deterministic system it is surrounded by a stable limit cycle for $c_o \gg c_o^{crit}$. The oscillations, however, lead to extremely small values of the concentration at the minima particularly in systems with large n. Then, the hypercycle is easily truncated in finite populations by extinction of one of its members and then the system converges inevitably towards the zero state P_o.

Stochastic simulations undertaken for examples of these various critical situations verified the extreme importance of fluctuations in regions of instability corresponding to the four points listed above.

Acknowledgements: One of us (A.M.R-V.) is indebted to Prof.Dr.H. Haken for the warm hospitality in his institute at Stuttgart university where most of the computer work was done, to Dr. Olmo for his advice in numerical problems and to F.Anzola, J.Bohorquez, D.Brooks, K.Gates and L.Lybbert for collaboration and stimulating discussions. We are indebted to Prof.Dr.K.Sigmund for critically reading the manuscript. The work performed in Wien was supported financially by the Austrian Fonds zur Förderung der wissenschaftlichen Forschung (Project no.4506). Technical assistence by Mag.M. Reithmaier is gratefully acknowledged.

1 J.Hofbauer, P.Schuster: "Dynamics of Linear and Nonlinear Autocatalysis and Competition", this volume.

2 M.Eigen, P.Schuster: "The Hypercycle-A Principle of Natural Self-Organization" (Springer, Berlin 1979)

3 K.Sigmund, P.Schuster: "Permanence and Univadability for Deterministic Population Models", this volume.
4 P.Schuster, K.Sigmund, R.Wolff: J.Math.Analysis and Applications 78, 88 (1980)
5 P.Schuster, K.Sigmund, R.Wolff: Bull.Math.Biol.40, 743 (1978)
6 P.Phillipson, P.Schuster: J.Chem.Phys.79,3807 (1983)
7 D.A.McQuarrie: J.Appl.Prob.4, 413 (1967)
8 C.W.Gardiner, S.Chaturvedi: J.Stat.Phys.17, 429 (1977)
9 N.G.van Kampen: "Stochastic Processes in Physics and Chemistry" (North-Holland, Amsterdam 1981)
10 D.T.Gillespie: J.Phys.Chem.81, 2340 (1977)
11 D.T.Gillespie: J.Compt.Phys.22, 403 (1976)
12 P.E.Phillipson, P.Schuster, F.Kemler: "Dynamical Machinery of a Biochemical Clock" (Bull.Math.Biol., in press)

Perturbation-Dependent Coexistence and Species Diversity in Ecosystems

M. Rejmánek

Centre of Biomathematics, Institute of Entomology, Czechoslovak Academy of Sciences, Na sádkách 702, 37005 České Budějovice, Czechoslovakia

A spatial simulation model is developed elucidating the effect of disturbance on species coexistence in competitive communities. In this context, the outstanding species diversity is interpreted in plant communities of snow avalanche areas. The crucial importance of the absolute size (spatial extent) for the persistence of perturbation-dependent ecosystems follows from the model. The implications for the strategy of biological conservation are discussed.

1. Introduction

One of the most fascinating questions in population and community ecology is how a large number of species utilizing common resources can coexist in one habitat [1], [2]. Several mechanisms have been proposed, including the microhabitat differentiation, openness to invasions, temporal partitioning, frequency dependent predation and harvesting (or rarefraction) by external disturbances.

It has recently been argued that, because of such mechanisms, few if any real biotic communities achieve a competitive equilibrium [3], [4]. In this perspective, all equilibrium models developed for solving the so-called "complexity-stability problem" in ecology, e.g. [5], [6], [7], seem to be of limited usefulness. Boreal and temperate forests disturbed by fires and windstorms [8], [9], tropical rain forests continuously influenced by tree-falls [10] and rocky intertidal communities exposed to destructive waves and floating logs [11] are examples of communities which have gradually been understood as non-equilibrium systems created and maintained by stochastic perturbations.

2. Example of an Ecological Problem: Species Diversity in Snow Avalanche Areas

The extraordinary species-richness of plant communities (colloquially called "gardens") situated on east-facing slopes of corries in the Sudeten Mountains (Krkonoše Mts., Hrubý Jeseník Mts., Kralický Sněžník) has occupied many botanists for more than 140 years. The general framework for the solution of this problem was provided by JENÍK [12], [13],[14]. According to his "Theory of Anemo-Orographical

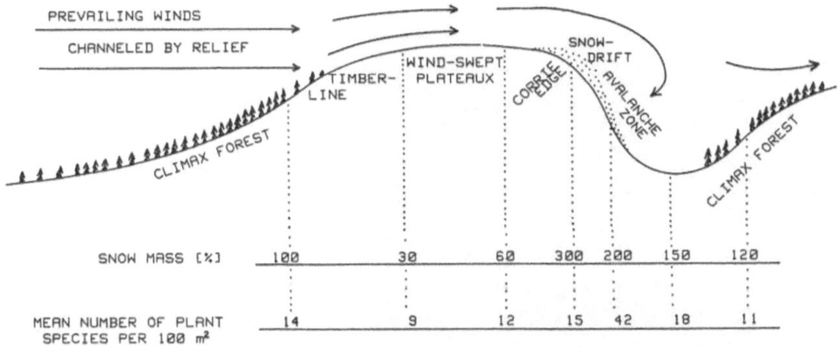

Fig. 1. The relationships between prevailing winds, geomorphology and vegetation along a typical Anemo-Orographical System. The distribution of snow mass is simplified from [14]. The mean numbers of vascular plant species are based on the author's results. The arrows mark wind directions and eddies

Systems", the interplay between relief and prevailing winds is responsible for highly uneven distribution of snow during winter. The prevailing west winds are chanelled by deeply incised valleys in the Sudeten Mountains upto wind-swept plateaux and the snow carried by the winds is deposited along the corrie edges and on east-facing leeward slopes (see Fig. 1). In long-term run, there is a feedback between snow cover and relief; the corrie edges are protected by long-lasting snow patches while the slopes situated below them are eroded by full-depth snow avalanches and landslides.

On the leeward slopes of the corries, the plant communities are not only protected by a thick snow cover during most of the winter but also regularly influenced by snow avalanches (Fig. 2). Allochthonous soil particles and plant propagules are also deposited by the winds in these leeward parts of the Anemo-Orographical Systems. Their resulting increased soil fertility and accesibility to species coming from other areas together with suitable microclimatic and hydrological conditions, represent the candidate factors explaining the species richness of these sites. However, regular snow avalanches appear to be the most powerful environmental factor of all.

The phenomenon of the Anemo-Orographical Systems was partly recognized by BILLINGS [15] in American mountains; he uses the term "mesotopographic gradient" in this connection. Likewise, VOGL [16], stressed the importance of periodic snow avalanches for the preservation of species-rich non-forest communities.

During the postglacial era, avalanches have played an important role in the succession of plant communities. Not even in warmer and

Fig. 2. The snow avalanche area in the leeward space of the Mumlava Anemo-Orographical System, western Krkonoše Mts.

wetter periods were the leeward slopes of the Sudeten cirques colonized by continuous dense forest [13]. This has saved numerous herbaceous species from extinction. The contemporary species-richness of avalanche slopes sharply contrasts with the species poverty of spruce forests at the same altitudes (see Fig. 1).

Although the importance of forest disturbance by avalanches for the maintenance of species diversity has been elucidated in many aspects, the mechanisms of coexistence of so many herbaceous species still largely remains an open question. In other words: there is still a link missing between disturbance and diversity. The following model is able to mimic some processes leading to the maintenance of high species diversity in snow avalanche areas or in similar perturbation-dependent ecosystems.

3. Model Development

There is now convincing evidence, from a variety of ecological systems that reduction of biomass by different perturbations (harvesting) can promote the coexistence within a set of competing species, some of which excludes one another without the perturbations. But the first attempts at incorporating indiscriminate harvesting into the Lotka-Volterra framework of population theory have been frustrating. Although the possibility of harvesting-mediated coexistence is easily shown [2], [17], the coefficients must be very preciesely balanced in the model. For example, in the system of two competing species

$$\frac{dN_1}{dt} = r_1N_1(K_1 - N_1 - \alpha_{12}N_2)/K_1 - hN_1$$
$$\frac{dN_2}{dt} = r_2N_2(K_2 - N_2 - \alpha_{21}N_1)/K_2 - hN_2, \qquad (1)$$

where N_1 and N_2 are the densities of species 1 and 2, r_1 and r_2 are their growth rates, K_1 and K_2 are the respective carrying capacities, α_{12} and α_{21} are the competition coefficients and h is the common harvest rate, species 2 is always excluded if $h = 0$ (the classical Lotka-Volterra system), $K_1 > \alpha_{12}K_2$ and $K_2 < \alpha_{21}K_1$. But the two species can coexist indefinitely if

$$\alpha_{12}\alpha_{21} < 1, \quad r_1 < r_2, \quad \text{and} \qquad (2)$$
$$(K_1 - \alpha_{12}K_2)/(K_1/r_1 - \alpha_{12}K_2/r_2) > h > (\alpha_{21}K_1 - K_2)/(\alpha_{21}K_1/r_1 - K_2/r_2).$$

It is clear that such delicate conditions as (2) are hardly maintained in real ecosystems. Similarly, the discrete and indiscriminate harvesting can only prolong but not maintain the species coexistence in real multispecies competition systems [18]. A spatio-temporal distribution of harvesting and species dispersion seems necessary for modelling perturbation-dependent diversity in real communities [19], [20].

Any model concerning the plant communities of snow avalanche areas should reflect at least four properties of ecological reality: interspecific competition, perturbation, an at least one-dimensional space and space-dependent dispersal. Perturbation is considered here as stochastic discrete harvesting in randomly chosen continuous parts of an area occupied by a community. In reality, this harvesting is performed by both full-depth (Bodenlawinen) and surface avalanches (Oberlawinen) [21]. Surface avalanches represent selective and partial harvesting: only trees are partly damaged along the avalanche path. Full-depth avalanches provide a heavy and indiscriminate harvesting; but they generally run down in narrower stripes than the surface avalanches. None of the published models can be simply adapted to the ecological requirements formulated above even if [18], [19] and [20] satisfy a number of them.

In a manageable model but without the loss of generality in its qualitative behaviour, the modelled community is reduced to three components obeying competitive hierarchy, i.e., a transitive relationship: (A) dicotyledonous herbs of the family Asteraceae (e.g. Crepis conyzifolia, Gnaphalium norvegicum, Hieracium prenanthoides, Hypochoeris uniflora), (B) grasses of the genus Calamagrostis (C. villosa, C. arundinacea) and (C) spruce (Picea alba). These three "species" represent the r-K selection continuum (see [23]) from the best colonizers but poor competitors to the better competitors but poor colonizers.

The dynamics of the system may be described by the following equations incorporating linear dispersion over a one-dimensional space $x \in \langle P;\ Q\rangle$

$$\left.\begin{aligned}
\frac{\partial N_1}{\partial t} &= D_1 \frac{\partial^2 N_1}{\partial x^2} + N_1 r_1 (K_1 - N_1 - \alpha_{12} N_2 - \alpha_{13} N_3)/K_1 \\
\frac{\partial N_2}{\partial t} &= D_2 \frac{\partial^2 N_2}{\partial x^2} + N_2 r_2 (K_2 - N_2 - \alpha_{21} N_1 - \alpha_{23} N_3)/K_2 \\
\frac{\partial N_3}{\partial t} &= D_3 \frac{\partial^2 N_3}{\partial x^2} + N_3 r_3 (K_3 - N_3 - \alpha_{31} N_1 - \alpha_{32} N_2)/K_3,
\end{aligned}\right\} \quad (3)$$

where $N_1 = N_1(t,x)$, $N_2 = N_2(t,x)$ and $N_3 = N_3(t,x)$ represent the cover percentages (horizontal projection) of the components A, B and C in time t and at point x; D_1, D_2 and D_3 are the respective dispersion (diffusion) coefficients. The equations are always assumed to satisfy the boundary conditions: $(\partial N_i/\partial x)_+(t,P) = (\partial N_i/\partial x)_-(t,Q) = 0$, $i = 1,2,3$.

The behaviour of this system in the absence of perturbations, with the parameter values $r_1 = 2$, $r_2 = 1.5$, $r_3 = 1.2$, $D_1 = 5$, $D_2 = 0.5$, $D_3 = 0.2$, $K_1 = K_2 = K_3 = 100\ \%$, $\alpha_{12} = 1.3$, $\alpha_{13} = 1.8$, $\alpha_{21} = 0.3$, $\alpha_{23} = 1.3$, $\alpha_{31} = 0.001$, $\alpha_{32} = 0.5$ and the intial cover values $N_1^0(x) = 24\ \%$, $N_2^0(x) = 19\ \%$ and $N_3^0(x) = 15\ \%$ for each $x \in \langle P;Q\rangle$ can be seen from Fig. 3. The values of the competition and diffusion coefficients have been chosen with regard to the bionomy of each component and to the competition coefficients estimated from models of vegetation succession based on Lotka-Volterra equations [24]. At these parameter values, the component C will always win in the competition in (3) and the component B will always win in (3) in the absence of component C, i.e., if $N_3(x) = 0$ for each $x \in \langle P;Q\rangle$. The components involved then satisfy the competitive hierarchy: $C > B > A$.

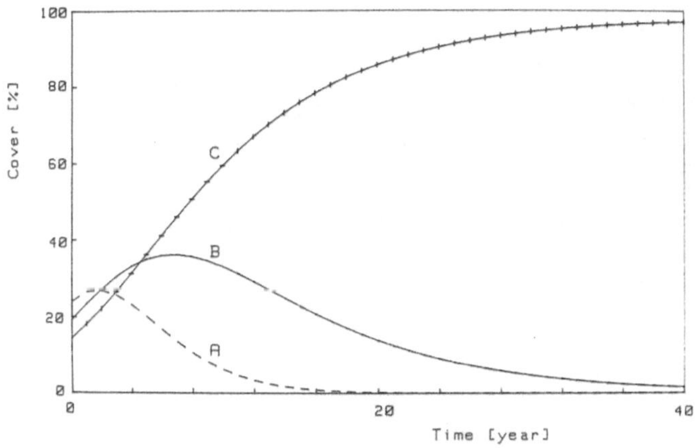

Fig. 3. Dynamics of the system (3) in the absence of perturbations. A - herbs of the family Asteraceae, B - grasses of the genus Calamagrostis, C - spruce (Picea alba)

We can now introduce spatially limited perturbations, i.e., discrete harvesting by avalanches in the system (3). We assume that two types of perturbations (avalanches) occur with defined independent mean frequencies in time and defined extensions in randomly chosen positions on the interval $\langle P;Q \rangle$ understood as a conture crossing an avalanche slope: (a) surface avalanches with the extension $(Q - P)/4$ and harvest fraction 0.5 with respect to the component C only (this means that C is reduced to 50 % of the cover achieved at each point of this interval) and (b) full-depth avalanches with the extension $(Q - P)/10$ and harvest fraction 1.0 (total harvest) with respect to all three components. In order to keep the mean frequency of perturbations equal at each point $x \in \langle P;Q \rangle$, the harvested intervals $(Q - P)/4$ and $(Q - P)/10$ are randomly distributed on the circular line obtained by connecting the two ends of the original straight line, i.e., by identifying the two end points. The resulting "species" diversity at time t is calculated according to the Shannon-Wiener information measure

$$H' = - \sum_{i=1}^{3} (\bar{N}_i/N)\ln(\bar{N}_i/N), \qquad (4)$$

where $\bar{N}_i$ is the mean value of $N_i(t,x)$ over the interval $\langle P;Q \rangle$ at time t and $N = \bar{N}_1 + \bar{N}_2 + \bar{N}_3$.

A computer program was written in BASIC HP 9845 to implement the model.

4. Results

Figure 4 shows the coexistence of the three plant community components mediated by 0.2 year^{-1} mean frequency of b-type perturbations (avalanches) in system(3). All parameters and initial conditions are the same as in Fig. 3. The development of each component during the same time interval is shown in Fig. 5 where the space interval ⟨P;Q⟩, P = 0, Q = 40, is displayed as well.

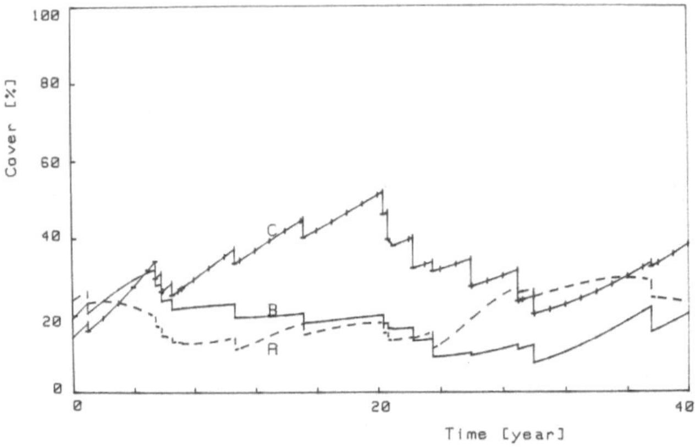

Fig. 4. Dynamics of the system (3) with perturbations: mean frequency of full-depth avalanches = 0.2 per year. Such conditions guarantee a long-term coexistence of all three components. Cover of each component is calculated as the mean value of $N_i(t,x)$ over the interval ⟨P;Q⟩ at time t

The effect of different combinations of mean frequency of surface and full-depth avalanches on mean long-term diversity of the modelled community is summarized in Fig. 6. This figure is based on 28 different perturbation régimes, each simulated for the time interval of 640 years. The rather narrow region of coexistence of the three components corresponds roughly to the mean diversity higher than 0.8. The results presented confirm a general hypothesis frequently found in the literature [3], [25], [26], namely that an intermediate frequency of disturbance will produce the highest species diversity if the diversity is a measure of both species abundance and number. A large region of coexistence of the two non-forest components is in agreement with the real situations in avalanche zones.

The result of the restriction of the total community space to 1/4 ,i.e., reduction of the interval ⟨P;Q⟩ to ⟨P';Q'⟩, keeping all other conditions of the system unchanged, is seen in Fig. 7. The

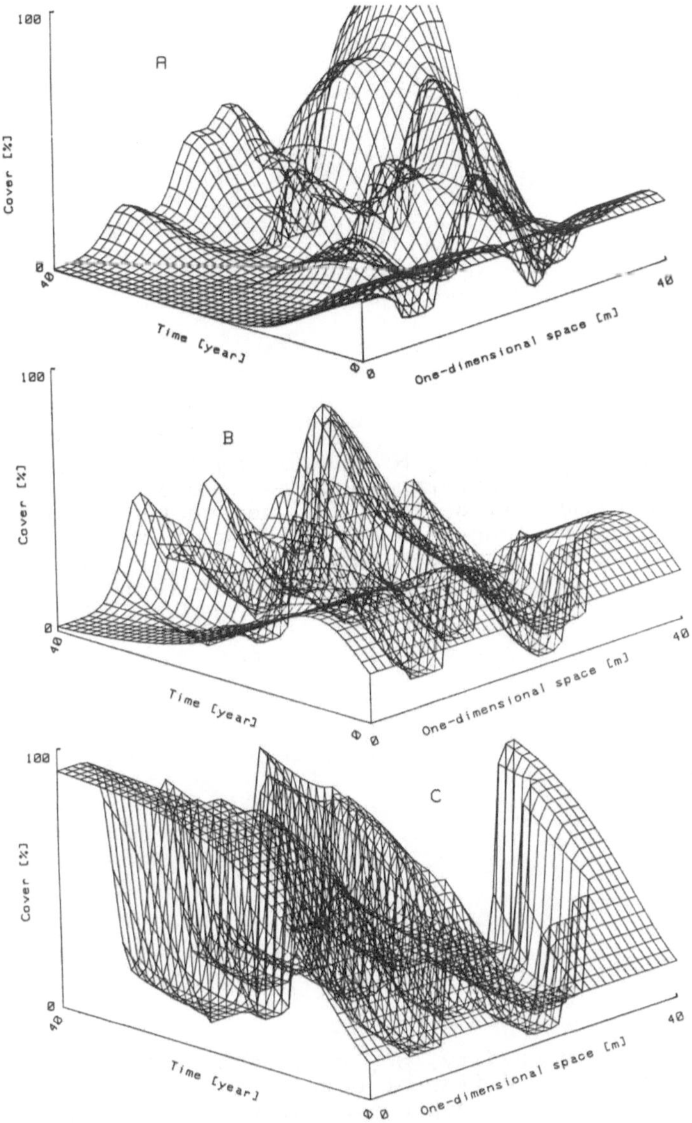

Fig. 5. Spatio-temporal dynamics of the system shown in Fig. 4. The different responses of each component to the perturbations as well as the competitive relations between the components can be recognized here

long-term maintenance of all three components is now impossible and the regions of two-component coexistence are much narrower than in the previous system, which is four times as large.

Figures 6 and 7 serve as an illustration of the crucial importance of the perturbation-dependent ecosystem size for ecosystem persistence. In a different theoretical framework, a similar conclusion was made by CROWLEY [27].

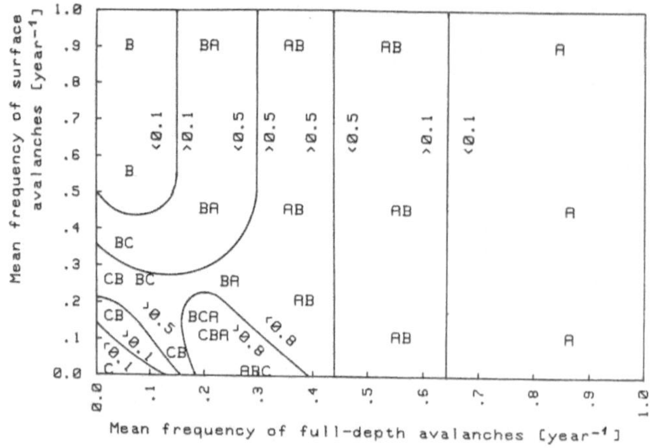

Fig. 6. Isopleths of the mean diversity, H', resulting from the long-term dynamics of (3) perturbed by different combinations of mean frequencies of both surface and full-depth avalanches on the interval ⟨P;Q⟩, P = 0, Q = 40. Regions of long-term coexistence of different combinations of A, B and C components are indicated

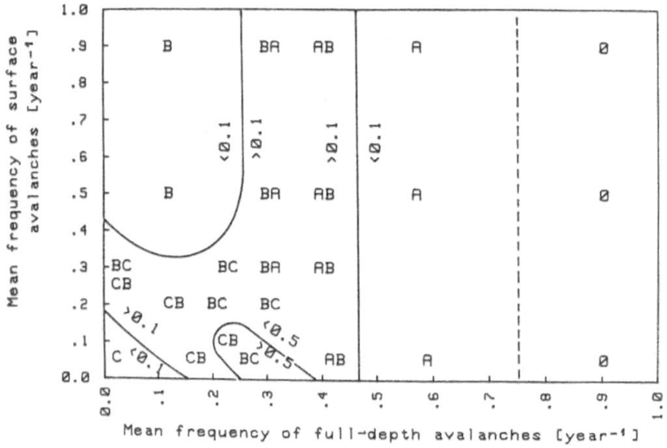

Fig. 7. Isopleths of the mean diversity, H', resulting from the long-term dynamics of (3) perturbed by different combinations of mean frequencies of both surface and full-depth avalanches on the reduced interval⟨P';Q'⟩, P' = 15, Q' = 25. Extensions of both types of avalanches are the same as in Fig. 6

Several other results have been obtained by means of the model. When larger disturbances are introduced (e.g., (Q - P)/5 for full-depth avalanches), the lower mean frequency (e.g., 0.05 year^{-1}) is sufficient for the coexistence of all three components (in the absence of surface avalanche). However, the corresponding diversity (4) is generally less than 0.8 owing to the dominance of component A (r-strategist). This result is in agreement with the prediction given by [28]. An ostencibly contraintuitive outcome has resulted from the introduction of the environmental gradient in the model (growth rates, now dependent on x, $r_i = r_i(x)$, differ in their maxima on $x \in \langle P;Q \rangle$ or $x \in \langle P'; Q' \rangle$). While in the larger system, $\langle P; Q \rangle$, the introduction of the environmental gradient supports diversity and persistence, it has an opposite effect on the interval $\langle P';Q' \rangle$ in the reduced system. This result will be discussed in more details elsewhere.

Finally, there seem to exist some relations between complexity and persistence (the ability of a system to maintain species abundances within acceptable ranges) of the analysed perturbation-dependent system, which resemble the results of the stability analysis in equilibrium linear multispecies systems [5]. Persistence generally decreases in the present study as the number of species increases, and as both the strength and number of inter-species connections increase. But this relation can be compensate by spatial extension of the system. In all simulations performed, the persistence has been positively correlated with diversity, H'. But there is no direct causal relationship here.

5. Discussion

The inclusion of space and perturbations into a three-component competition model, even in a crude way, results in a behaviour which is significantly different from that of non-spatial equilibrium models. A number of improvements of the model could be made (introduction of more realistic modes of dispersal, age-dependent competition coefficients, space-dependent immigration, discrete environmental gradients, etc.) but the basic qualitative relations elucidating the causal connection between perturbance and diversity will surely be maintained or even fortified.

The dependence of the persistence of perturbation-dependent ecosystems on their spatial extension is to be interpreted in terms of nature conservation. In concordance with the theory of island biogeography [29], [30], the persistence of many natural ecosystems in small preserves often becomes extremely expensive or impossible. For example, the partial spatial restriction of the snow avalanche zone by the introduction of some avalanche-resistant shrubs (e.g., Pinus mugho) will finally result in the extinction of some (rare) species, even if all species are present during the first, say 10 or 20 years. Truly natural ecosystems can only be preserved on scales large enough to ensure their persistence.

Reformulation of the "complexity-stability problem" in the framework of the non-equilibrium theory is badly needed. In fact, many difficulties arise in this field [31]. Estimation of parameters in systems that are far from an equilibrium and the definition of "normal system behaviour" represent some of them. It should be stressed that systems like (3), superimposed by temporaly and spatialy stochastic discrete harvesting, are not tractable analytically any more. Systematic numerical simulations then seem to be the only way out. However, the generality of the results obtained remains an open question.

Acknowledgements

I wish to thank Drs. J. Jeník, P. Kindlmann, J. Štursa, J. Lepš, J. Květ, I. Dostálková, and V. Křivan for many helpful discussions related to this paper.

1 G.E. Hutchinson: Am.Nat. 93, 145 (1959)
2 S.A. Levin: in The Mathematical Theory of the Dynamics of Biological Populations II (ed. R.W. Hiorns and D. Cooke), pp. 137-194 (Academic, London 1981)
3 J.H. Connell: Science 199, 1302 (1978)
4 H. Caswell: Am. Nat. 112, 127 (1978)
5 R.M. May: Stability and Complexity in Model Ecosystems (Princeton Univ. Press, Princeton 1973)
6 P. Kindlmann and M. Rejmánek: Ecol. Model. 16, 85 (1982)
7 M. Rejmánek, P. Kindlmann and J. Lepš: J. Theor. Biol. 101, 649 (1983)
8 O.L. Loucks: Am. Zool. 10, 17 (1970)
9 D. Sprungel: J. Ecol. 64, 889 (1976)
10 P.Richards and G.B. Williamson: Ecology 56, 1226 (1975)
11 S.A. Levin and R.T. Paine: Proc. Natn. Acad. Sci, U.S.A. 71, 227 (1974)
12 J. Jeník: Preslia 31, 337 (1959)
13 J. Jeník: Alpinská vegetace Krkonoš, Kralického Sněžníku a Hrubého Jeseníku (Nakl. ČSAV, Praha 1961)
14 J. Štursa, J. Jeník, J. Kubíková, M. Rejmánek and T. Sýkora: Opera Corcont. 10, 111 (1973)
15 W.D. Billings: BioScience 23, 697 (1973)
16 R.J. Vogl: in The Recovery Process in Damaged Ecosystems (ed. J. Cairns), pp. 63-94 (Ann Arbor Science, Ann Arbor 1980)
17 J.N. Kapur: Math. Biosci. 51, 175 (1980)
18 M. Huston: Am. Nat. 113, 81 (1979)
19 P. Yodzis: Competition for Space and the Structure of Ecological Communities (Springer, Berlin 1978)
20 A. Hastings: Theor. Popul. Biol. 18, 363 (1980)
21 M. Mellor: in Rockslides and Avalanches I (ed. B. Voight), pp. 753-792 (Elsevier, Amsterdam 1978)
22 J. Weiner and P.T. Conte: Ecol. Model. 13, 131 (1981)
23 E.R. Pianka: Evolutionary Ecology (Harper and Row, New York 1978)
24 J. Lepš and K. Prach: Folia Geobot. Phytotax. 16, 61 (1981)
25 R.T. Paine and R.L. Vadas: Limnol. Oceanogr. 14, 710 (1969)
26 R. Abugov: Ecology 63, 289 (1983)
27 P.H. Crowley: Bull. Math. Biol. 39, 157 (1977)
28 T.E. Miller: Am. Nat. 120, 533 (1982)
29 J.M. Diamond: Biol. Conserv. 7, 129 (1975)
30 M.E. Soulé and B.A. Wilcox, eds.: Conservation Biology (Sinauer, Sunderland 1980)
31 R. van Hulst: Vegetatio 43, 147 (1980)

Chance and Necessity in Urban Systems

P.M. Allen, M. Sanglier, and G. Engelen
Chimie-Physisque II, C.P. 231, Université Libre de Bruxelles,
Bruxelles 1050, Belgium

The conceptual difficulties of modelling human systems are briefly discussed, contrasting the new methods emerging from the theory of self-organizing systems with the descriptive methods usually used. A generic model of an urban system is described and its behaviour explored.

1. Introduction

Over recent years several different authors have begun to lay the foundations of truly dynamic models of urban phenomena (1)(2)(3)(4). These attempts are varied responses to the limited success that planners have often experienced in executing projects decided on the basis of hoped for results calculated using existing models. The reasons for this lack of success are deeply rooted in the modelling methodology, and indeed in the scientific method itself when applied to complex human systems.

For the scientific method is that of induction and intuition. We define a class of situations which are analogous (this is where intuition plays a role) and having observed certain behaviour in given circumstances, we suppose that in a seemingly similar case the behaviour will again be observed. If so, we keep our rule. If not we must try something else. We learn that what we thought similar in fact was not and we learn of essential differences.

How can we decide that two cases are similar, since this is what we must attempt to do in order to 'predict' the future evolution of a system on the basis of the behaviour of the class of systems to which it belongs. It is of course possible in the case of simple particles without internal structure to construct replicas of the system, or equivalently to perform repeatable experiments. But for a neighbourhood, a city or a region, how can we ever be sure that each case is not unique and science therefore impotent?

The question is one concerning 'description' as opposed to 'understanding', where this latter term implies some 'generic ' or 'transferable' law. Let us examine closer the problem of modelling dynamically a complex system. An equation is always about 'accounting', about a conservation between left and right hand sides, and in the case of a system changing over time, a differential equation relies on a conservation of the 'changes' in each side. The variable x changes at precisely the rate that the processes of growth, decline, aggregation and dilution impose.

$$\frac{dx}{dt} = \text{rates of(creation - destruction + arrival - exit)}$$

These mechanisms written explicitly, contain 'parameters' that must be 'calibrated' in any particular application. But there are two sources of change in this equation. The first is the disequilibrium that may exist at time t ($dx/dt \neq 0$), which will lead to a change in x over time, even when the parameters remain constant. The second is the evoution of the parameters which, even if dx/dt is initially zero, will generate a change in x. We have in one limit the evolution of a differential equation under fixed boundary conditions, in the other that of a 'solution' to the differential equation under changing boundary conditions. In general, a real system will exhibit a subtle mixture of both, but in fact most 'models' in the human sciences retain only this second mechanism of change. Systems are calibrated assuming that $dx/dt = 0$ (that meaningful simultaneous relations exist between the variables) corresponding, it is supposed, either to a maximised utility or entropy, in the circumstances. This assumes that the time scale of the dynamic equation is much more rapid than that which is of interest for the study. Typically, therefore, the dynamics of an urban model results from the exogenous change of population, of income or of economic structure,which is distributed over space by means of the model according to the structure corresponding to the 'solution' present initially. It is for this reason that these models are essentially descriptive and of fixed structure, a structure which is extrapolated into the future in a quasi-equilibrium manner.

The question of course is, firstly, whether the initial structure corresponds to a 'solution', and secondly whether this 'solution' will remain stable over time. As is now well known, complex systems involving non-linearities and feedback may possess multiple solutions, and the bifurcation of a particular solution is a common phenomenon. In order to be able to comment on the stability or instability of the structure which is present initially, it is necessary to use the dynamic equation itself to explore system behaviour around the existing solution.

In this way the system can probe the 'validity' of its own state of organization, and either retain it nevertheless (stability) or move away to some other branch of solution, some other state of organization and perhaps of complexity.

In the models which we have developed we propose to retain both sources of change in the differential equations, in order to be able to discuss the stability of our system over time, and to explore the future in a manner which includes the possibility of structural change.

A key point that arises, once we admit that a complex system may evolve both through the progressive displacement of a particular branch of solution under exogenous changes and also by jumping to a new branch, is that branches can differ qualitatively in their nature, possessing different characteristic pleasures and problems. We are faced with a fundamental non-conservation of emergent properties involved in such an evolution.

This is well illustrated by the amusing example of 'origami' which is the subject of figure (1), where an initially uninteresting piece of paper can be folded into many striking forms according to a 'tree' of evolution, where characteristic traits emerge over time. We have discussed this analogy in detail elsewhere (5), and we shall not therefore discuss it further here.

Returning to our intial reflections concerning the sometimes disappointing performance of past models in real applications, we perhaps can now see good reasons for this lack of success, in that

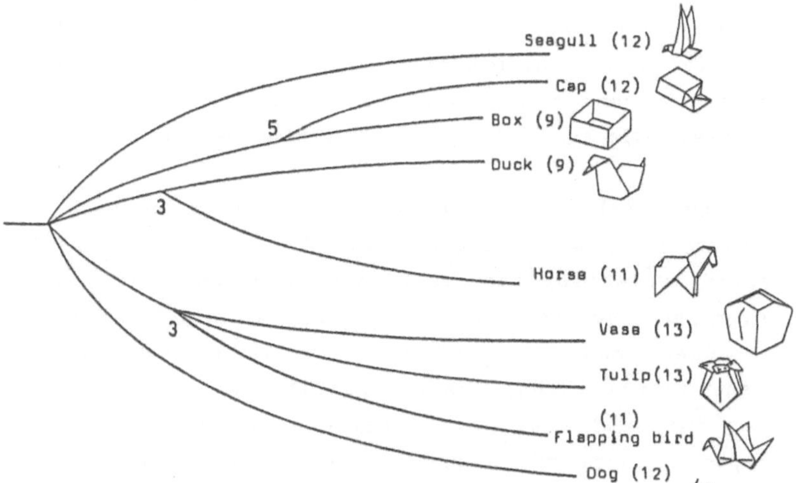

Figure 1. A bifurcating tree of Origami forms

they are essentially statistically fitted descriptions (calibration) which are extrapolated into the future under various scenarious for exogenous events.

The dynamic models which we propose pose the invariance necessary to any attempt at modelization, not at the level of urban structures, but instead at that of actors preferences which are fashioned by their professional and private roles, and by the beliefs and perceptions which condition them. It is through successive spatial instabilities that complex behaviour evolves in a system of interacting actors, each of which may have very simple preferences and criteria. It is in the occurrence of these instabilities, in their probabile frequency and type that the stochastic aspects of the world play a vital role as we shall discuss. Also we shall show how the simple 'generic' preferences of actors can give rise to quite different urban forms, where the actors exhibit different behaviour. In this way our approach attempts to go beyond the 'particular', which must inevitably be only descriptive, and to attain some degree of general applicability.

2. An Urban Model

Over recent years several examples of these ideas have been studied in fields as varied as ecology (6), regional geography (7), oil exploration (8), economic markets (8) and, the case we shall develop here, intra-urban evolution (10). This has already been described in some detail elsewhere and so we shall simply give a brief summary here. It is based on the dynamic interaction of the different urban actors, in their constant cooperative and competitive efforts to survive and to function successfully, which lead to behaviour which expresses not only the 'needs' (utility) of each actor, but also his limited means of action and of perception.

Each variable of the model represents the density at a particular location of a certain type of actor, each type having its own locational criteria reflecting the 'role' the actor has been assigned (or believes he has been assigned) by society. The model expresses the changes induced in these densities by the mutual interaction of the actors through their conflicting or complementary criteria. In this model we have considered 7 variables consisting of 5 types of employment and 2 types of resident. Other disaggregations are of course possible, and may be found to be more suitable for different circumstances, the choice depends on the study requirements, and more fundamentally on the existence of distinctive locational criteria which may characterize an actor.

The 5 types of employment are:

- heavy industry (Steel, Automobile,Paper...)
- light industry (electronics,)
- exporting tertiary(business, finance .....)
- infrequent specialized tertiary
- elementary tertiary

The schema of interaction is shown in figure (2), and each of the 5 variables obeys an equation of the form:

$$\frac{dS_J^L}{dt} = \epsilon^L S_J^L \left(1 - \frac{S_J^L}{M_J^L}\right) \qquad (1)$$

where M_J^L is the 'potential market' for L if available at J in quantity S_J^L.

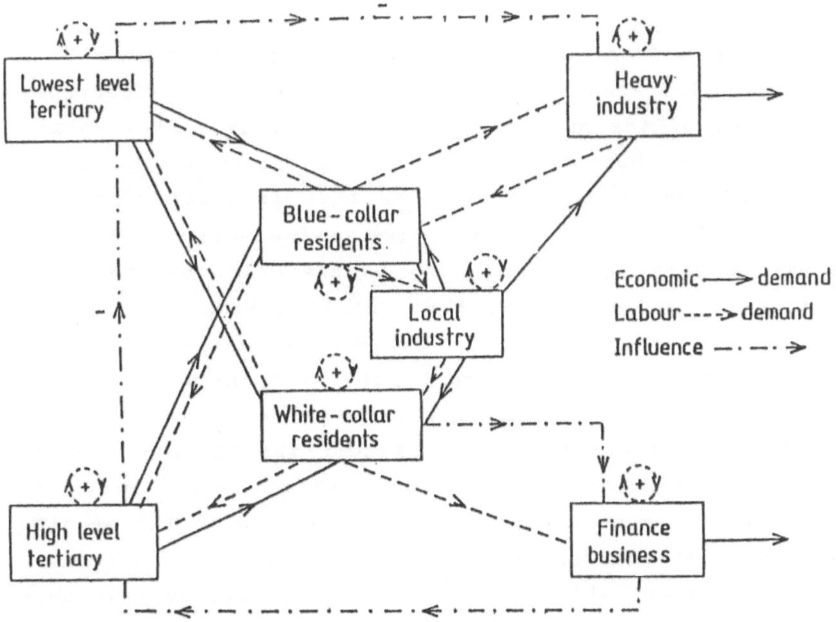

Figure 2. The interaction scheme for our intra-urban model

$$M_J^L = D_{ext}^L \frac{A_{Jext}}{\sum A_{J'ext} + A_{ext2}} + \sum_{J'} \left[\sum_{L'=1,5} B^{L'L} \frac{S_{J'}^{L'} A_{JJ'}^L}{\sum_{J''} A_{J''J'}^L + A_{extJ'}^L} + \sum_{L'=6,7} B^{L'L} \frac{X_{J'}^{L'} A_J^L}{\sum_{J''} A_{J''J'}^L + A_{extJ'}^L} \right]$$

External demand falling on city — Demand from city employment — Demand from urban consumers

The 2 types of population are:

- blue collar
- white collar

The equation obeyed by each of these is,

$$\frac{dX_J^L}{dt} = \eta^L X_J^L \left(1 - \frac{X_J^L}{M_J^L}\right) \qquad \text{where}$$

$$M_J^L = \frac{X_{ext}^L \, R_{jext}}{\sum_{J'} R_{J'ext} + R_{ext}{}^2} + \sum_{J'} \sum_{L'=1,5} \zeta^{L'L} \frac{S_{J'}^{L'} \, R_{JJ'}}{\sum_{J''} R_{J''J'} + R_{extJ'}}$$

The criteria of location of these different actors are those which are usually supposed, taking into account their particular needs. In the scheme of figure (2), we see how the external demand produces jobs, the jobs attract residents, and the residents concentrate further economic demand generating more jobs, which if the tertiary centre becomes sufficiently important becomes an attractor of regional economic demand, complementing the original economic base of the city.

Our model, a basic set of 'urban mechanisms', is represented by a set of non-linear differential equations each of which describes the time evolution of the number of jobs or residents af a particular type at a given point. In a homogeneous space one possible 'solution' of such equations would be to have an equal distribution of all variables on all points. Such a 'non-city', although theoretically possible, corresponds to an unstable solution, and any fluctuations by actors around this solution will result in a 'higher pay-off', and this will drive the system to some structured distribution of actors, with varying amounts of aggregation and decentralization. However, in reality, there are two reasons for the structure of the system: the first is due to the non-linear interaction mechanisms which give rise to instabilities as mentioned above; the second is due to the spatial inhomogeneities of the terrain and of the transportation network.

In the model simulations which we shall present here, we have supposed two transportation networks on which interactions and accessibilities are calculated. One is a road network for private transportation taking into account 3 different qualities of road, the other is a set of 4 public transport networks, describing the possibilities of travel by train, bus, metro and tram. We have therefore, a dynamic land use/transportation model which permits the multiple repercussions involved in the various decisions concerning land use or transportation to be explored as the effects

are propagated, damped or amplified around our interactive scheme of figure (2).
The simulations described here have been based broadly on the evolution of a city ressembling Brussels. The global characteristics of employment and population are shown in table (1), and the spatial evolution of urban structure is shown in figures (5),(6), (7); and (8) for successive times of the simulation.

VARIABLE	T = 10	T = 20	T = 30
Total Employment	729,600	669,500	674,300
Total number of Active Residents	462,670	411,560	414,200
Coefficient of Employment	1.58	1.63	1.63
Employment Structure:			
- Industry	25%	22%	22%
- Tertiary	75%	78%	78%
Structure of commuter flows from outside the urban centre			
- Blue Collar	40%	33%	33%
- White Collar	33%	44%	44%

Table 1. Global figures generated by our model simulation, in order to explore the evolution of a city which resembles Brussels

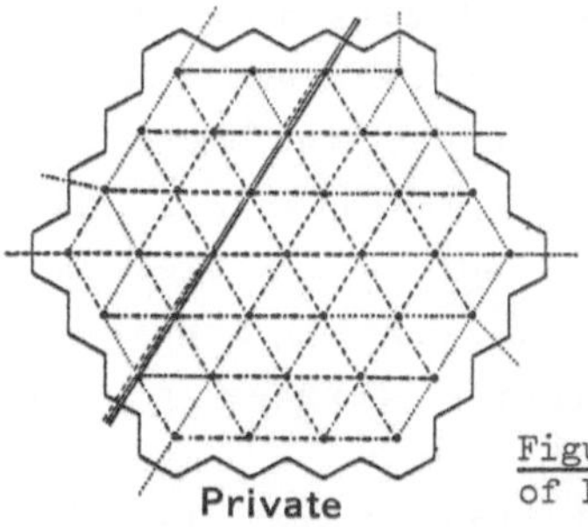

Figure 3. The road network of our city, based on that of Brussels

The simulation times of 0, 10, 20 and 30 are of course somewhat unreal but they are supposed to describe changes of urban structure which could occur over some 40 or 50 years. The initial condition of our simulation, which of course affects the structure which evolves is in fact taken from figures resulting from a previous simulation made earlier without a transportation network.

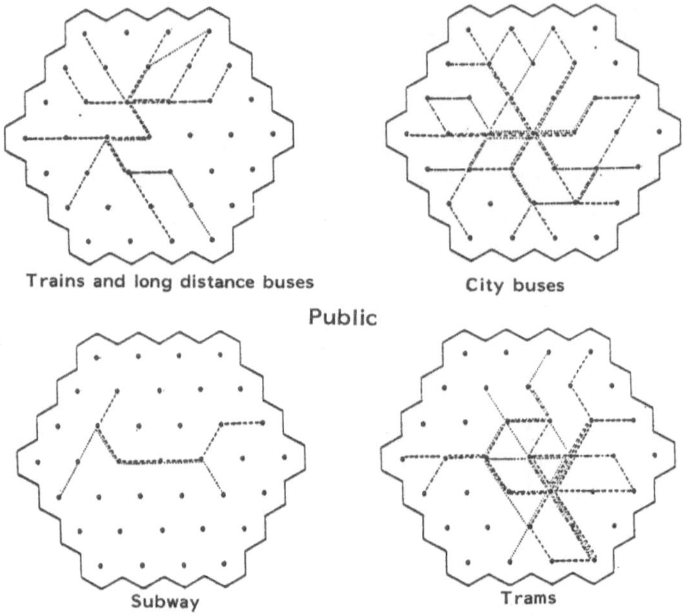

Each link of each network is characterized by a "time and money" cost. Travel behaviour depends on the relative sensitivity of an actor to these

Figure 4. The four public transport networks of our city

We see that our urban system evolves to a complex interlocked structure of mutually dependent concentrations. We have two poles of heavy industry, and a distribution of blue collar residents reflecting this. Light industry, after remaining diffuse for some time concentrates in one pole in the north east. Financial and business employment in the city centre, at around t = ,
begins to spread through the urban space. Then, it exceeds a threshold on the point adjacent to the centre and grows dramatically there, causing the decentralized locationsthoughout the city to decrease. The white collar and blue collar residents spread out (many live outside the system) according to the accessibilities of the networks, and a spatial hierarchy of shopping centres appears, serving the suburban population and encouraging further urban sprawl.

This completes our description of the evolution of our system according to the deterministic equations of our model, and starting from the particular initial conditions that we have used. In the next section we shall discuss why such a simulation would be insufficient as a basis for decision making and city planning. In reality such a system will not run its course undisturbed by external and internal disturbances. This is the point that we shall examine in the next section.

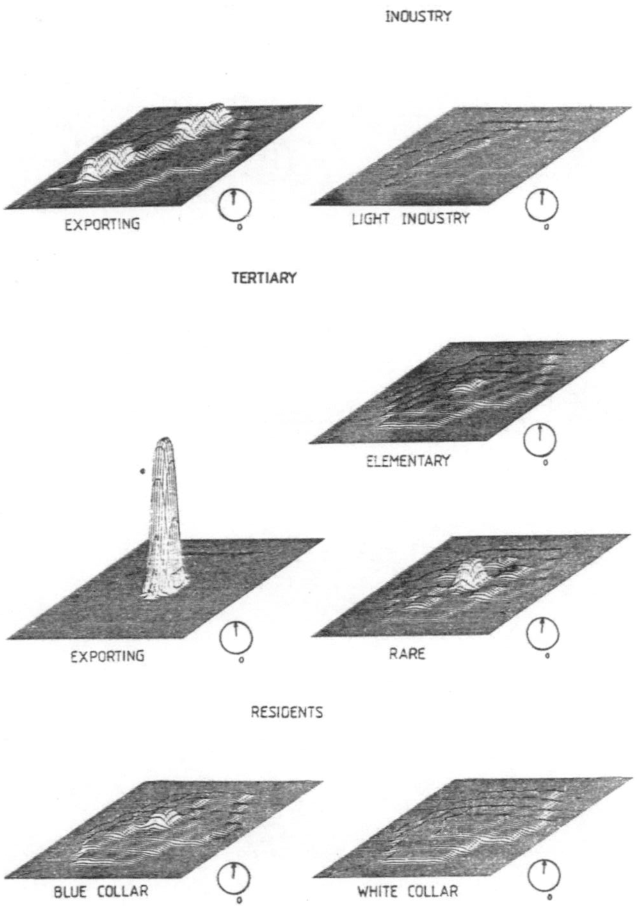

Figure 5. The initial distribution of employment and residences in our urban space

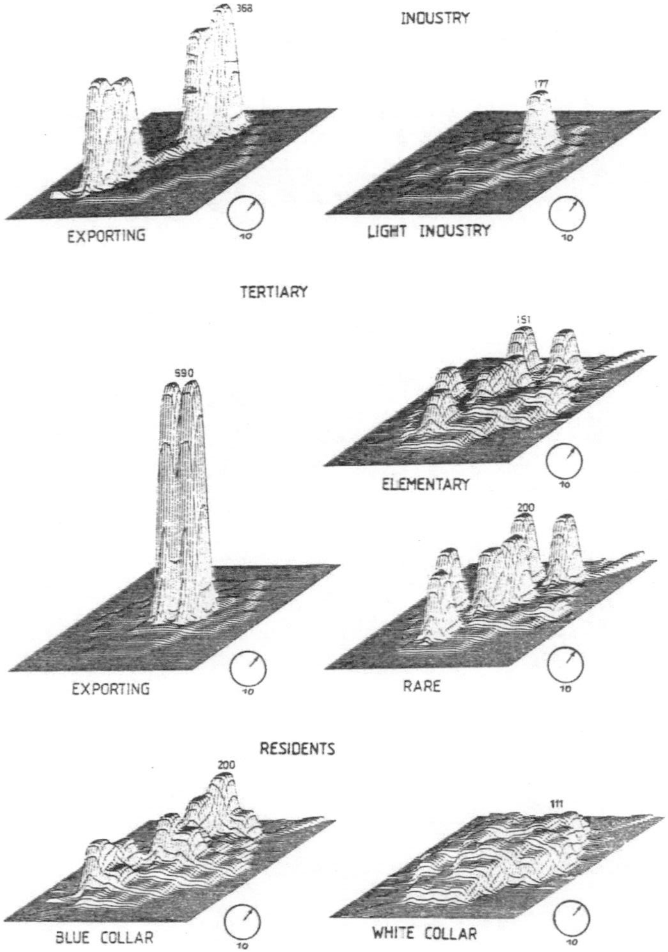

Figure 6. The distribution of employment and residences at t = 10

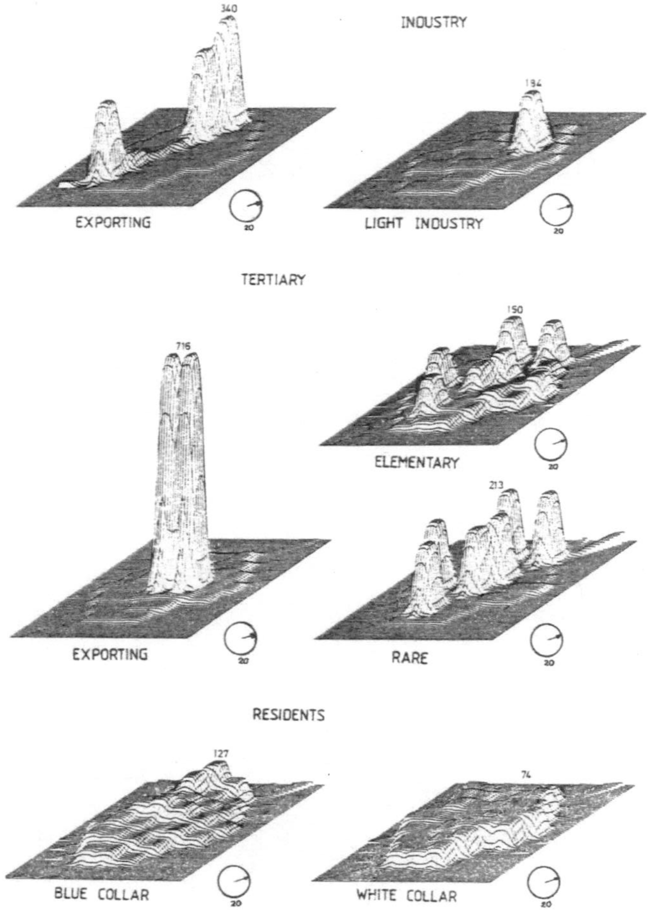

Figure 7. The distribution of employment and residences at t = 20

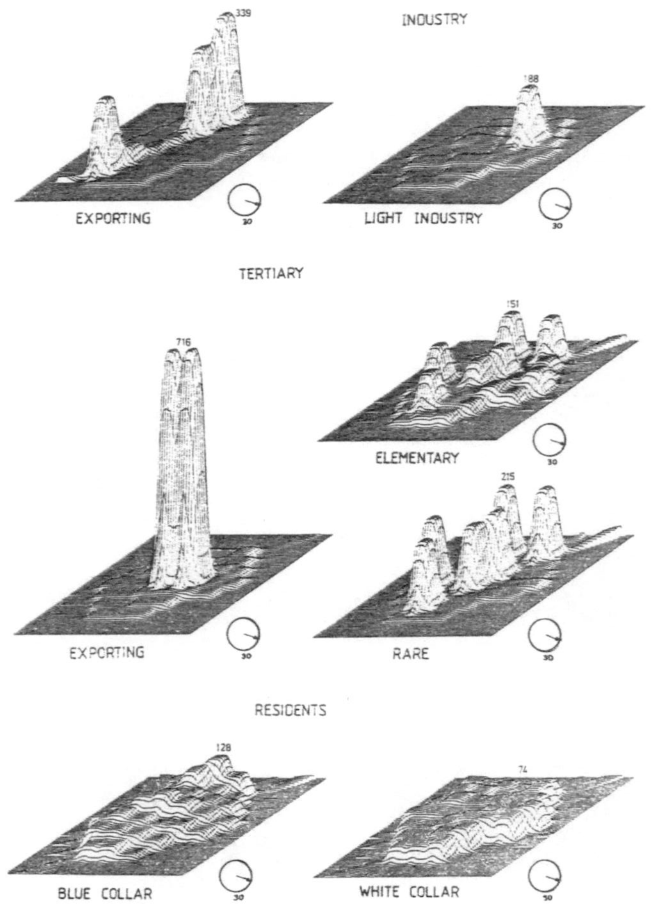

Figure 8. The distribution of employment and residences at t = 30

3. Stochastic Effects and a Reduced Description

The ideas sketched out in our first section tell us that the deterministic equations governing the average behaviour of the elements of a complex system are in fact insufficient to determine precisely the state of the system and even its nature, because of the existence of many branches of possible solution. Only the effects of factors and events <u>not</u> included in the differential equations break this ambiguity and 'decide' which branch precisely the system will really be on ! In this way an event of historical significance is one which is not contained in the average behaviour of the elements. Just as the events which are important in origami are not the mechanical properties of the paper, but the intervention of human hands which decide at times and places when it is possible, where and when to fold the paper.

A 'solution' is simply a situation in which the processes of growth and decline at each point cancel out, either constantly or perhaps over some cyclical change. With non-linear equations it is possible for the cancellation to occur for different situations as successive mechanisms play the important role in this balance. This is why multiple solutions can occur, and why, in that case the effects of elements and factors not contained in the model are of importance in discussing its possible future behaviour.

The most obvious contribution to system behaviour described by deterministic equations containing average behaviour, is the <u>non-average</u> behaviour. If our equations decree that exactly 7,623.225 residents will inhabit a certain zone at a particular time, then clearly, since the residents live in ignorance of our extremely exacting equations, they are likely as few as 7,500, or as many as 7,800 at that time , even though on average the rates of growth and migration we have used are correct.

So, in any model of a complex system we must recognize that the differential equations form a reduced description of a much 'richer' reality, and that one way of capturing its possible evolution is to admit the presence of fluctuations around the average values of the variables of the system. This constitutes a constant 'probing' of the stability of any particular spatial organization, and can precipitate structural changes at certain times. Because the precise nature of this 'non-average' behaviour is not part of our model it is reasonable to represent this by the 'least controversial' choice, that is by stochastic noise. The interplay of this with the deterministic, but non-linear, equations generates a structural evolution of the system in which different stages are separated by instabilities.

In the figure (9) we see how very small stochastic differences in our variables can be amplified and lead to other spatial distributions of the variables and to other flow patterns on the networks. In this case small gaussian errors, with mean 3%, are added to the employment variables of our reference simulation each unit of time. Heavy industry, and exporting tertiary are barely affected, their implantation being strongly influenced by factors such as accessibility which are not changed by these small fluctuations. However, as we see, light industry, and rare tertiary activities are affected and, if these simulations were continued we may well find at some point that these differences play a vital role in some further structural changes.

Another important contribution to evolution is that due to parametric fluctuations, expressing changes in the 'environment' of the system. In an urban system, for example, changing weather conditions can seriously affect traffic patterns and also the types and quantities of goods consumed, and obviously breakdowns, accidents, failures and repairs in the public and private transportation networks can all have considerable influence on urban evolution. Fluctuating economic conditions outside the city, in agriculture, in the external demands or supplies of the various products and raw materials used by the city can all play a critical role in its development, and it is important to explore any future policies or plans within the context of this 'noisy' environment. Also the long term success of an urban system may well be best assured by

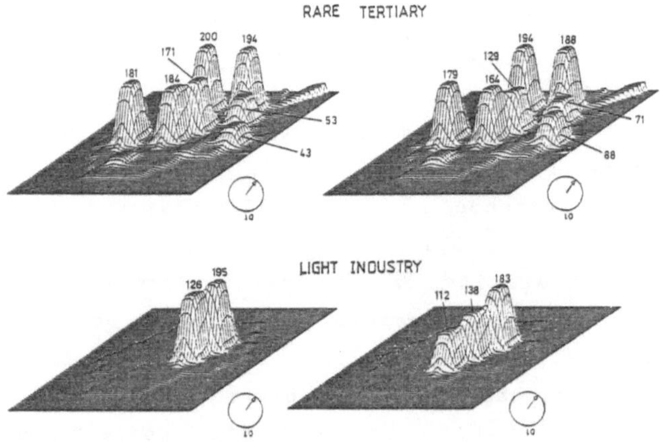

Figure 9. The effects of small (3%) fluctuations on the distribution of rare tertiary and light industry

taking measures which retain its resilience, but of course allow it to be sufficiently competitive in the short term.
Another type of effect, which has a certain underlying stochastic characteris that of 'happenstance' and of external intervention.
One particular example of 'happenstance' is shown in figure (10), where an urban centre grows from exactly the same intial conditions as before, with identical locational criteria of the actors and parameter values, except that instead of the city developing with a canal/river/railway crossing it as before, we have in its place a line of hills. The only affect of this on the model is that accessibility for the supplies of raw materials and for the export of finished products of heavy industry, instead of being favoured along this axis, is slightly worse than elsewhere.
We see that the city that evolves, figure (10), is totally different from that of our reference simulation shown in figures (5) to

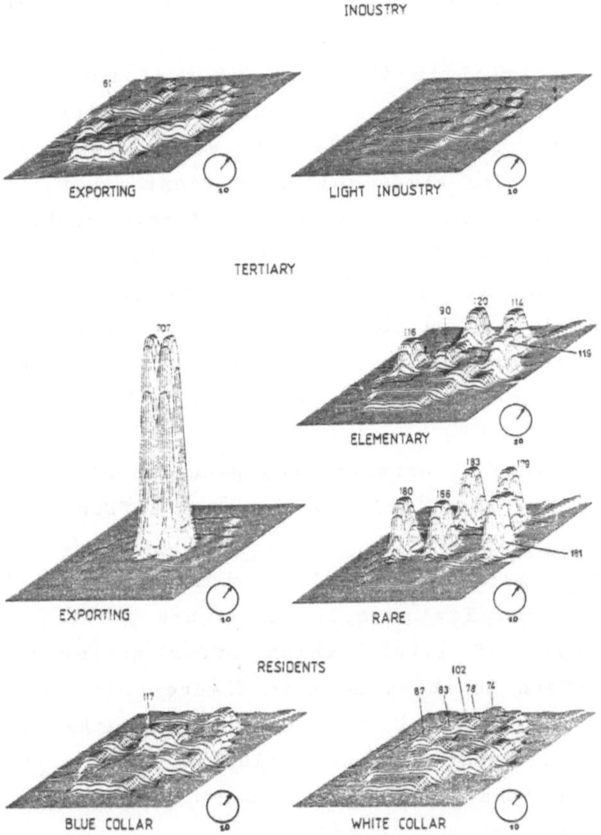

Figure 10. The distribution of employment and residences in another city, where, instead of a river there is a chain of hills

(8). Industry is dispersed throughout the rest of the city, and with it the blue collar residents. The white collar, instead of concentrating in the south-east aggregate instead along the line of hills which have replaced our 'canal'.The distances travelled to work are not the same as before, nor are the 'costs' of shopping trips or the distribution of tertiary activity. Furthermore, not only are the spatial distributions of the variables different, but also the global quantities of industrial and tertiary employment, and of white and blue collar residents are modified. This is because the absense of an axis of good accessibility for industry lowers the 'attractivity' of the whole city for this type of investment and our model takes into account this change in its global performance.

This underlines our remarks concerning the fact that global quantities are not constraints on an evolution, but on the contrary are observables which are generated by all the local events in the system.

Again, our approach is 'generic' in the sense that it should be contrasted with one based on 'observed behaviour' of a particular system. For example, a model based on this city would suppose as part of the 'utility' function of white collar residents that they 'wish' to live 'on the hills'. This means that in future explorations made with that model the 'hills' would play the role of attractor. However, as we can see from our model the interaction of the location criteria of the actors of our system can produce white collar neighbourhoods along the line of hills without any such factor appearing in the preferences of white collar residents.

Our simple, interacting locational criteria can generate many different cities. In the figures that follow we see the influence of other 'happenstances'.

A city which starts to grow along a shore line where port facilities are practicable is shown for example in figure (11). The effect of a different path of rivers which offer preferenial access to the outside markets is shown next in figure (12). Here, industry is in competition with the C.B.D. for space in the very centre of the city, producing yet another evolution, marked once again by different observed behaviour of the actors.

Having discussed these types of chance effects of the effects of interventions of various kinds. In other recent publications we have explored the evolution resulting from a whole series of pos-

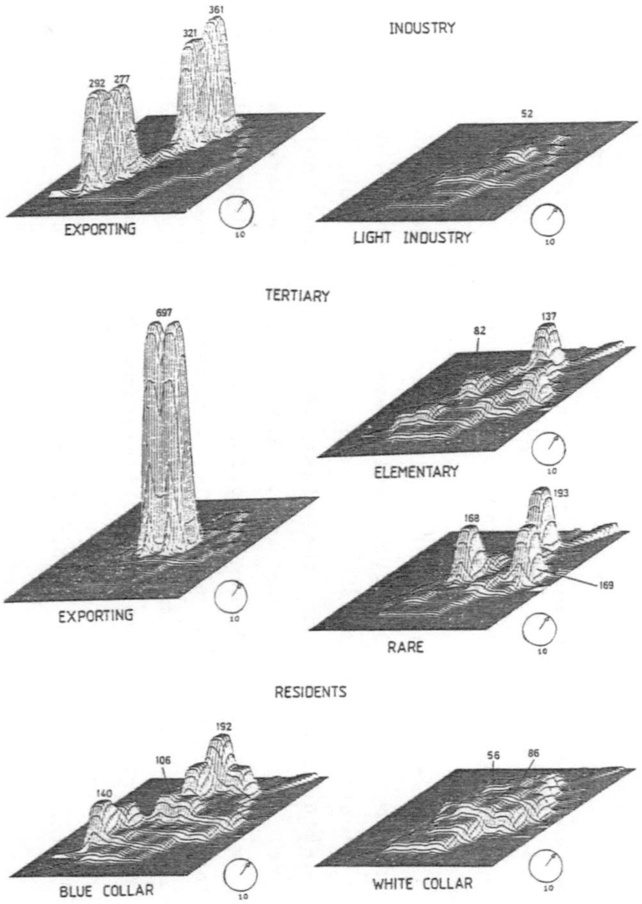

Figure 11. The distribution of employment and residences in another city which has grown up on a lake or sea shore

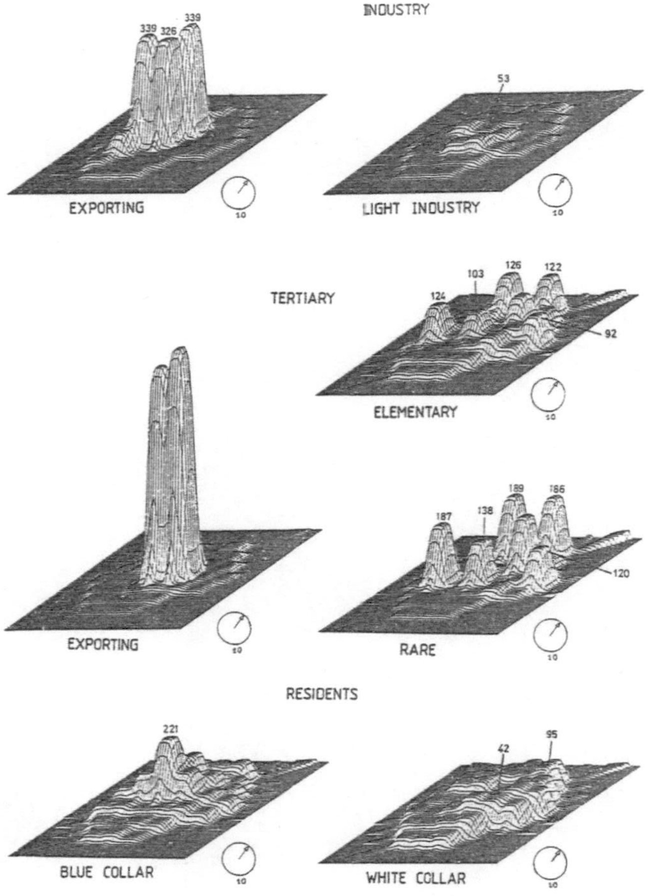

Figure 12. The distribution of employment and residences in a city which develops around a different river pattern

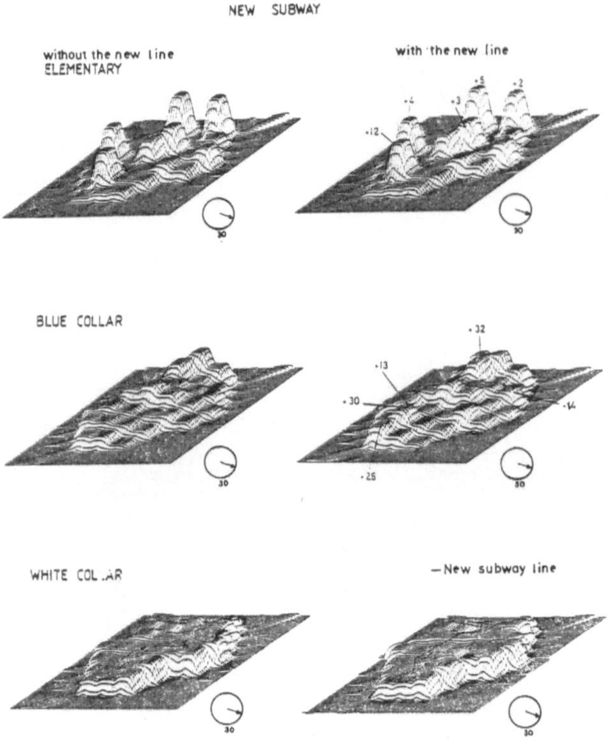

Figure 13. Some of the effects of the implantation of a new subway line between the north and the south of our theoretical city

sible decisions in different sectors. Here, we shall simply show three which illustrate some important aspects of these.
In figure (13) we see the differences that are produced if in our reference simulation a new subway line is added, with trains running of course with a certain frequency and speed. Clearly, we could produce a whole series of such results, for comparison, with different possible routes, frequencies and costs, and in this way a 'cost/benefit' estimation could be made of the various possibilities taking into account the wider context of the whole urban structure, and also the long term. Another possible case is that of the implantation of a new shopping centre. In figure (14) we show the time place and size of investment can be critical in determining its probability of success. Again a more detailed study of this particular problem could examine the advantages of different temporal strategies of profit reduction which can aid the implantation even if the resources available initially are limited.

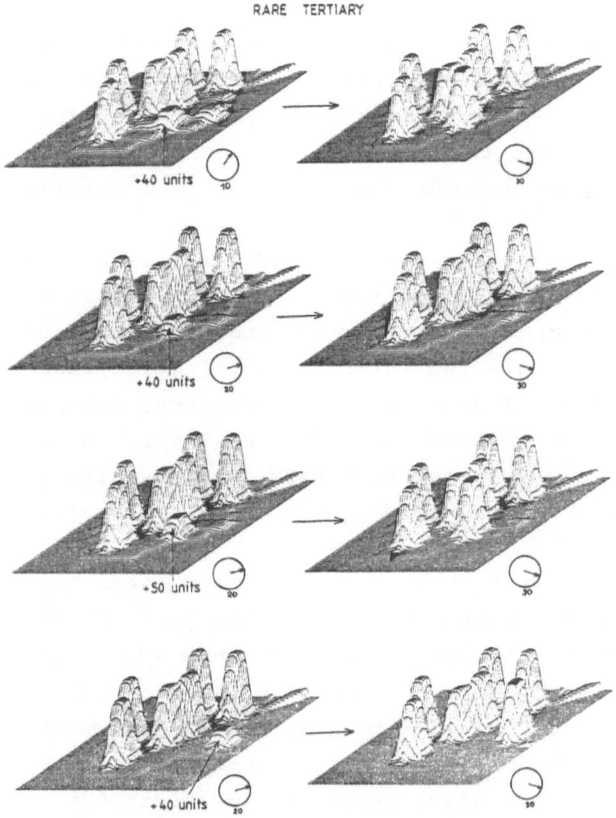

Figure 14. A series of simulations showing the importance of the size, the moment and the location for a successful investment in the retail structure

Finally, we see different possible spatial distributions of financial and business employment, which of course imply quite different distributions for the other 6 variables, each one resulting from the possible effects of the growth of telematics. The use of decentralized office computer systems leads to a reduction in the need for aggregation of firms, and as we see at a certain critical value result in the disappearance of the Central Business District - at present perhaps the most striking feature of most large cities at present. Clearly, also, daily traffic patterns are totally changed and such a shift in the spatial distributions corresponds in reality to a real revolution in city structure.

4. Conclusions

In our models of human systems based on the concepts emerging from the study of dissipative structures in physics and chemistry, we adopt a quite new approach to the question of anticipating the future, and evaluating policies. We recognize the fact that evolution results from the dialogue between the deterministic equations of change expressing the average behaviour of actors, and a whole series of 'perturbations' from outside the model, or from outside this level of description.

Instead of 'stochastic' effects being simply minor irritations for the modeller, producing a deviation from the 'mean' of his prediction, we see that in fact they drive the system from one state of organization to another - they are the vital force of evolutionary change. It is under the effect of stochastic fluctuations, arising in human systems from limited information, from 'arational' behaviour, from unusual local circumstances or simply from a desire to be different, that structural instabilities occur and new average behaviours arise.

Such changes can also be precipitated by a fluctuating environment, in which case they represent system 'adaptation' to the uncertainy and extreme possibilities of its environment. Other vital factors which a modeller may represent as a 'stochastic' or 'random' element in his simulations are the probabilities of 'external' intervention in the system, by macro-actors not included in the equations of the model which describe microdecisions.

Instead of viewing a model as 'closed' and capable of predicting the future course of events, merely reflecting a scenario of exogenous changes, in a mechanical fashion, we are led another view. Our model provides a set of dynamic relationships which can be which can be used to explore the many possible futures, and alternative structures, which could evolve as a result of various perturbations and shocks which it may experience.

This new view of the importance of stochasticity, uncertainty and adaptation in the evolutionary process suggests that in a very broad range of contexts, long term survival may be related to the presence of stochasticity in a system. Just as in biological evolution, the existence and even the rate of genetic mutation may be itself a result of that evolution, so in social and cultural systems long term survival may depend on the presence of 'arational', 'chaotic' and 'original' members of the social group probing the efficacity of possible 'other behaviour'.

Such a viewpoint stresses the dangers of short term, narrow 'optimization' procedures often used to make decisions in the economic, social and political spheres. These methods, often of great apparent computer aided sophistication, threaten society with a pro-

gramme for its own fossilization at best, or more usually with collapse. The adaptive possibilities of societies allow them to survive and to compete in the real world, and these adaptive possibilities are related to the presence of 'original thought' and this must ultimately be based on mental models which are also capable of structural reorganizations.
Perhaps, in stochasticity, we see the source of man's creative intelligence.

References

1) R.White,(1979) Geographical Analysis, 9 , 226-243
2) A.Wilson, (1981) "Catastrophe Theory and Bifurcation Analysis in Urban and Regional Geography" Croom Helm, London.
3) A.Wilson, (1982) "Evolution of Urban Spatial Structure" Review of Progress and research problems using S.I.A.Models. Bulletin of the Inst.Math. and its applications, 18, 90-100.
4) P.M.Allen and M.Sanglier, "A Dynamic Model of a Central Place System", Geographical Analysis, (1979) Volume 11, No.3.
5) P.M.Allen, "Evolution and Design in Human Systems", Environment and Planning B.Vol.9 (1982)
6) P.M.Allen, "The Evolutionary Paradigm of Dissipative Structures", AAAS Selected Symposium, Editor E.Jantsch. Westview Press, Boulder, Colorado. (1980)
7) P.M.Allen and M.Sanglier, "Urban Evolution, Self-Organization and Decision Making", Environment and Planning A, (1981), Vol.13.
8) B.Braunschweig, Proceedings of the 7th International Systems Dynamics Conference, Brussels, 1982, Special Edition, Large Scale Systems, North Holland, 1983.
9) P.M.Allen,M.Sanglier, G.Engelen and F.Boon, Proceedings of the 7th International Systems Dynamics Conference, Brussels, 1982, Special Edition, Large Scale Systems, North Holland, 1983.
10) P.M.Allen, M.Sanglier, G.Engelen and F.Boon. "Evolutionary Spatial Models of Urban and Regional systems", Systemi Urbani.

Random Phenomena in Nonlinear Systems in Connection with the Volterra Approach

K.F. Albrecht: Academy of Sciences GDR, Institute for Cybernetics and Information Processes, Kurstraße 33
DDR-1086 Berlin, German Democratic Republic

V. Chetverikov and W. Ebeling: Humboldt-University, Section of Physics
Unter den Linden 6, DDR-1086 Berlin, German Democratic Republic

R. Funke: Academy of Sciences GDR, Institute for Mathematics, Mohrenstraße 39, DDR-1086 Berlin, German Democratic Republic

W. Mende: Academy of Sciences GDR, Institute for Geography and Geoecology, Rudower Chaussee 5, DDR-1199 Berlin-Adlershof
German Democratic Republic

M. Peschel: Scientific Department Mathematics/Cybernetics, Academy of Sciences GDR, Rudower Chaussee 5, DDR-1199 Berlin-Adlershof
German Democratic Republic

EXPONENTIAL CHAINS are a base for modelling growth processes, they include the HYPERCYCLE CONCEPT. The robustness of EXPONENTIAL CHAINS under amplitude bounded stochastic influences is studied analytically, by computer simulation and with imbedding into stochastic differential equations of Ito type. For the EVOLON model of growth processes probability distributions for different kinds of stochastic disturbances are studied.

1. Introduction

In MENDE, PESCHEL [1] the concept of EXPONENTIAL CHAINS was introduced as a possibility to understand the reality of hyperbolic growth processes. In greater detail this concept was studied in MENDE, PESCHEL [2,3].

An EXPONENTIAL CHAIN is determined by

$$dx_i/dt = K_i x_i x_{i+1} \qquad i = 0,1,2,,,\ldots.$$

with the normalized initial conditions

$$x_i(0) = 1$$

This is a signal expansion concept similar to wellknown Taylor expansion.

Very often a truncated EXPONENTIAL CHAIN is considered restricting i on $i = 0,1,\ldots,N$ and putting $x_{N+1} = 1$.

In this case the bottom signal x_o will be designated by x_{ON}. An EXPONENTIAL CHAIN with periodically changing coefficients

$$K_o, K_1, \ldots, K_{r-1}$$

with r being a basic period corresponds to a HYPERCYCLE of order r after EIGEN, SCHUSTER [4].

For $K_i = K$ an EXPONENTIAL CHAIN generates hyperbolic growth, because then we have

$$x_o(t) = 1/(1 - Kt)$$

We have called this the Law of Large Numbers in Ecology. In this paper the behaviour of EXPONENTIAL CHAINS will be studied under the influence of stochastic perturbations of the chain parameters K_i.

This is of importance, because EXPONENTIAL CHAINS can reflect real behaviour only under the condition that the qualitative phenomena are stable under some kind of stochastic disturbances. We first show convergence and stochastic stability of EXPONENTIAL CHAINS under some conditions which are not too restrictive. Roughly speaking we have to distinguish the 3 cases $\bar{K} > 0$, $\bar{K} < 0$ and $\bar{K} = 0$ with $\bar{K}$ being the mean value of K_i, $i = 0,1,\ldots$ The most difficult case is $\bar{K} = 0$.

Then we try to verify the results by computer simulation. For $\bar{K} = 0$ we always meet a tendency to an explosive process. Afterwards we imbed the hyperbolic growth process

$$dx/dt = K\,x^2$$

into an Ito-like stochastic process and study its behaviour in dependence on the strength of the stochastic influence.

These results are consistent to what we have found by computer simulation and show, that only when the 'negative' stochastic parameter perturbation passes a certain threshold we meet a stochastic trajectory with extinction.

There is a strong connection between the concept of EXPONENTIAL CHAINS and the EVELON model following the hyperlogistic differential equation PESCHEL, MENDE [3,5]

$$dx/dt = K\,x^k(B - x^w)^l$$

This is a generalisation of a lot of existing growth models as for example logistic growth for $k = w = 1$ or Bertalanffy growth model for $l = 1$ BERTALANFFY [6].

In the last part of the paper we are interested in distributions for x under the condition that the parameter of the EVOLON model are stochastically disturbed.

For some examples it is shown that quite different distributions arise depending on the strength and the character of the stochastic influence. At the end some possibilities for generalisation of these considerations on general Lotka-Volterra systems are sketched.

2. Robustness of EXPONENTIAL CHAINS against Amplitude-limited Stochastic Parameter Disturbances

We consider signal representations by finite EXPONENTIAL CHAINS

$$F\, x_i = K_i x_{i+1} \quad i=0,1,\ldots,N \quad \text{and} \quad x_{N+1} = 1$$

$$x_i(0) = 1$$

with $F = d\ \ln/dt$ and $F^{-1} = \exp\,(\int_0^t)$

We are interested in the properties of

$$x_{ON}(t) = F^{-1}(K_o F^{-1}(K_1 \ldots F^{-1}(K_N\ 1(t))\ldots)$$

We assume

$$K_i = k_i + \varepsilon_i$$

with ε_i beeing independent stochastic variables with amplitude restrictions of the form

$$-\,\varepsilon \leqslant \varepsilon_i \leqslant \varepsilon$$

We distinguish 3 different cases.

I. Strictly positive parameters $K_i > 0$.

The operator

$$F^{-1}(K_i .)$$

is a monotoneous operator.

We get the following inequalities:

For $w \leqslant u \leqslant v$ we have

$$F^{-1}\ ((k_i - \varepsilon)w) \leqslant F^{-1}(k_i u) \leqslant F^{-1}((k_i + \varepsilon)v)$$

We introduce the following notations

$x_{O,N}$, $x_{O,N;-\varepsilon}$, $x_{O,N;\varepsilon}$ are the bottom signals of the EXPONENTIAL CHAINS with the parameters k_i, resp. $k_i - \varepsilon$, resp. $k_i + \varepsilon$.

If we apply the inequality above for $i=N,N-1,\ldots 0$ we get obviously

$$x_{O,N,-\varepsilon} \leqslant x_{O,N} \leqslant x_{O,N;+\varepsilon}$$

Because of the monotony property of the generating operators all these EXPONENTIAL CHAINS produce an increase of their bottom signals if we adjoin a next level N+1.

Therefore we have now

$$\begin{array}{ccccc} x_{0,N;-\varepsilon} & \leq & x_{0,N} & \leq & x_{0,N;+\varepsilon} \\ \wedge & & \wedge & & \wedge \\ x_{0,N+1;-\varepsilon} & \leq & x_{0,N+1} & \leq & x_{0,N+1;+\varepsilon} \end{array}$$

Sufficient for the convergence of an EXPONENTIAL CHAIN with nonnegative parameters is the condition

$K_i \leq K$ for all i

In such a case obviously the EXPONENTIAL CHAIN with $K_i = K$ is a majorant.

Therefore we get now the inclusion

$$x_{0,\infty;-\varepsilon} \leq x_{0,\infty} \leq x_{0,\infty;+\varepsilon}$$

The result is: Also under the influence of stochastic disturbances with amplitudes in the restricted ε- interval the corresponding EXPONENTIAL CHAIN is converging and is included in the area built up by the deterministic boundary curves

$x_{0,\infty;-\varepsilon}$ and $x_{0,\infty;+\varepsilon}$

In strong analytical sense the convergence holds only in a time-interval restricted right-hands by the pole of $x_{0,\infty},_{-\varepsilon}$.

Most interesting is the special case with $k_i = k$ (Homogenous EXPONENTIAL CHAIN).
In this case (Law of large numbers in Ecology) we get

$$1/(1-(k-\varepsilon)t) < x_{0,\infty} < 1/(1-(k+\varepsilon)t)$$

II. Strictly negative parameters $K_i < 0$.
Now $F^{-1}(K_i.\ )$ is not a monotoneous operator, because multiplication with K_i changes the sign of the inequality.
Maybe

$w \leq u \leq v$

Then we have

$$F^{-1}(k_iw) \geq F^{-1}(k_iu) \geq F^{-1}(k_iv)$$

Moreover holds

$$F^{-1}((k_i+\varepsilon)w) \geq F^{-1}(k_iu) \geq F^{-1}((k_i-\varepsilon)v)$$

If we now apply the operator of the next level i+1 we get again the monotony property

$$F^{-1}((k_{i+1}-\varepsilon)F^{-1}(k_i+\varepsilon)w)) \leqslant F^{-1}(k_{i+1}\,F^{-1}(k_i u)) \leqslant$$

$$\leqslant F^{-1}((k_{i+1}+\varepsilon)\,F^{-1}((k_i-\varepsilon)v)$$

That means: Always a pair of consecutive operators again generate together a monotoneous operator.

If we restrict on EXPONENTIAL ÇHAINS with an even number of levels and combine the corresponding operators pairwise we get the following inclusion

$$x_{0,2N;-\varepsilon,+\varepsilon} \leqslant x_{0,2N} \leqslant x_{0,2N;+\varepsilon,-\varepsilon}$$

If we now adjoin another pair of levels we get the result

$$\begin{array}{ccccc} x_{0,2N;-\varepsilon,+\varepsilon} & \leqslant & x_{0,2N} & \leqslant & x_{0,2N;+\varepsilon,-\varepsilon} \\ \vee & & \vee & & \vee \\ x_{0,2(N+1);-\varepsilon,+\varepsilon} & \leqslant & x_{0,2(N+1)} & \leqslant & x_{0,2(N+1);+\varepsilon,-\varepsilon} \end{array}$$

Now obviously the corresponding bottom signals $x_{0,2N}$ form a monotoneously decreasing sequence bounded from below by zero. Therefore these sequences are necessarily converging.

For the limits we get the following inclusion

$$x_{0,\infty;-\varepsilon,+\varepsilon} \leqslant x_{0,\infty} \leqslant x_{0,\infty;+\varepsilon,-\varepsilon}$$

For the special case $k_i=k$ (Homogenous EXPONENTIAL CHAIN), holds

$$F\,\xi_0 = (k-\varepsilon)\,\xi_1, \quad F\,\xi_1 = (k+\varepsilon)\,\xi_0$$

$$\xi_0 = x_{0,\infty;-\varepsilon,+\varepsilon} \quad \text{and} \quad \xi_1 = x_{0,\infty;+\varepsilon,-\varepsilon}$$

This is a Hypercycle of order r = 2.

This Hypercycle allows closed analytical solutions, namely

$$\xi_0 = 2\,\varepsilon/(\,|k|\,(\exp(2\,\varepsilon\,t)-1)+\,\varepsilon\,(\exp(2\,\varepsilon\,t)+1))$$

$$\xi_1 = 2\,\varepsilon/(\,\varepsilon\,(\exp(-2\,\varepsilon\,t)+1)-\,|k|\,(\exp(-2\,\varepsilon\,t-1))$$

For small ε both signals differ only a little from $1/(1+|k|\,t)$, a hyperbolically decreasing signal.

Therefore we get for the case I (all parameters K_i strictly positive) and for the case II (all parameters K_i strictly negative) robust convergence, namely for case I an exploding signal and for the case II an extincting signal.

III. EXPONENTIAL CHAINS with sign-changing parameters K_i

Especially the case $\bar{K} = 0$ belongs to this type of behaviour. In this more difficult case we got up to now no finite result for convergence or robustness.

The following consideration using linearisation gives only an impression what is going on in this case.
Because

$$\exp(ax) > 1 + ax \quad \text{and} \quad F^{-1}(K.) > 1 + K\,I.$$

we have

$$x_{0,N} > (1+IK_o(1+IK_1(\ldots(1+IK_N\,1(t)) =$$

$$= 1 + K_o t + K_o K_1 t^2/2! + \ldots + K_o K_1 \ldots K_N\, t^{N+1}/(N+1)!$$

For small values of the corresponding arguments we have

$$\bar{x}_{0,N} \sim \sum_{i=0}^{N+1} a_i\, t^i/i! \qquad \bar{a}_i = \bar{K}_o \bar{K}_1 \ldots \bar{K}_i$$

$$\sigma^2_{x_{0,N}} \sim \sum_{i=0}^{N+1} \sigma^2_{a_i} (t^i/i!)^2$$

For the case $\bar{a}_i = \bar{a}$ for all i, that means $\bar{K}_i = \bar{K}$ $\bar{x}_{0,N}$ for small arguments differs not much from exp(Kt).
That means for $\bar{K} > 0$ after some time the condition of small arguments will be violated, for $\bar{K} < 0$ nothing can be concluded, because we have

$$x_{0,N} > \sum_{i=0}^{N+1} a_i t^i/i!$$

For $\bar{K} = 0$ we see, that in the mean $\bar{x}_{0,N}$ will stay at the value 1.
Interesting is the behaviour of the variance $\sigma^2_{x_{0,N}}$.
In all 3 cases the variance is monotoneously and quickly increasing. This means in the case $\bar{K} < 0$ the possibility of an exploding signal and in the case $\bar{K} = 0$ the instability of the state $x_{0,N} = 1$.

We see that at the beginning of this process for small values of arguments the process is very similar to a stochastic process with independent ingrements or something like a Brownian motion. It is possible to derive inclusion properties also for an EXPONENTIAL CHAIN with sign changing coefficients for example if the signs of the coefficients occur periodically.

At first we have to seek a sign-period with an even number of negative signs. For example for the sign-sequence $-,+,-,+,....$ we have to combine always four consecutive coefficients (K_o,K_1,K_2,K_3), (K_4,K_5,K_6,K_7)... to get a monotoneous operator. With this operator we can make the upper and lower estimations for a given length $N = 4k$ for the chain.

Then we adjoin another quadruple of coefficients $(K_N,K_{N+1},K_{N+2},K_{N+3})$. Depending on the sign of the first coefficient K_N we get either a decreasing sequence or an increasing sequence of x_{ON} (decreasing for $K_N < 0$, and increasing for $K_N > 0$).

3. Simulation Experiments with Stochastically Disturbed EXPONENTIAL CHAINS.

With an EXPONENTIAL CHAIN

$$F\, x_i = K_i x_{i+1} \qquad i = 0,1,\ldots,N$$

$$x_{N+1} = 1 \quad \text{and} \quad x_i(0) = 1$$

simulation experiments were performed with parameters K_i being stochastic variables of the form

$$K_i = \bar{K} + \sigma \xi_i$$

with stochastic variables ξ_i stochastically independent and uniformly distributed in $(-1,+1)$.

This is obviously a stochastically disturbed homogenous EXPONENTIAL CHAIN.

The corresponding deterministic EXPONENTIAL CHAIN has the hyperbolic solution

$$x_o(t) = 1/(1-\bar{K}\, t)$$

Because of the nonlinear stochastic influence we can expect a certain bias, therefore we use in the parameter estimation process a biased process of the form

$$x_{o\delta}(t) = 1/(1- (K + \delta)t)$$

belonging to a reference EXPONENTIAL CHAIN with the parameters $K_i = \bar{K} + \delta$ and corresponding $x_{o\delta,N}(t)$.

In the simulation experiments always a sequence of random parameters K_i was chosen (f - number of repeated experiments) and the corresponding EXPONENTIAL CHAIN was integrated in the interval $(0,t_u)$ with the step-width Δt and using a numerical integration method INT (RK-Runge-Kutta of 4.th order, RKM-Runge-Kutta-Merson).

Evaluated in the simulation experiments were the mean value estimation

$$\bar{x}_{0,N}(k\ \Delta t) = \frac{1}{f} \sum x_{0,N;j}\ (k\ \Delta t)$$

and the following deviation measure (relative standard deviation)

$$s_H = \text{squr}\ \left(\frac{1}{f} \sum \left(\frac{x_{0,\sigma}\ (k\ \Delta t) - x_{0,N;,j}\ (k\ \Delta t)}{x_{0,\sigma}\ (k\ \Delta t)} \right)^2 \right)$$

to the reference hyperbola
and

$$s_T = \text{squr}\ \left(\frac{1}{f} \sum \left(\frac{x_{0,\sigma;N}(k\ \Delta t) - x_{0,N;j}(k\ \Delta t)}{x_{0,\sigma;N}(k\ t)} \right)^2 \right)$$

to the reference EXPONENTIAL CHAIN.

After apriori-tests to choose a suitable step-width Δt, and an appropriate integration method and good approximation of the hyperbola by a finite EXPONENTIAL CHAIN it was decided to work with

$f = 1000$, $N = 10$, $\Delta t = 0{,}01$ and Runge-Kutta method of 4-th order.

Under these conditions we must not differentiate between s_H and s_T, we put $s = s_H = s_T$ in this case.

It turns out, that there is a certain but small influence of σ on s, it is not necessary to take it into account.

Now we discuss the main results for the 3 cases $\bar{K} = 1$, $\bar{K} = -1$ and $\bar{K} = 0$. $\bar{K} = 1$ Fig. 1 shows the effect on s.

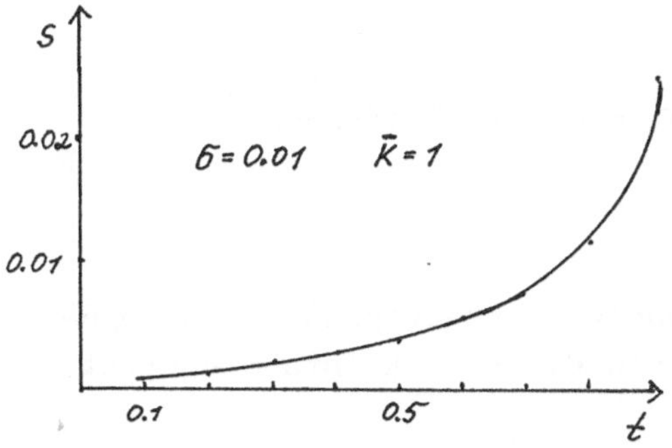

Fig.1

$\overline{K} = -1.$

Fig. 2 shows the effect on s in dependence on $k\Delta t$ and σ. In this case we know that the reference curves describe hyperbolic decrease. Nevertheless the relative deviation over time is increasing but almost linearly.

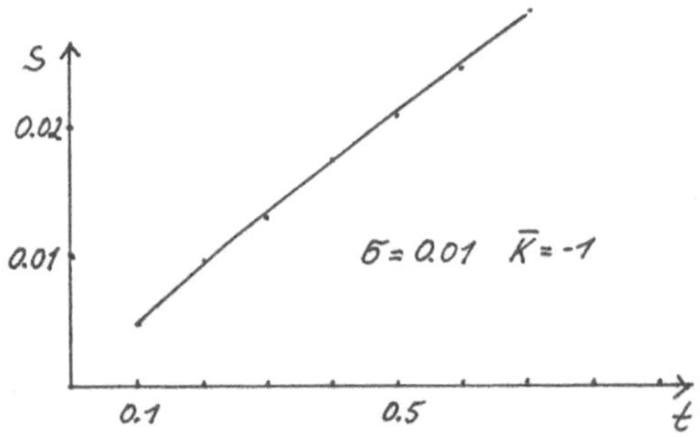

Fig.2

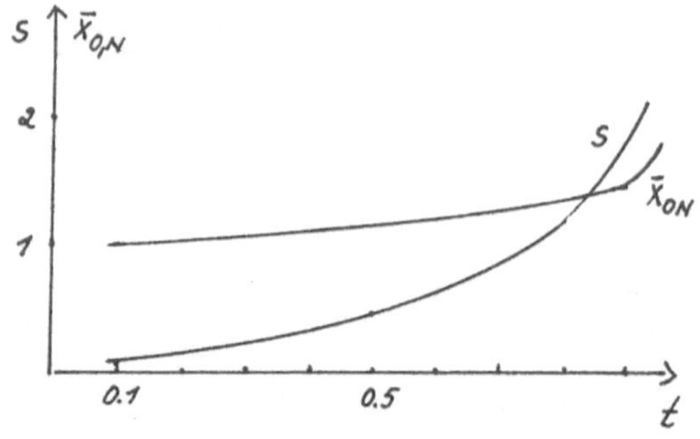

Fig.3

$\overline{K} = 0$

Fig. 3 shows the effect on s in dependence on t and σ . At the beginning s always increases almost linearly but for larger values of k Δ t we then always meet an exploding process. This also holds for $\bar{x}_{0,N}(k\ t)$, Remarkably the pole to be expected for $\bar{x}_{0,N}$ is later in comparison with the pole for s.

As small as σ might be we always meet this exploding process with an empirically found relation of the form

$$t_p\ \sigma \sim 1.35$$

where t_p is the pole for s.

If we change stochastically the parameters K_i also after every integration step Δt, we get similar results.
In this case we get an empirical relation of the form

$$t_p\ \sigma \sim 8.1$$

4. Stochastic Imbedding of Hyperbolic Growth Processes

We imbed the differential equation

$$dx/dt = K\ x^2 \qquad x(0) = 1$$

representing a hyperbolic growth process corresponding to a homogenous infinite EXPONENTIAL CHAIN with all $K_i = K$ into a context of stochastic disturbances.

At the same time we could consider $y(t) = 1/x(t)$ with

$dy/dt = -K$ and $y(0) = 1$

For the corresponding stochastic processes under the stochastic influence we introduce the notations $\xi(t)$ and $\eta(t)$.
We consider the stochastic differential equation

(1) $d\,\eta(t) = -K\,dt + \sqrt{h\;\eta(t)}\;d\,w(t)$ with

$\eta(0) = 1$, K and h are constants, $h > 0$, $w(t)$ Wiener-process.

Applying the Ito-formula on the transformation $\xi(t) = 1/\eta(t)$ we get

(2) $d\,\xi(t) = \xi^2(t)\,(K+h)\,dt - \sqrt{h\,\xi^3(t)}\;d\,w(t), \quad \xi(0) = 1$

We now study the phenomena of explosion and extinction with the help of absorption properties of $\eta(t)$ at the boundary values 0 respective ∞. For this purpose we restrict the values of $\eta(t)$ on the interval $[\varepsilon, M]$ with a small ε and a large M putting for $s > t$ $\eta(s) = \varepsilon$ if $\eta(t) = \varepsilon$ resp. $\eta(s) = M$, if $\eta(t) = M$. $a(t,y) = -K$ and $b(t,y) = \sqrt{h\,y}$ fulfil in respect of y a Lipschitz-condition

$$|a(t,y_1)-a(t,y_2)| + |b(t,y_1)-b(t,y_2)| \leq L\,|y_1-y_2|$$

Therefore a unique solution of (1) with the absorption condition exists. Now we transform state-variable η and time t in such a way that we get absorption in 0 respective in 1.

$$\tilde{\eta}(t) = \frac{\sqrt{\eta\,(4t\,(\sqrt{M}-\sqrt{\varepsilon})^2/h)} - \sqrt{\varepsilon}}{\sqrt{M}-\sqrt{\varepsilon}}$$

Instead of (1) we get now

$$d\,\tilde{\eta}(t) = \tilde{a}\,(\tilde{\eta}(t))\,dt + d\,\tilde{w}\,(t),$$

$\tilde{\eta}(0) = y_0 = \dfrac{1-\sqrt{\varepsilon}}{\sqrt{M}-\sqrt{\varepsilon}}$, $\tilde{w}\,(t)$ is again a Wiener-process,

$\tilde{a}\,(\tilde{\eta}) = -\dfrac{\alpha}{\tilde{\eta}(t)+\beta}$ with $\alpha = 2K/h + 1/2$, $\beta = \sqrt{\varepsilon}/(\sqrt{M}-\sqrt{\varepsilon})$

May be $\tilde{\eta}_{y,s}(t)$ in $y \in (0,1)$ and for $t > s$ the solution of the stochastic differential equation

$$d\,\tilde{\eta}_{y,s}(t) = \tilde{a}\,(\tilde{\eta}_{y,s}(t))dt + d\,\tilde{w}(t), \qquad \tilde{\eta}_{y,s}(s) = y$$

with absorption in 0 or 1 respectively, because $\tilde{a}$ is independent on time t. This is a Markov-process and its finite-dimensional distributions only depend on t-s. Therefore we can restrict on $\tilde{\eta}_y(t) = \tilde{\eta}_{y,0}(t)$ and $t > 0$.

May be

$$p_i(t,y) = P(\tilde{\eta}_y(t) = i), \quad y \in (0,1), \; t > 0 \text{ and } i = 0,1$$

(probability of absorption up to time t in 0 resp. in 1)

$$q(t,y) = 1 - (p_0(t,y) + p_1(t,y))$$

For these probability-functions the following important theorem holds GICHMAN, SKOROCHOD [7]

<u>Theorem:</u> For the solution $\tilde{\eta}_y(t)$ of the stochastic differential equation $d\,\tilde{\eta}_y(t) = \tilde{a}(\tilde{\eta}_y(t))dt + d\,\tilde{w}(t)$ for $t > 0$ with absorption in 0 resp. 1 and a function $\tilde{a}$ twice continuously differentiable the absorption probability-functions $p_i(t,y)$ are solutions of the following Kolmogorov partial differential equations

(3) $$\partial p_i(t,y) / \partial t = \tilde{a}(y)\, \partial p_i(t,y)/\partial y + \frac{1}{2}\, \partial^2 p_i(t,y)/\partial y^2$$

for i = 0,1 in the sense of generalized functions with the boundary conditions

$$\lim_{t \to 0} p_i(t,y) = 0 \qquad \lim_{y \to 0} p_1(t,y) = \lim_{y \to 1} p_0(t,y) = 0$$

$$\lim_{y \to 0} p_0(t,y) = \lim_{y \to 1} p_1(t,y) = 1$$

In our case the Kolmogorov-equation has the form

(4) $$\partial p_i(t,y)/\partial t = -\frac{\alpha}{(y + \beta)}\, \partial p_i(t,y)/\partial y + \frac{1}{2}\, \partial^2 p_i(t,y)/\partial y^2$$

A solution in terms of an infinite sum can be find easily in the special case $\alpha = 0$. In this case we get the differential equations

(5) $$\partial p_i(t,y)/\partial t = \frac{1}{2}\, \partial^2 p_i(t,y)/\partial y^2, \quad i = 1,2.$$

The solutions are

$$p_0(t,y) = 1 - y - 2\sum_{n=1}^{\infty} e^{-n^2\pi^2 t/2} (\sin n\pi y)/(n\pi)$$

$$p_1(t,y) = y + 2\sum_{n=1}^{\infty} e^{-n^2\pi^2 t/2} (-1)^n (\sin n\pi y)/(n\pi)$$

Considering these solutions for $y = y_0$ and the time transformation

$$t = \tau \frac{h}{4(\sqrt{M} - \sqrt{\varepsilon})^2} \qquad (\tau \text{ is here the real time})$$

we get the following result for large but finite M.

$$\lim_{\tau \to \infty} p_0\left(\tau \frac{h}{4(\sqrt{M}-\sqrt{\varepsilon})^2}, y_0\right) = \frac{1 - 1/\sqrt{M}}{1 - \sqrt{\varepsilon}/\sqrt{M}} \approx 1,$$

$$\lim p_1\left(\tau \frac{h}{4(\sqrt{M}-\sqrt{\varepsilon})^2}, y_0\right) = \frac{1 - \sqrt{\varepsilon}}{\sqrt{M} - \sqrt{\varepsilon}} \approx 0.$$

Obviously holds $\varrho(\infty, y_0) = 1 - (p_0(\infty, y_0) + p_1(\infty, y_0)) = 0.$ The limits are continuous in ε and we can put $\varepsilon = 0$. In regard of the time transformation the result can be formulated as follows. For every $\delta > 0$ there exists a M_δ and for all $M > M_\delta$ holds

$$- p_0(\infty, y_0) = p_1(\infty, y_0) < \delta$$

In case of $\alpha \neq 0$ something can be said about the asymptotic behavior of the solution.

If $p_i(t,y)$ converge for $t \to \infty$ with a very small slope

$$\lim_{t \to \infty} \partial p_i(t,y)/\partial t = 0$$

asymptotically $p_i(\infty, y)$ follow the ordinary differential equation

$$\frac{1}{2} \overset{\circ\circ}{p}_i(\infty, y) - \frac{\alpha}{y+\beta} \overset{\circ}{p}_i(\infty, y) = 0$$

with the following analytically closed solutions

$$p_0(\infty, y) = 1 - p_1(\infty, y) = ((1+\beta)^{2\alpha+1} - (y+\beta)^{2\alpha+1})$$

$$/((1+\beta)^{2\alpha+1} - \beta^{2\alpha+1}) \quad \text{for} \quad \alpha \neq -1/2$$

$$p_0(\infty, y) = 1 - p_1(\infty, y) = (\ln(1+\beta) - \ln(y+\beta))/(\ln(1+\beta) -$$

$$- \ln \beta) \quad \text{for} \quad \alpha = -1/2$$

Consequently the process $\tilde{\eta}_y(t)$ will be absorbed with probability 1 for all $y \in (0,1)$.

By a more detailed discussion of this formula we get the following results: $\tilde{\eta}(0) = y_0 = \beta(\varepsilon^{-1/2} - 1)$

1. $\alpha > -1/2$, that means $K > -h/2$

$$p_0(\infty, y_0) = (1 - M^{-(\alpha+1/2)})/(1-(\varepsilon/M)^{\alpha+1/2}) \approx 1$$

$$p_1(\infty, y_0) = 1 - p_0(\infty, y_0) \approx 0.$$

That means, the process $\xi(t)$ is exploding even in the case where K is negative but above a certain threshold -h/2. This corresponds completely to the findings in section 2 and 3.

2. $\alpha = -1/2$, that means $K = -h/2$
Now the behaviour is completely indefinite, because we get now for example

$$p_0(\infty, y_0) = \frac{1}{1 - \ln M/\ln \varepsilon} \rightarrow \begin{cases} 1 & \text{for } \varepsilon \to 0 \\ 1/2 & \text{for } \varepsilon = 1/M \text{ and } M \to \infty \\ 1/(1+q) & \text{for } \varepsilon = 1/M^q \text{ and } M \to \infty \end{cases}$$

$$p_1(\infty, y_0) = 1 - p_0(\infty, y_0)$$

3. $\alpha < -1/2$, that means $K < -h/2$
In this case the inhibitation of the system is high and is still supported by the stochastic influence.
Now we get

$$p_0(\infty, y_0) = \varepsilon^{-(\alpha+1/2)}(M^{-(\alpha+1/2)}-1)/(M^{-(\alpha+1/2)} - \varepsilon^{-(\alpha+1/2)})$$
$$\rightarrow 0 \quad \text{for } \varepsilon \to 0$$

$$p_1(\infty, y_0) = 1 - p_0(\infty, y_0)$$

In this case we get extinction with probability 1 by putting $\varepsilon = 0$. Also this result is consistent with the findings in section 2 and 3. The probabilities in the limit $t \to \infty$ depend in case 1 on M. In case 1 the probability for extinction of $\xi(t)$ is

$$p_1(\infty, y_0) = \left(\frac{1}{M}\right)^{\alpha + 1/2} \qquad \text{with } \varepsilon = 0.$$

That means, the answer on the question, if $\xi(t)$ will explode or will be extincted depends on the value of the 'minimum for existence', the lower absorption level $\frac{1}{M}$ of $\xi(t)$.

In case 3 we meet extinction independent from the 'minimum for existence' (M finite, $\varepsilon \to 0$).

5. Probability Distributions for the Stochastically Disturbed EVOLON

Onedimensional problems

Let us first consider the simplest case of an Verhulst-Pearl-EVOLON where the birth rates fluctuate according to a Gaussian white noise process

$$dx/dt = (\alpha + \sigma \xi(t))\, x - \beta x^2$$

The deterministic trajectory is well known. In linear approximation one gets for the deviations

$$\delta x(t) = x(t) - \bar{x}(t)$$

from the deterministic trajectory the equation

$$\frac{d}{dt} \langle \delta x^2 \rangle = 2 \langle \delta x^2 \rangle (\alpha - 2\beta \bar{x}(t)) + 2 \sigma^2 \bar{x}^2(t)$$

This linear equation with time dependent coefficients can easily be solved and we obtain in this way the time dependence of the relative deviation

$$\Delta(t) = \langle \delta x^2 \rangle^{1/2} / \bar{x}(t)$$

Fig. 4 shows that the time developement depends on the initial condition. It is interesting to note that the deviation remains finite only in the case $\alpha > 0$, $\beta > 0$. In that case we may give the final distribution explicitely

$$P^0(x) = \frac{(\beta / \sigma^2)^{(\alpha / \sigma^2)}}{\Gamma(\alpha/\sigma^2)} \, x^{(\alpha/\sigma^2)-1} \exp\left(-\frac{\beta x}{\sigma^2}\right), \; x > 0$$

The case which is of special interest for our consideration

$$\alpha = 0, \quad \beta = -K < 0$$

i.e. hyperbolic growth, shows a monotonous increase of the relative deviation in time and therefore a stationary distribution does not

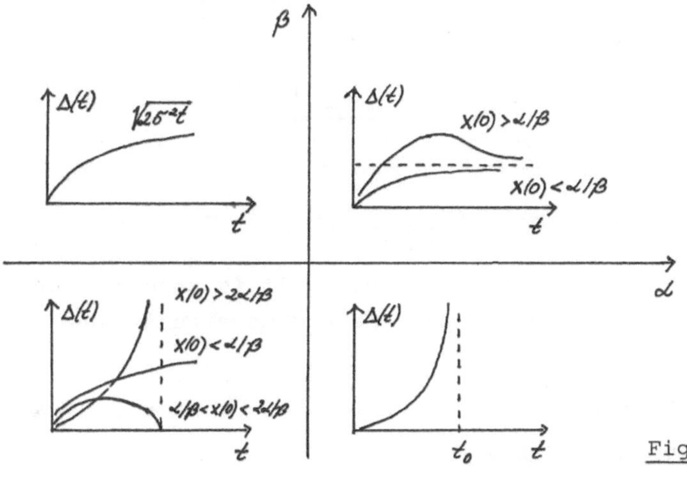

Fig.4

exist. An integration of the linear approximation with $K = t_o^{-1}$ yields

$$\langle \delta x^2 \rangle_t = \frac{1}{(1-Kt)^4} \langle \delta x^2 \rangle_{t=0} + \int_0^t dt' \frac{2\sigma^2}{(1-Kt')^2(1-Kt+Kt')^4}$$

We see that the mean square deviation rapidly increases; we have to note however that the linear theory describes only the very beginning of this process. Further let us consider the more general case of an stochastically disturbed EVOLON given by the equation

$$dx/dt = K\, x^k\, (B-x^w)^l + \sum_{i=0}^{r} g_i(x)\, \xi_i(t)$$

We assume an additive noise source and two noise sources arising from stochastic parameter disturbances

$$K \rightarrow K + \sigma_1\, \xi_1(t) \quad \text{and} \quad B \rightarrow B + \sigma_2\, \xi_2(t)$$

This leads to

$$g_o = g_o(x), \quad g_1 = \sigma_1 x^k (B-x^w)^l, \quad g_2 = \sigma_2 K\, x^k (B-x^w)^{l-1}$$

Without changing the character of the process we can get rid of w (Putting $y = x^w$) and thus we can assume $w = 1$.

In order to preserve the condition $x(t) \geq 0$ the noise sources should satisfy the conditions

$$g_i(0) = 0 \text{ and } g_o(x) \geq 0 \quad \text{for} \quad x \geq 0.$$

In the case of Gaussian white noise

$$\langle \xi_i(t).\ \xi_j(t) \rangle = \delta_{ij}\ \delta(t-t')$$

we can determine closed analytical formulas for the stationary distribution, namely EBELING [8]

$$P^o(x) \sim D(x)^{-1/2} \quad \exp\left(2 \int_0^x d\xi\, \frac{K\, \xi^k (B-\xi)^l}{D(\xi)}\right)$$

with

$$D(x) = g_o^2(x) + \sigma_1^2\, x^{2k}(B-x)^{2l} + \sigma_2^2\, K^2 x^{2k} (B-x)^{2(l-1)}$$

Now let us consider the case of coloured noise. It is possible to find the stationary distribution for any one - dimensional stochastic equation

$$dx/dt = v(x) + g(x)\, \tau(t)$$

with so called telegraph noise (see Fig. 5)

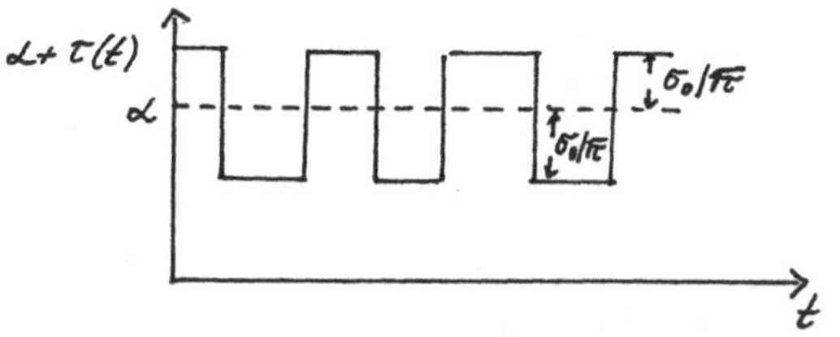

Fig.5

$$\tau(t) = \frac{\sigma_0}{\sqrt{\tau}} (-1)^{n(0,t)}$$

where n(0,t) is the Poisson stationary stream with characteristic time τ. We do not know the exact form of the characteristic functional for this process, but know all mean values. For example

$$\langle \tau(t) \rangle = 0, \quad \langle \tau(t)\, \tau(t') \rangle = \frac{\sigma_0^2}{\tau} \exp\left(- \frac{|t-t'|}{\tau}\right)$$

The stationary density P can be obtained as a solution of the system of two different equations

$$\frac{d}{dx} v P + \frac{d}{dx} g \Psi = 0, \quad \frac{d}{dx} v \Psi + \frac{\sigma_0^2}{\tau} g P + \frac{1}{\tau} \Psi = 0$$

with additional conditions

$$P \geq 0, \quad \int P\, dx = 1, \quad \int \Psi\, dx = 0$$

If we are interested in the solution with zero probability stream, we can write down $\Psi = -v g^{-1} P$ and substitute this value into the second equation to abtain the formal solution KLJACKIN [9]

$$P(x) = N g(x) \left[\frac{\sigma_0^2}{\tau} g^2(x) - v^2(x) \right]^{-1} \cdot$$
$$\exp\left\{ \frac{1}{\tau} \int^{x} dy \left[\frac{\sigma_0^2}{\tau} g^2(y) - v^2(y) \right]^{-1} v(y) \right.$$

However we must be attentive using this formula because it is necessary to carry out all additional demands.

Finally let us study another case which may be solved explicitely: the EVOLON equation with k = 1, l = 1, w -1 (Bernoulli model) disturbed by a stationary Ornstein-Uhlenbeck noise u(t):

$$dx/dt = x\,(\alpha - \beta x^{w}) + x^{w+1}\, u(t)$$

$$\langle u(t) \rangle = 0$$

$$\langle u(t)\, u(t') \rangle = \frac{\sigma_0^2}{\tau} \exp\left[- |t-t'| / \tau \right]$$

Following the paper [10] one can obtain an exact equation for the probability density. If the inequality $(1 + w \alpha \tau) > 0$ is satisfied, we find the stationary distribution corresponding to the zero probability stream:

$$P^o(x) = Nx^{-w-1} \exp\left(- \frac{\alpha}{2w\sigma^2} x^{-2w} + \frac{\beta}{w\sigma^2} x^{-w} \right)$$

$$\sigma^2 = \sigma_o^2 (1 + w \alpha \tau)^{-1}$$

Multidimensional problems

We consider the general Lotka-Volterra system with stochastical birth rates which are represented by a Gaussian white noise process

$$dx_i/dt = x_i (e_i + \sum G_{im} \xi_m(t)) - x_i \sum g_{ij} x_j + \partial_i$$

with $\partial_i = +0$. In a similar way as above we obtain in linear approximation a closed equation for the correlation function of the deviations from the means

$$\frac{d}{dt} \langle \delta x_i(t) \delta x_j(t) \rangle = \left[(e_i + e_j) - \sum_e (g_{ie} + g_{je}) \bar{x}_e(t) \right] \langle \delta x_i(t) \delta x_j(t) \rangle$$

$$- \sum_e \left[\bar{x}_i(t) g_{ie} \langle \delta x_e(t) \delta x_j(t) \rangle + \bar{x}_j(t) g_{je} \langle \delta x_e(t) \delta x_i(t) \rangle \right]$$

$$+ 2 \sum G_{im} G_{jm} \bar{x}_i(t) \bar{x}_j(t)$$

This non-autonomous linear system depends on the deterministic solution, it may be solved by standard methods. Depending on the vector a_i and the matrices g_{ij} and G_{ij}, many different cases are possible and only for a restricted set of coefficients the mean quadratic deviation remains finite.

Of special interest of course is the case that there exists exactly one stationary state with

$$e = g\, x^o$$

being a stable attractor point. In this case the final probability distribution may be calculated at least approximately. We have to distinguish the following cases:

a) Conservative system $g_{ij} = -g_{ji}$

In this case it is wellknown, that

$$H = \sum (x_i - x_i^o - x_i^o \ln (x_i / x_i^o))$$

is a first integral.

Then

$$P^o(x) \sim \frac{1}{\prod x_i} f(H)$$

is a possible stationary distribution, where f is any differentiable function EBELING, FEISTEL [11]

Of more interest is the following case

b) Canonical-dissipative system

We introduce $D_{ij} = \frac{1}{2} \sum G_{im} G_{jm}$

g can be decomposed uniquely into

$$g_{ij} = g_{ij}^{A} + g_{ij}^{S} \qquad g_{ij}^{A} = - g_{ji}^{A} \qquad g_{ij}^{S} = g_{ji}^{S}$$

May be $g_{ij}^{S} \neq 0$ and fulfil the condition

$$g_{ij}^{S} = \beta D_{ij}$$

In physics relations of the type given above between dissipative and fluctuative terms are called fluctuation-dissipation relations; they are satisfied for a big class of systems (Nyquist-relation, Einstein-relation etc.). In ecology fluctuation-dissipation relations hold only for a certain subclass of the set of possible networks. A special important case is that non-diagonal losses and correlations of fluctuations may be neglected

$$g_{ij} = \beta D_{ij} = \lambda \delta_{ij}$$

Then a closed analytical stationary distribution can be determined

$$P^{o}(x) \sim \frac{1}{\prod x_i} \exp (- \beta H)$$

c) Systems near to the case b

"near" cannot be precisely specified.

Here approximately the following formula for the stationary distribution holds

$$P^{o}(x) \sim \frac{1}{\prod x_i \; Q(H)^{1/2}} \exp \left(- \int_0^{H} dw \; G(w)/Q(w) \right)$$

with G, Q being of the form

$$G(H) = \sum_{i,j} g_{ij}^{S} I_{ij}(H) \; ; \quad Q(H) = \sum_{i,j} D_{ij} I_{ij}(H)$$

$$I_{ij}(H) = \int \frac{dx_1}{x_1} \cdots \frac{dx_n}{x_n} \delta(H-H(x)) (x_i - x_i^o) (x_j - x_j^o)$$

$$\approx \delta_{ij} \, c_n \left[x_i^o x_j^o \, H^{n+1} \right]^{1/2}$$

Simple examplesand results of computation

a) Stationary distribution of the logistic model with white noise. The deterministic stationary value is

$$x_o = \alpha / \beta$$

We introduce the parameter

$$\delta = \langle (x-x_o)^2 \rangle / x_o^2 = \sigma^2 / \alpha$$

Fig. 6 shows which types of stationary distributions are possible under these circumstances.

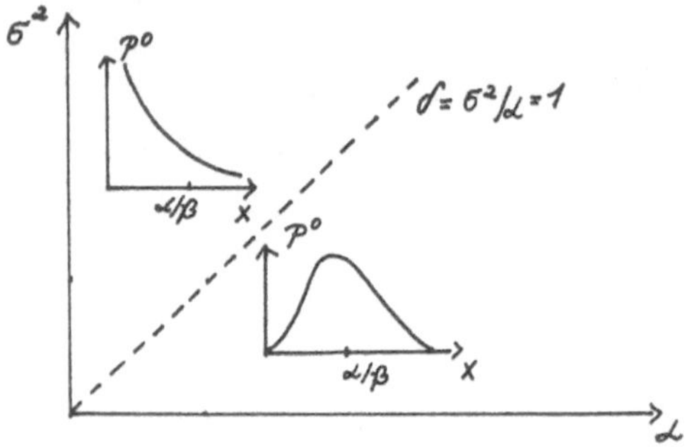

Fig.6

For $\delta < 1$, that means a moderate noise influence, we get what we should expect, an unimodular similar to Gaussian distribution with a certain shift of the maximum to the left of the stationary point. For large noise influence $\delta > 1$ we get a concentration of the distribution in the neighbourhood of zero, that means an increasing tendency to extinction.

b) Stationary distribution of the logistic model with telegraph noise. Based on a Poisson point process in time a random telegraph signal according to Fi. 5 is used for the parameter disturbance. Depending on the time-constant τ now a definite shift of the stationary point with the maximum amount $\delta\alpha = \sigma_0 / \sqrt{\tau}$ can take place. Fig. 7 shows that besides the distributions we already now other monostable and bistable distributions can occur.

c) stationary distribution of the Bernoulli model with Ornstein - Uhlenbeck noise.
All possible forms of distributions for different values α, β we have shown in Fig. 8, where the full line shows the case $w > o$, and the broken line - the case $-1 < w < 0$.

In conclusion we want to underline that Volterra-systems are quite sensitive to the influence of external noise. The class of examples studied here shows that the noise may give rise to drastic qualitative changes. Looking e.g. at Fig. 6-8 one sees that in some cases the deterministic prediction by mean values looses any sense. Further we see that even the information on the quadratic deviations is in some cases not sufficient to specify the form of the distribution function.

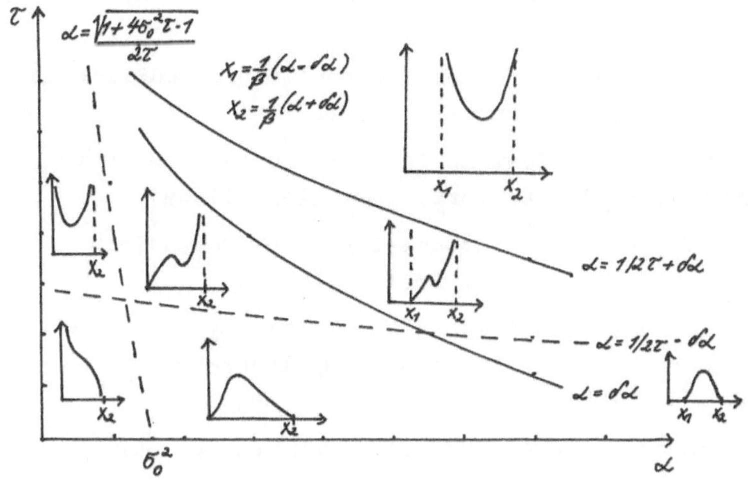

Fig.7

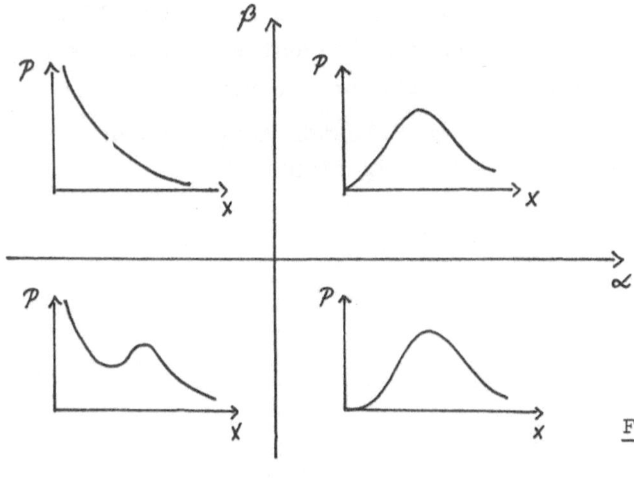

Fig.8

References

[1] W. Mende, M. Peschel: Problems of fuzzy modelling, control and forecasting of time-series and some aspects of evolution. IFAC Symposium on Controm Mechanisms in Bio- and Ecosystems, Plenary Paper, Leipzig, (1977)

[2] W. Mende, M. Peschel: Structure-Building Phenomena in Systems with Power-Product Forces, In: Haken H.: Chaos and Order in Nature, 196-206, Springer-Verlag, Berlin, (1981)

[3] M. Peschel, W. Mende: Leben wir in einer Volterra-Welt? - Ein ökologischer Zugang zur angewandten Systemanalyse. Mathematical Research 14, Akademie-Verlag, Berlin, (1983)

[4] M. Eigen, P. Schuster: The Hypercycle. Springer-Verlag, Berlin, (1979)

[5] M. Peschel, W. Mende: Probleme der mathematischen Modellierung von Evolutionsprozessen, msr 12, (1981)

[6] L.K. Bertalanffy: General Systems Theory: Foundations, Development, Applications, New York: Braziller, (1968)

[7] I.I. Gichman, A.W. Skorochod: Stochastische Differentialgleichungen, Akademie-Verlag, Berlin, (1971)

[8] W. Ebeling: Structural Stability of Stochastic Systems. In: Haken H.: Chaos and Order in Nature, Springer-Verlag, Berlin, 188-195, (1981

[9] V.J. Kljackin, Izvestij Vuzov. Radiofisika, tom 20, 562, (1977)

[10] S.E. Pitovranov, V.M. Chetverikov: Corrections to the Diffusion Approximation in Stochastic Differential Equations. Theoret. i Matemat. Fizika 35, 211-223, (1978)

[11] W. Ebeling, R. Feistel: Physik der Selbstorganisation und Evolution, Akademie-Verlag, Berlin, (1982)

List of Contributors

Springer Series in Synergetics

Editor: **H. Haken**

Volume 23
Synergetics of the Brain
Proceedings of the International Symposium on Synergetics at Schloß Elmau, Bavaria, May 2-7, 1983
Editors: **E. Başar, H. Flohr, H. Haken, A. J. Mandell**
1983. 192 figures. VIII, 377 pages. ISBN 3-540-12960-X

Volume 20
H. Haken
Advanced Synergetics
Instability Hierarchies of Self-Organizing Systems and Devices
1983. 105 figures. XV, 356 pages. ISBN 3-540-12162-5

Volume 17
Evolution of Order and Chaos
in Physics, Chemistry, and Biology
Proceedings of the International Symposium on Synergetics at Schloß Elmau, Bavaria, April 26-May 1, 1982
Editor: **H. Haken**
1982. 189 figures. VIII, 287 pages. ISBN 3-540-11904-3

Volume 16
L. A. Blumenfeld
Physics of Bioenergetic Processes
Volume Editor: H. Haken
1983. 30 figures. X, 132 pages. ISBN 3-540-11417-3

Volume 15
W. Horsthemke, R. Lefever
Noise-Induced Transitions
Theory and Applications in Physics, Chemistry, and Biology
1984. 56 figures. XV, 318 pages. ISBN 3-540-11359-2

Volume 14
W. Weidlich, G. Haag
Concepts and Models of a Quantitative Sociology
The Dynamics of Interacting Populations
1983. 71 figures. XII, 217 pages. ISBN 3-540-11358-4

Volume 13
C. W. Gardiner
Handbook of Stochastic Methods
for Physics, Chemistry and the Natural Sciences
1983. 29 figures. XIX, 442 pages. ISBN 3-540-11357-6

Volume 12
Nonlinear Phenomena in Chemical Dynamics
Proceedings of an International Conference, Bordeaux, France, September 7-11, 1981
Editors: **C. Vidal, A. Pacault**
1981. 124 figures. X, 280 pages. ISBN 3-540-11294-4

Volume 11
Chaos and Order in Nature
Proceedings of the International Symposium on Synergetics at Schloß Elmau, Bavaria, April 27-May 2, 1981
Editor: **H. Haken**
1981. 134 figures. VIII, 275 pages. ISBN 3-540-11101-8

Volume 10
Y. L. Klimontovich
The Kinetic Theory of Electromagnetic Processes
Translated from the Russian by A. Dobroslavsky
1983. XI, 364 pages. ISBN 3-540-11458-0

Volume 9
Numerical Methods in the Study of Critical Phenomena
Proceedings of a Colloquium, Carry-le-Rouet, France, June 2-4, 1980
Editors: **J. Della Dora, J. Demongeot, B. Lacolle**
1981. 83 figures. IX, 269 pages. ISBN 3-540-11009-7

Volume 8
Stochastic Nonlinear Systems
in Physics, Chemistry, and Biology
Proceedings of the Workshop, Bielefeld, Federal Republic of Germany, October 5-11, 1980
Editors: **L. Arnold, R. Lefever**
1981. 48 figures. VIII, 237 pages. ISBN 3-540-10713-4

Volume 7
L. A. Blumenfeld
Problems of Biological Physics
1981. 38 figures. IX, 224 pages. ISBN 3-540-10401-1

Springer-Verlag
Berlin
Heidelberg
New York
Tokyo